Proceedings
of the
Steklov Institute of Mathematics

edited by

S. M. Nikol'skiĭ

Number 127 (1975)

Boundary Value Problems of Mathematical Physics. IX

edited by

O. A. Ladyženskaja

Providence, Rhode Island

AMERICAN MATHEMATICAL SOCIETY

1977

Академия Наук
Союза Советских Социалистических Республик

ТУРДЫ

ордена Ленина
МАТЕМАТИЧЕСКОГО ИНСТИТУТА
имени В. А. СТЕКЛОВА

CXXVII

Ответственный редактор
академик С. М. НИКОЛЬСКИЙ

КРАЕВЫЕ ЗАДАЧИ
МАТЕМАТИЧЕСКОЙ ФИЗИКИ. 9

под редакцией О. А. ЛАДЫЖЕНСКОЙ

Издательство "Наука"
Ленинград 1975

AMS (MOS) subject classifications (1970). Primary 35A05, 35B05, 35B15, 35B45, 35D99, 35J25, 35J40, 35J45, 35J60, 35K50, 35M05, 35N05, 35N10, 35Q10, 46E15, 65N10, 65N20, 76D05, 76W05; Secondary 35Q10, 41A30, 46E30, 47H15, 76B15, 76D05, 92A05.

Library of Congress Cataloging in Publication Data CIP

Main entry under title: (Revised)

Boundary value problems of mathematical physics.

Translation of (romanized: Kraevye zadachi mathematicheskoĭ fiziki)
Includes bibliographies.
1. Boundary value problems. 2. Differential equations, Partial. 3. Mathematical physics.
I. Ladyshenskai͡a, Olga Aleksandrovna, ed. II. Series: Akademii͡a nauk SSR. Matematicheskiĭ institut. Proceedings, no. 83 [etc.]
QA1.A413 no. 83, etc. 515'.353 67-6187
ISBN 0-8218-3027-9

PROCEEDINGS OF THE STEKLOV INSTITUTE OF MATHEMATICS
IN THE ACADEMY OF SCIENCES OF THE USSR

(Труды математического института им. В. А. Стеклова, т. CXXVII, 1975)

TABLE OF CONTENTS

Proc. Steklov Inst. Math.
127 (1975)

PREFACE

In this collection V. A. Solonnikov continues his investigations on overdetermined boundary value problems of elliptic type. In his paper the solvability of such problems is established in a broad scale of function spaces introduced by K. K. Golovkin (the so-called fractional spaces). Along the same lines, the paper by Š. Sahaev studies an overdetermined problem of parabolic type arising in magnetohydrodynamics. Sharp estimates for its solutions are obtained in the norms of the spaces $W_p^{2,1}(Q_T)$ and $C^{1+2\alpha,\frac{1}{2}+\alpha}(Q_T)$, where $Q_T = \Omega \times [0, T]$ and Ω is a bounded domain in three-dimensional Euclidean space. In the paper by Sahaev and Solonnikov unique local solvability in the class of smooth functions is established for an initial-boundary value problem in magnetohydrodynamics. Unique solvability of this and two other problems in magnetohydrodynamics in certain classes of generalized solutions was proved earlier by Ladyženskaja and Solonnikov. Some generalizations of the system of Navier-Stokes equations are given by A. P. Oskolkov, establishing a number of estimates for solutions of these systems.

Quasilinear equations of elliptic type are studied in the papers of A. V. Ivanov and A. L. Treskunov. More precisely, Ivanov presents a class of elliptic systems of the form

$$\sum_{i,j=1}^{n} A_{ij}(x, \mathbf{u}, |\mathbf{u}_x|)\frac{\partial^2 u^s}{\partial x_i \partial x_j} = B^s(x, \mathbf{u}, \mathbf{u}_x), \quad s = 1, \ldots, N,$$

where $\mathbf{u}(x) = (u^1(x), \ldots, u^N(x))$, $x = (x_1, \ldots, x_n)$, for which any solution $\mathbf{u}(x)$ bounded in the whole space R^n (or not increasing too rapidly as $|x| \to \infty$) is a constant. Other criteria distinguishing a constant from the rest of the solutions are also established.

Treskunov proves the classical solvability of a boundary value problem with oblique derivative for the equation $\Sigma_{i,j=1}^2 a_{ij}(x, u, u_x)u_{x_i x_j} = a(x, u, u_x)$ under very mild smoothness of the functions $a_{ij}(x, u, p)$ and $a(x, u, p)$.

L. Stupjalis investigates unique solvability of initial-boundary value problems for equations of mixed type. In one part Ω_1 of the domain $\Omega \subset R^n$ this equation is of elliptic or parabolic type, and in the other Ω_2 it is hyperbolic.

On the surface where the equation changes type, conjunction conditions arising in certain problems of mathematical biology are imposed. These conditions turn out to be "better" than the conjunction conditions arising in diffraction theory: unique solvability holds under weaker assumptions on the structure of the equations in Ω_1 and Ω_2.

The papers by N. K. Korenev concern the grid method. In the first the simplest known difference approximation u_h of elements $u(x)$ of the space $W_p^1(\Omega)$ is analyzed for the case where the u_h are equal to zero on the boundary of the grid domain $\Omega_h \subset \Omega$. For domains Ω with smooth boundary it is known that the piecewise constant completions $\tilde{u}_h(x)$ and $\tilde{u}_{x_i}$ for the approximation u_k converge in $L_p(\Omega)$ to $u(x)$ and $\partial u(x)/\partial x_i$ respectively. The author shows that the same is also true for a broad class of domains Ω with nonsmooth boundaries. In the second paper one of the difference schemes proposed by Ladyženskaja for a steady state nonlinear boundary value problem for the Navier-Stokes equations is analyzed. It is proved that if the nonlinear problem has a finite number of solutions $\mathbf{u}^{(i)}$, $i = 1, \ldots, N$, and the linear boundary value problem obtained by Oseen linearization of the original problem on each of the solutions $\mathbf{u}^{(i)}$ is uniquely solvable, then for all sufficiently fine grids the difference equations mentioned above have exactly N solutions $\mathbf{u}_h^{(i)}$, which converge to $\mathbf{u}^{(i)}$ as $h \to 0$.

Translated by B. SILVER

Trudy Mat. Inst. Steklov
127 (1975)

Proc. Steklov Inst. Math.
127 (1975)

THEOREMS OF LIOUVILLE TYPE FOR SOLUTIONS OF QUASILINEAR ELLIPTIC SYSTEMS

UDC 517.919

A. V. IVANOV

Abstract. Theorems of "Liouville type" are proved for certain classes of elliptic systems of nondivergent form. These theorems state that every linear function which is a solution defined on the whole space and which does not grow too rapidly at infinity is constant. For systems of a more general form (depending explicitly on the independent variables), it is proved that every solution which is defined on the whole space R^n and which has sufficiently small oscillation in R^n is constant. These results follow from a priori estimates obtained in the paper for the gradients of solutions of the systems under consideration.

Bibliography: 6 titles.

In an arbitrary (not necessarily bounded) region $\Omega \subset R^n$ we consider a system of the form

$$\left.\begin{array}{c} A_{ij}(x, u, |u_x|)\, u^s_{ij} = B^s(x, u, u_x), \\ s = 1, 2, \dots, N, \end{array}\right\}^{(1)} \qquad (0.1)$$

where

$$x = (x_1, \dots, x_n) \in \Omega, \quad u = (u^1, \dots, u^t, \dots, u^N), u_x = (u^1_x, \dots, u^t_x, \dots, u^N_x),$$

$$u^t_x = (u^t_1, \dots, u^t_k, \dots, u^t_n), \quad u^t_k = \frac{\partial u^t}{\partial x_k}, \quad u^s_{ij} = \frac{\partial^2 u^s}{\partial x_i \partial x_j}, \quad |u_x| = \sqrt{\sum_{t=1}^{N}\sum_{k=1}^{n} (u^t_k)^2}.$$

We assume that the coefficients A_{ij} and B^s, $s = 1, \dots, N$, belong to the class C^1. We denote the smallest eigenvalue of the matrix $\|A_{ij}(x, u, |p|)\|$ by $\lambda = \lambda(x, u, |p|)$. The ellipticity of the system (0.1) follows from the fact that $\lambda(x, u, |p|) > 0$. Here and in what follows,

$$p = (p^1, \dots, p^t, \dots, p^N), \quad p^t = (p^t_1, \dots, p^t_n),$$

$$|p| = \sqrt{\sum_{t=1}^{N}\sum_{k=1}^{n} (p^t_k)^2}.$$

It is the purpose of this paper to prove theorems of Liouville type for solutions of systems of the form (0.1). Such theorems (which consist of statements to the effect that every solution which is defined on the whole space and either

AMS (MOS) subject classifications (1970). Primary 35B05, 35B45, 35J45, 35J60.

(1) Here and in what follows, repeated subscripts indicate summation over those subscripts. However, summation over superscripts will be denoted by a Σ.

is bounded or does not grow too rapidly as $|p| \to \infty$ is constant) were established for quasilinear elliptic equations of the form

$$A_{ij}(x, u, u_x) u_{ij} = B(x, u, u_x)$$

in an earlier work of the author [1]. For equations of the form

$$A_{ij}(u_x) u_{ij} = B(u, u_x), \ \partial B/\partial u \geqslant 0,$$

analogous theorems were obtained earlier by Serrin [2]. We note also the paper of Redheffer [3], which contains a bibliography of earlier works on theorems of Liouville type.

The proof of these Liouville theorems for solutions of systems of the form (0.1) is based on local a priori bounds for the gradients of solutions of these systems. In this paper, we identify certain classes of systems of the form (0.1) for which we can establish these local bounds. These classes have nonempty intersection with those considered in [4] and include the class of elliptic systems of the form

$$A_{ij}(x, u) u^s_{ij} = B^s(x, u, u_x), \quad s = 1, 2, \dots, N,$$

which were first introduced by Ladyženskaja and Ural'ceva (cf. [5]) and which have been studied especially from the point of view of the availability of local bounds for the gradients of the solutions.

As an example, we state one of the simplest results in this paper. It is proved that every solution of the elliptic system of the form

$$\left. \begin{gathered} A_{ij}(|u_x|) u^s_{ij} = B^s(u^s, |u_x|), \quad s = 1, 2, \dots, N, \\ \frac{\partial B^s}{\partial u^s} \geqslant 0, \quad |p| \, |B^s_{|p|}| \leqslant \text{const} \, |B^s|, \end{gathered} \right\} \tag{0.2}$$

which is sufficiently smooth on the whole space R^n and which satisfies the condition

$$u = o(|x|), \quad |u_x| = O(1), \quad |x| \to \infty, \tag{0.3}$$

is constant (or, more precisely, all of its components are constant).

The above result continues to hold if certain restrictions are imposed on the matrix $\|A_{ij}\|$ in (0.2), and the condition (0.3) is replaced by

$$u = o(|x|), \quad |x| \to \infty. \tag{0.4}$$

We note that the condition (0.4) is the best possible in the sense that, even for an elliptic equation–Laplace's equation (which is included in the class of systems of the form (0.2))—there exist solutions which grow like $O(|x|)$. In fact, every linear function is a solution of Laplace's equation.

Some of the results obtained in this paper for systems having the general form (0.1) are of a new Liouville type; it is shown, for example, that every solution which is sufficiently smooth on the whole space R^n and whose oscillation in R^n does not exceed a certain quantity determined by the structure of the system (0.1) is constant.

§1. Local bounds for the gradients of solutions of elliptic systems

We introduce some notation. Let τ be an arbitrary unit vector ($|\tau| = 1$). We let $A^\tau = A_{ij}(x, u, |p|)\tau_i\tau_j$. Let $D = D(x, u, p)$. We introduce the notation

$$\delta^s D = \sum_{t=1}^{N} \frac{\partial D}{\partial u^t} u_k^t \frac{u_k^s}{|u_x^s|} + \frac{\partial D}{\partial x_k} \frac{u_k^s}{|u_x^s|}.$$

We suppose that for $x \in \Omega$, $m_1 \leq u^s \leq m_2$ ($s = 1, \ldots, N$) and $|p| > L$, where m_1, $m_2 = \text{const}$ and $L = \text{const} > 0$, the coefficients $A_{ij}(x, u, |p|)$ and $B^s(x, u, p)$ of the system (0.1) satisfy the following conditions:

$$\|A\| \leqslant \mu \frac{\mathcal{E}_1}{|p|^{2-r}}, \quad 0 \leqslant r < 2, \tag{1.1}$$

$$|p|\,|A^\tau_{|p|}| \leqslant \sqrt{\frac{\mu_0}{Nn}\frac{A^\tau \mathcal{E}_1}{|p|^2}}, \quad |\delta^s A^\tau| \leqslant \sqrt{\frac{\mu_1}{n} A^\tau \mathcal{E}_1}, \tag{1.2}$$

$$|B^s| \leqslant \frac{\mu_2}{N}\mathcal{E}_1, \quad |\delta^s B^s| \leqslant \frac{\mu_3}{N}\mathcal{E}_1 |p|, \quad |B^s_p| \leqslant \sqrt{\frac{\mu_4}{Nn}\mathcal{E}_1 \lambda}, \tag{1.3}$$

where μ, $\mu_0 = \text{const} > 0$, $\mu_i = \text{const} \geq 0$ ($i = 1, 2, 3, 4$), $s = 1, \ldots, N$, and

$$\|A\| = \sqrt{\sum_{i,j=1}^{n} A_{ij}^2}, \quad \mathcal{E}_1 = \sum_{s=1}^{N} A_{ij} p_i^s p_j^s.$$

Theorem 1.1. *Let the coefficients of the elliptical system* (0.1) *satisfy the conditions* (1.1)–(1.3), *and let* $u(x) \in C^3(\Omega)$ *be any solution of this system such that*

$$m_1 \leqslant u^s(x) \leqslant m_2, \quad x \in \Omega, \quad s = 1, 2, \ldots, N. \tag{1.4}$$

Suppose that the constants μ_1, μ_2, μ_3, μ_4 *and the quantity* $m = m_2 - m_1$ *satisfy the conditions*

$$\mu_2 \leqslant \frac{\varkappa}{4N(\varkappa+1)^2 m}, \quad \mu_1 + \mu_3 + \mu_4 \leqslant \frac{1}{32N(\varkappa+1)^2 m^2 \ln(\varkappa+1) m}, \tag{1.5}$$

where $\kappa = \exp\{4(\mu_0 + 1)N\}$. *Then at any point* x_0 *which, together with the sphere* $K_\rho(x_0) \equiv \{x \in R^n\colon |x - x_0| < \rho\}$, *is contained in* Ω, *we have*

$$\left.\begin{aligned} &|u_x(x_0)| \leqslant 2\max(L, \Theta), \\ &\Theta = \max\left\{\frac{\sqrt{2^9 \mu_0}\,(\varkappa+1)\,m}{\sigma\rho},\ \left[\frac{16\sqrt{(8N+3\sqrt{n}+10)\,\mu\,(\varkappa+1)\,m}}{\sigma\rho}\right]^{1/\sigma}\right\}, \end{aligned}\right\} \tag{1.6}$$

where $\sigma = 1 - r/2$.

Proof. We apply the operator $u_k^s\,\partial/\partial x_k$ to the kth equation of (0.1). Then for the function $v^s = \sum_{k=1}^n (u_k^s)^2$ we have, at any point of Ω,

$$\frac{1}{2}A_{ij}v_{ij}^s = A_{ij}u_{ki}^s u_{kj}^s - \frac{1}{2}\sum_{t=1}^N \frac{\partial A_{ij}}{\partial |p|} v_k^t \frac{u_k^s}{|u_x|} + \sum_{t=1}^N \frac{\partial B^s}{\partial p_l^t} u_{lk}^t u_k^s + \sqrt{v^s}\,\delta^s (B^s - A_{ij}u_{ij}^s). \tag{1.7}$$

Summing (1.7) over $s = 1, \ldots, N$, and letting $v = \sum_1^n v^s$, we get

$$\left.\begin{aligned} \frac{1}{2}A_{ij}v_{ij} = {} & \sum_{s=1}^N A_{ij}u_{ki}^s u_{kj}^s - \frac{1}{2}\sum_{s=1}^N \frac{\partial A_{ij}}{\partial |p|}\frac{u_k^s}{|u_x|} u_{ij}^s v_k \\ & + \sum_{s,t=1}^N \frac{\partial B^s}{\partial p_l^t} u_{lk}^t u_k^s + \sum_{s=1}^N \sqrt{v^s}\,\delta^s (B^s - A_{ij}u_{ij}^s). \end{aligned}\right. \tag{1.8}$$

Further, let $w = v\zeta$, where $\zeta = \zeta(\xi(x))$, $\zeta(\cdot)$ is defined and increasing in $[0, 1]$, $\zeta(0) = 0$, $\zeta(1) = 1$, and

$$\xi(x) = \begin{cases} \dfrac{\rho^2 - |x - x_0|^2}{\rho^2} & \text{in } K_\rho(x_0) \\ 0 & \text{outside of } K_\rho(x_0). \end{cases} \tag{1.9}$$

It is clear that

$$|\xi_x| \leqslant \frac{2}{\rho}, \quad \|\xi_{xx}\| \leqslant \frac{2\sqrt{n}}{\rho^2}, \tag{1.10}$$

where

$$|\xi_x| = \sqrt{\sum_{i=1}^n \xi_i^2}, \quad \|\xi_{xx}\| = \sqrt{\sum_{i,j=1}^n \xi_{ij}^2}.$$

Taking into account the facts that

$$v = \frac{w}{\zeta}, \quad v_i = \frac{w_i}{\zeta} - \frac{\zeta_i}{\zeta^2}w, \quad v_{ij} = \frac{w_{ij}}{\zeta} - \frac{w_i\zeta_j + w_j\zeta_i}{\zeta^2} - \left(\frac{\zeta_{ij}}{\zeta^2} - \frac{2\zeta_i\zeta_j}{\zeta^3}\right)w,$$

we get, in place of (1.8),

$$A_{ij}w_{ij}=2\zeta\sum_{s=1}^{N}A_{ij}u^s_{ki}u^s_{kj}-\sum_{s=1}^{N}\frac{\partial A_{ij}}{\partial|p|}\frac{u^s_k}{|u_x|}u^s_{ij}\Big(w_k-\frac{\zeta_k}{\zeta}w\Big)+2\sum_{s,t=1}^{N}\frac{\partial B^s}{\partial p^t_l}u^t_{lk}u^s_k\zeta$$
$$+2\zeta\sum_{s=1}^{N}\sqrt{v^s}\,\delta^s(B^s-A_{ij}u^s_{ij})+2A_{ij}w_i\frac{\zeta_j}{\zeta}+\Big(A_{ij}\frac{\zeta_{ij}}{\zeta}-2A_{ij}\frac{\zeta_i}{\zeta}\frac{\zeta_j}{\zeta}\Big)w. \tag{1.11}$$

Now we put $w = \Sigma_{s=1}^{N} z(u^s)\overline{w}$, where $z(\cdot) > 0$. Then, taking into account the facts that

$$w=\sum_{s=1}^{N}z(u^s)\,\overline{w},\qquad\text{where}\qquad z(\cdot)>0.$$
$$w_i=\sum_{s=1}^{N}[z'(u^s)u^s_i\overline{w}+z(u^s)\overline{w}_i],$$
$$w_{ij}=\sum_{s=1}^{N}[z''(u^s)u^s_iu^s_j\overline{w}+z'(u^s)u^s_{ij}\overline{w}+z'(u^s)u^s_i\overline{w}_j+z'(u^s)u^s_j\overline{w}_i+z(u^s)\overline{w}_{ij}],$$

we have, in place of (1.11),

$$\sum_{s=1}^{N}z(u^s)A_{ij}\overline{w}_{ij}+b_k\overline{w}_k=-\sum_{s=1}^{N}z''(u^s)A_{ij}u^s_iu^s_j\overline{w}+2\zeta\sum_{s=1}^{N}A_{ij}u^s_{ki}u^s_{kj}$$
$$-\sum_{s=1}^{N}\frac{\partial A_{ij}}{\partial|p|}u^s_{ij}\frac{u^s_k}{|u_x|}\left[\sum_{t=1}^{N}z'(u^t)u^t_k-\frac{\zeta_k}{\zeta}\sum_{t=1}^{N}z(u^t)\right]\overline{w}+2\sum_{s,t=1}^{N}\frac{\partial B^s}{\partial p^t_l}u^t_{lk}u^s_k\zeta$$
$$+2\zeta\sum_{s=1}^{N}\sqrt{v^s}\,\delta^s(B^s-A_{ij}u^s_{ij})-\sum_{s=1}^{N}z'(u^s)B^s\overline{w}+2A_{ij}\sum_{s=1}^{N}z'(u^s)u^s_i\frac{\zeta_j}{\zeta}\overline{w}$$
$$+\Big(A_{ij}\frac{\zeta_{ij}}{\zeta}-2A_{ij}\frac{\zeta_i}{\zeta}\frac{\zeta_j}{\zeta}\Big)\sum_{s=1}^{N}z(u^s)\overline{w}, \tag{1.12}$$

where we need not know the form of the b_k for what follows. We note that, in arriving at (1.12), we used the fact that $A_{ij}u^s_{ij} = B^s$.

Let x_* be an absolute maximum point of $\overline{w}(x)$. Since $\overline{w}(x) = 0$ outside of $K_\rho(x_0)$, we may assume that $x_* \in K_\rho(x_0)$. Let $\theta > 0$ be an arbitrary fixed number, and let $L > 0$ be the number in the hypothesis of the theorem (i.e. a lower bound for $|p|$ for all p for which (1.1)–(1.3) are satisfied). Then we have the following alternative:

$$\text{either } v(x_*) > L^2 \text{ and } w(x_*) > \theta^2, \tag{1.13}$$

$$\text{or at least one of the inequalities (1.13) is not true.} \tag{1.14}$$

First of all, we consider the second possibility. If $v(x_*) \le L^2$, then from the relations

$$w=\sum_{s=1}^{N} z(u^s)\bar{w},\quad w=v\zeta,\quad \zeta(x_0)=1 \tag{1.15}$$

it follows that

$$v(x_0)=w(x_0)\leqslant \max_{\Omega} w\leqslant \max\sum_{s=1}^{N} z(u^s)\bar{w}(x_*)\leqslant \frac{\max\sum_{s=1}^{N} z(u^s)}{\min\sum_{s=1}^{N} z(u^s)}\, w(x_*)$$

$$\leqslant \frac{\max\sum_{s=1}^{N} z(u^s)}{\min\sum_{s=1}^{N} z(u^s)}\, v(x_*)\leqslant \frac{\max\sum_{s=1}^{N} z(u^s)}{\min\sum_{s=1}^{N} z(u^s)}\, L^2.$$

If $w(x_*)\leq \theta^2$, then, with the help of (1.15), we get in precisely the same way that

$$v(x_0)\leqslant \frac{\max\sum_{s=1}^{N} z(u^s)}{\min\sum_{s=1}^{N} z(u^s)}\,\theta^2.$$

Thus either both of the inequalities in (1.13) hold, or

$$v(x_0)\leqslant \frac{\max\sum_{s=1}^{N} z(u^s)}{\min\sum_{s=1}^{N} z(u^s)}\max(L^2,\ \theta^2). \tag{1.16}$$

Now we show that, if we choose $z(\cdot)$ and θ properly, the first of the above alternatives cannot hold. We suppose the contrary: let both of the inequalities of (1.13) hold (for a value of θ which will be described below). Making use of (1.1)–(1.3), we transform (1.12) at the point x_*. We note that we can use these conditions at x_* because $|u_x(x_0)|>L$, which follows from the first of the conditions in (1.13). We have the following estimates (which can be proved in the same way as the analogous estimates in [1] and [6]):

$$\left|\sum_{s=1}^{N}\frac{\partial A_{ij}}{\partial|p|}u^s_{ij}\frac{u^s_k}{|u_x|}\sum_{t=1}^{N} z'(u^t)u^t_k\bar{w}\right|\leqslant\frac{\varepsilon}{2}\zeta\sum_{s=1}^{N}A_{ij}u^s_{ki}u^s_{kj}+\frac{\mu_0}{2\varepsilon}\frac{\left(\sum_{s=1}^{N}|z'(u^t)|\right)^2}{\sum_{s=1}^{N} z(u^t)}\mathscr{E}_1\bar{w}; \tag{1.17}$$

$$\left|\sum_{s=1}^{N}\frac{\partial A_{ij}}{\partial|p|}u^s_{ij}\frac{u^s_k}{|u_x|}\sum_{t=1}^{N} z(u^t)\frac{\zeta_k}{\zeta}\bar{w}\right|\leqslant\frac{\varepsilon}{2}\zeta\sum_{s=1}^{N}A_{ij}u^s_{ki}u^s_{kj}+\frac{2\mu_0}{\varepsilon}\frac{1}{v\rho^2}\left(\frac{\zeta'}{\zeta}\right)^2\sum_{s=1}^{N} z(u^s)\mathscr{E}_1\bar{w}; \tag{1.18}$$

$$2\zeta\left|\sum_{s=1}^{N}\sqrt{v^s}\,\delta^s(A_{ij}u^s_{ij})\right|\leqslant\frac{\varepsilon}{2}\zeta\sum_{s=1}^{N}A_{ij}u^s_{ki}u^s_{kj}+\frac{2}{\varepsilon}\mu_1\sum_{s=1}^{N}z(u^s)\,\mathcal{E}_1\overline{w}; \qquad (1.19)$$

$$2\left|\sum_{s=1}^{N}\frac{\partial B^s}{\partial p^t_l}u^t_{lk}u^s_k\zeta\right|\leqslant\frac{\varepsilon}{2}\zeta\lambda\sum_{s=1}^{N}(u^s_{xx})^2+\frac{2\mu_4}{\varepsilon}\sum_{s=1}^{N}z(u^s)\,\mathcal{E}_1\overline{w}; \qquad (1.20)$$

where $(u^s_{xx})^2=\sum_{i,j=1}^{n}(u^s_{ij})^2$;

$$2\zeta\left|\sum_{s=1}^{N}\sqrt{v^s}\,\delta^sB^s\right|\leqslant 2\mu_3\sum_{s=1}^{N}z(u^s)\,\mathcal{E}_1\overline{w}; \qquad (1.21)$$

$$\left|\sum_{s=1}^{N}z'(u^s)\,B^s\overline{w}\right|\leqslant\mu_2\sum_{s=1}^{N}|z'(u^s)|\,\mathcal{E}_1\overline{w}; \qquad (1.22)$$

$$2\left|A_{ij}\sum_{s=1}^{N}z'(u^s)\,u^s_i\frac{\zeta_j}{\zeta}w\right|\leqslant\frac{\left(\sum_{s=1}^{N}|z'(u^s)|\right)^2}{\sum_{s=1}^{N}z(u^s)}\mathcal{E}_1\overline{w}+4NA_{ij}\frac{\zeta_i}{\zeta}\frac{\zeta_j}{\zeta}\sum_{s=1}^{N}z(u^s)\,\overline{w}; \qquad (1.23)$$

$$(4N+2)\,A_{ij}\frac{\zeta_i}{\zeta}\frac{\zeta_j}{\zeta}\sum_{s=1}^{N}z(u^s)\,\overline{w}\leqslant 4\,(4N+2)\,\mu\frac{1}{v^\sigma\rho^2}\left(\frac{\zeta'}{\zeta}\right)^2\sum_{s=1}^{N}z(u^s)\,\mathcal{E}_1\overline{w}, \qquad (1.24)$$

where $\sigma=1-r/2\in[0,1]$;

$$2A_{ij}\frac{\zeta_{ij}}{\zeta}\sum_{s=1}^{N}z(u^s)\,\overline{w}\leqslant 2\,(4+2\sqrt{n})\,\mu\frac{|\zeta''|+|\zeta'|}{\zeta}\frac{1}{v^\sigma\rho^2}\sum_{s=1}^{N}z(u^s)\,\mathcal{E}_1\overline{w}. \qquad (1.25)$$

We put $\epsilon=1/2$ in (1.17)–(1.20). Then from (1.12) and (1.17)–(1.25) it follows that, at the point x_*,

$$\sum_{s=1}^{N}z(u^s)A_{ij}\overline{w}_{ij}\geqslant-\sum_{s=1}^{N}z''(u^s)\,A_{ij}u^s_iu^s_j\overline{w}-\left\{(\mu_0+1)\frac{\left(\sum_{s=1}^{N}|z'(u^s)|\right)^2}{\sum_{s=1}^{N}z(u^s)}+\mu_2\sum_{s=1}^{N}|z'(u^s)|\right.$$
$$+\left[4\,(\mu_1+\mu_3+\mu_4)+\frac{4\mu_0}{v\rho^2}\left(\frac{\zeta'}{\zeta}\right)^2+\frac{4\,(4N+2)\,\mu}{v^\sigma\rho^2}\left(\frac{\zeta'}{\zeta}\right)^2\right. \qquad (1.26)$$
$$\left.\left.+\frac{2\,(4+2\sqrt{n})\,\mu}{v^\sigma\rho^2}\frac{|\zeta''|+|\zeta'|}{\zeta}\right]\sum_{s=1}^{N}z(u^s)\right\}\mathcal{E}_1\overline{w}.$$

Let

$$\zeta(\xi)=\xi^{2/\sigma} \quad \text{for} \quad \xi\in[0,\ 1]. \tag{1.27}$$

Then

$$\zeta'=\frac{2}{\sigma}\xi^{2/\sigma-1},\quad \zeta''=\frac{2}{\sigma}\left(\frac{2}{\sigma}-1\right)\xi^{2/\sigma-2},\quad \frac{|\zeta'|}{\zeta}=\frac{2}{\sigma}\cdot\frac{1}{\zeta^{\sigma/2}},\quad \left(\frac{\zeta'}{\zeta}\right)^2=\frac{4}{\sigma^2}\cdot\frac{1}{\zeta^{\sigma}},$$

$$\frac{|\zeta''|+|\zeta'|}{\zeta}\leqslant\frac{6}{\sigma^2}\cdot\frac{1}{\zeta^{\sigma}}.$$

Consequently,

$$\left(\frac{\zeta'}{\zeta}\right)^2\frac{1}{v}\leqslant\frac{4}{\sigma^2 w},\quad \left(\frac{\zeta'}{\zeta}\right)^2\frac{1}{v^{\sigma}}\leqslant\frac{4}{\sigma^2 w^{\sigma}},\quad \frac{|\zeta''|+|\zeta'|}{\zeta v^{\sigma}}\leqslant\frac{6}{\sigma^2 w^{\sigma}}. \tag{1.28}$$

We also put

$$z(u)=\ln\eta,\quad \eta=\varkappa m+u-m_1, \tag{1.29}$$

where $\kappa = \text{const} \geq e$, $m_1 \leq u \leq m_2$ and $m = 1 + m_2 - m_1$, m_1 and m_2 being the constants in (1.4). We note that $\kappa m \leq \eta \leq (\kappa + 1)m$. Clearly, $z'(u) = 1/\eta$ and $z''(u) = 1/\eta^2$. The following inequalities are easily verified:

$$\begin{gathered}
-z''(u^s)\geqslant\frac{1}{(\varkappa+1)^2 m^2},\quad \sum_{s=1}^{N}|z'(u^s)|\leqslant\frac{N}{\varkappa m},\\
N\ln\varkappa m\leqslant\sum_{s=1}^{N}z(u^s)\leqslant N\ln(\varkappa+1)m,\\
\frac{\left(\sum_{s=1}^{N}|z'(u^s)|\right)^2}{\sum_{s=1}^{N}z(u^s)}\leqslant\frac{N\left(\frac{\varkappa+1}{\varkappa}\right)^2}{\ln\varkappa}\cdot\frac{1}{(\varkappa+1)^2 m^2}.
\end{gathered} \tag{1.30}$$

Now from (1.26), (1.28), (1.30) and the fact that $w(x_*) > \theta^2$ at the point x_* (cf. (1.13)), we get, at this point,

$$\begin{aligned}
\sum_{s=1}^{N}z(u^s)A_{ij}\bar{w}_{ij}\geqslant\Bigg[&\frac{1}{(\varkappa+1)^2 m^2}-\frac{(\mu_0+1)N\left(\frac{\varkappa+1}{\varkappa}\right)^2}{\ln\varkappa}\frac{1}{(\varkappa+1)^2 m^2}-\frac{\mu_2 N}{\varkappa m}-\\
&-4N(\mu_1+\mu_3+\mu_4)\ln(\varkappa+1)m-\frac{16\mu_0}{\rho^2\sigma^2\Theta^2}-\frac{(64N+24\sqrt{n}+80)\mu}{\rho^2\sigma^2\Theta^{2\sigma}}\Bigg]\mathcal{E}_1\bar{w}.
\end{aligned} \tag{1.31}$$

Let $\kappa = e^{4(\mu_0+1)N}$. Then

$$\frac{(\mu_0+1)N\left(\frac{\varkappa+1}{\varkappa}\right)^2}{\ln\varkappa}\leqslant\frac{1}{2}.$$

Now we impose on the quantities m, μ_1, μ_2, μ_3 and μ_4 the conditions

$$\mu_2 \leqslant \frac{\varkappa}{4N(\varkappa+1)^2 m}, \tag{1.32}$$

$$\mu_1+\mu_3+\mu_4 \leqslant \frac{1}{32N(\varkappa+1)^2 m^2 \ln(\varkappa+1) m}. \tag{1.33}$$

Finally, we choose θ as follows:

$$\theta = \max\left\{\frac{\sqrt{2^9\mu_0}\,(\varkappa+1)\,m}{\sigma\rho},\ \left[\frac{2^4\sqrt{(8N+3\sqrt{n}+10)\,\mu}\ (\varkappa+1)\mathrm{m}}{\sigma\rho}\right]^{1/\sigma}\right\}. \tag{1.34}$$

Then in place of (1.31) we get, at the point x_*,

$$\sum_{s=1}^{N} z(u^s)\, A_{ij}\bar{w}_{ij} \geqslant \frac{1}{16(\varkappa+1)^2 m^2}\,\mathcal{E}_1\bar{w},$$

which clearly contradicts the assumption that $\bar{w}(x)$ has a maximum at $x_* \in K_\rho(x_0)$. Consequently the bound (1.16) holds, and we have that

$$|u_x^s(x_0)| \leqslant \frac{\ln(\varkappa+1)}{\ln \varkappa}\max(L,\ \theta) \leqslant 2\max(L,\ \theta), \tag{1.35}$$

because, taking into account (1.29) and recalling our choice of κ, we have

$$\frac{\max \sum_{s=1}^{N} z(u^s)}{\min \sum_{s=1}^{N} z(u^s)} \leqslant \frac{\ln(\varkappa+1)+\ln m}{\ln\varkappa+\ln m} \leqslant \frac{\ln(\varkappa+1)}{\ln\varkappa} \leqslant \ln 2+1 \leqslant 2.$$

The value of θ in (1.35) is defined by (1.34). This proves Theorem 1.1.

We now give several other forms of Theorem 1.1, taking into account the critical role of the argument u^s for the function B^s.

We introduce the new notation

$$\tilde{\delta}^s B^s = \sum_{t\neq s} \frac{\partial B^s}{\partial u^t}\frac{u_k^t u_k^s}{|u_x|} + \frac{\partial B^s}{\partial x_k}\frac{u_k^s}{|u_x^s|}.$$

Then

$$\delta^s B^s = \tilde{\delta}^s B^s + \frac{\partial B^s}{\partial u^s}|u_x^s|. \tag{1.36}$$

Replacing $\delta^s B^s$ in (1.12) by the sum (1.36) and taking into account the fact that $\partial B^s/\partial u^s$ is multiplied by a positive factor, we find that we need only impose a one-sided restriction on the growth of $\partial B^s/\partial u^s$. In fact, we have the following theorem.

Theorem 1.1′. *If, in the hypotheses of Theorem 1.1, (1.3) is replaced by the conditions*

$$|B^s|\leqslant\frac{\mu_2}{N}\mathcal{E}_1,\quad |\delta^s B^s|\leqslant\frac{\mu_3}{2N}\mathcal{E}_1|p|,\quad \frac{\partial B^s}{\partial u^s}\geqslant-\frac{\mu_3}{2N}\mathcal{E}_1,\quad |B^s_p|\leqslant\sqrt{\frac{\mu_4}{Nn}\mathcal{E}_1\lambda},\tag{1.37}$$

then the conclusion of Theorem 1.1 still holds.

We consider now a system of the form

$$\left.\begin{array}{c}A_{ij}(x,\ u,\ |u_x|)u^s_{ij}=B^s(x,\ u,\ |u_x|),\\ s=1,\ 2,\ \ldots,\ N.\end{array}\right\}\tag{1.38}$$

Let the coefficients $A_{ij}(x, u, |p|)$, $B^s(x, u, |p|)$, $s = 1, 2, \ldots, N$, for $x \in \Omega$, $m_1 \leq u^s \leq m_2$, $s = 1, 2, \ldots, N$ and $|p| > L$ satisfy the conditions (1.1) and (1.2) and the conditions

$$\operatorname{tr}\|A_{ij}\|\cdot|p|^2\leqslant\nu\mathcal{E}_1,\tag{1.39}$$

$$B^s(x,\ u,\ |p|)=B^s_1(x,\ u,\ |p|)+B^s_2(x,\ u,\ |p|),\quad \frac{\partial B^s}{\partial u^s}\geqslant-\frac{\mu_3}{2N}\mathcal{E}_1,\tag{1.40}$$

$$|B^s_1|\leqslant\frac{\mu_2}{N}\mathcal{E}_1,\quad |\delta^s B^s_1|\leqslant\frac{\mu_3}{2N}\mathcal{E}_1|p|,\quad |p||(B^s_1)_{|p|}|\leqslant\frac{\mu_4}{N}\mathcal{E}_1,\tag{1.41}$$

$$|p||(B^s_2)_{|p|}|\leqslant\nu_1|B^s_2|,\quad |\delta B^s_2|\leqslant\nu_2|B^s_2||p|,\tag{1.42}$$

where $\mu_2, \mu_3, \mu_4, \nu, \nu_1$ and ν_2 are nonnegative constants.

Theorem 1.2. *Let the coefficients of the elliptic system* (1.38) *satisfy the conditions* (1.1), (1.2), *and* (1.40)–(1.42), *and let* $u(x) \in C^3(\Omega)$ *be an arbitrary solution of this system such that* (1.4) *holds. Suppose that the constants* μ_1, μ_2, μ_3, μ_4, ν_2 *and the quantity* $m = m_2 - m_1$ *satisfy the conditions*

$$\left.\begin{array}{c}\nu_1\mu_2+\mu_4\leqslant\dfrac{\varkappa}{4N(\varkappa+1)^2m},\\ 4\mu_1+2\mu_3+2\nu_2\mu_2+16\nu\nu_2^2\leqslant\dfrac{1}{8N(\varkappa+1)^2m^2\ln(\varkappa+1)m},\end{array}\right.\tag{1.43}$$

where $\kappa = \exp\{4(\mu_0 + 1 + 4\nu_0\nu_1^2 + 8\nu_0)N\}$. *Then at any point* x_0 *which, together with the sphere* $K_\rho(x_0)$, *is contained in* Ω, *we have*

$$\left.\begin{array}{l}|u_x(x_0)|\leqslant 2\max(L,\ \theta),\\ \theta=\max\left\{\dfrac{16\sqrt{3}(\mu_0+\nu\nu_1^2)(\varkappa+1)m}{\sigma\rho},\ \left[\dfrac{16\sqrt{3}(8N+3\sqrt{n}+10)\mu(\varkappa+1)m}{\sigma\rho}\right]^{1/\sigma},\right.\\ \qquad\left.\left[\dfrac{3\cdot 2^6(\nu_1\mu_2+\mu_4)(\varkappa+1)^2m^2}{\sigma\rho}\right]^{1/\sigma}\right\},\end{array}\right\}\tag{1.44}$$

where $\sigma = 1 - r/2$.

Proof. We make use of the proof of Theorem 1.1 (remembering the alternate version in Theorem 1.1$'$) and modify it where necessary. In (1.12), instead of

$$2 \sum_{s,\, t=1}^{N} \frac{\partial B^s}{\partial p_l^t} u_{lk}^t u_k^s \zeta,$$

we will have

$$\sum_{s=1}^{N} \frac{\partial B^s}{\partial |p|} \frac{u_k^s}{|u_x|} v_k.$$

As a result of the substitutions (1.15), this expression takes the form

$$\sum_{s=1}^{N} \frac{\partial B^s}{\partial |p|} \frac{u_k^s}{|u_x|} \sum_{t=1}^{N} z'(u^t) u_k^t \overline{w} - \sum_{s=1}^{N} \frac{\partial B^s}{\partial |p|} \frac{u_k^s}{|u_x|} \frac{\zeta_k}{\zeta} \sum_{t=1}^{N} z(u^t) \overline{w} + \tilde{b}_k \overline{w}_k, \tag{1.45}$$

where we will not need to know the form of the $\tilde{b}_k$ for what follows. We estimate (1.45) at a point x_* where $\overline{w}(x)$ has a maximum, assuming that (1.13) is satisfied there. Taking (1.40) and (1.41) into account, we get

$$\left| \sum_{s=1}^{N} \frac{\partial B_1^s}{\partial |p|} \frac{u_k^s}{|u_x|} \sum_{t=1}^{N} z'(u^t) u_k^t \overline{w} \right| \leqslant \mu_4 \sum_{s=1}^{N} |z'(u^s)| \mathcal{E}_1 \overline{w}, \tag{1.46}$$

$$\left| \sum_{s=1}^{N} \frac{\partial B_1^s}{\partial |p|} \frac{u_k^s}{|u_x|} \frac{\zeta_k}{\zeta} \sum_{t=1}^{N} z(u^t) \overline{w} \right| \leqslant 2\mu_4 \frac{1}{\sqrt{v}\,\rho} \frac{|\zeta'|}{\zeta} \sum_{s=1}^{N} z(u^s) \overline{w}. \tag{1.47}$$

To estimate the remaining expression

$$\sum_{s=1}^{N} \frac{\partial B_2^s}{\partial |p|} \frac{u_k^s}{|u_x|} \sum_{t=1}^{N} z'(u^t) u_k^t \overline{w} - \sum_{s=1}^{N} \frac{\partial B_2^s}{\partial |p|} \frac{u_k^s}{|u_x|} \frac{\zeta_k}{\zeta} \sum_{t=1}^{N} z(u^t) \overline{w} \tag{1.48}$$

we make use of the fact that

$$\frac{\partial B_2^s}{\partial |p|} = \frac{\dfrac{\partial B_2^s}{\partial |p|} \sqrt{v}}{B_2^s} \cdot \frac{A_{ij} u_{ij}^s - B_1^s}{\sqrt{v}}, \tag{1.49}$$

which is true because $B^s = B_1^s + B_2^s$ and $B^s = A_{ij} u_{ij}^s$. We assume that the denominator B_2^s in (1.49) is different from zero, because, if it were not, it would follow from (1.42) that $\partial B_2^s / \partial |p| = 0$, and therefore the expression (1.48) would vanish. Now taking into account (1.41) and the inequality

$$\frac{\left| \dfrac{\partial B_2^s}{\partial |p|} \right| \sqrt{v}}{|B_2^s|} \leqslant \nu_1, \tag{1.50}$$

which follows from (1.42), we get

$$\left|\sum_{s=1}^{N}\frac{\partial B_2^s}{\partial|p|}\frac{u_k^s}{|u_x|}\sum_{t=1}^{N}z'(u^t)u_k^t\bar{w}\right|\leqslant\nu_1\mu_2\sum_{s=1}^{N}|z'(u^s)|\mathcal{E}_1\bar{w}+\frac{\varepsilon}{8}\zeta\sum_{s=1}^{N}A_{ij}u_{ki}^su_{kj}^s+\frac{2\nu_1^2}{\varepsilon}\frac{\left(\sum_{s=1}^{N}|z'(u^s)|\right)^2}{\sum_{s=1}^{N}z(u^s)}\operatorname{tr}\|A_{ij}\|v\bar{w}, \tag{1.51}$$

$$\left|\sum_{s=1}^{N}\frac{\partial B_2^s}{\partial|p|}\frac{u_k^s}{|u_x|}\frac{\zeta_k}{\zeta}\sum_{t=1}^{N}z(u^t)\bar{w}\right|\leqslant 2\nu_1\mu_2\frac{1}{\sqrt{v}\rho}\frac{|\zeta'|}{\zeta}\sum_{s=1}^{N}z(u^s)\bar{w}+\frac{\varepsilon}{8}\zeta\sum_{s=1}^{N}A_{ij}u_{ki}^su_{kj}^s+\frac{2\nu_1^2}{\varepsilon}\left(\frac{\zeta'}{\zeta}\right)^2\frac{1}{v\rho^2}\sum_{s=1}^{N}z(u^s)\operatorname{tr}\|A_{ij}\|v\bar{w}. \tag{1.52}$$

In a similar way, we estimate the term

$$2\zeta\sum_{s=1}^{N}\sqrt{v^s}\tilde{\delta}^sB^s=2\zeta\sum_{s=1}^{N}\sqrt{v^s}\tilde{\delta}^sB_1^s+2\zeta\sum_{s=1}^{N}\sqrt{v^s}\tilde{\delta}^sB_2^s.$$

From (1.41) it follows that

$$2\zeta\left|\sum_{s=1}^{N}\sqrt{v^s}\tilde{\delta}^sB_1^s\right|\leqslant\mu_3\sum_{s=1}^{N}z(u^s)\mathcal{E}_1\bar{w}. \tag{1.53}$$

Writing

$$\tilde{\delta}^sB_2^s=\frac{\tilde{\delta}^sB_2^s}{B_2^s\sqrt{v}}(A_{ij}u_{ij}^s-B_1^s)\sqrt{v},$$

at x_*, where we may assume that B_2^s is different from zero, taking into account the fact that

$$\left|\frac{\tilde{\delta}^sB_2^s}{B_2^s\sqrt{v}}\right|\leqslant\nu_2$$

by virtue of (1.42), and using (1.41), we get

$$2\zeta\left|\sum_{s=1}^{N}\sqrt{v^s}\tilde{\delta}^sB_2^s\right|\leqslant 2\nu_2\mu_2\sum_{s=1}^{N}z(u^s)\mathcal{E}_1\bar{w}+\frac{\varepsilon}{2}\zeta\sum_{s=1}^{N}A_{ij}u_{ki}^su_{kj}^s+\frac{8\nu_2^2}{\varepsilon}\operatorname{tr}\|A_{ij}\|v\sum_{s=1}^{N}z(u^s)\bar{w}. \tag{1.54}$$

The term

$$2\zeta \sum_{s=1}^{N} \frac{\partial B^s}{\partial u^s} v$$

will also be present in (1.12) (cf. (1.36)). By virtue of (1.40), we can estimate it from below:

$$2\zeta \sum_{s=1}^{N} \frac{\partial B^s}{\partial u^s} v \geqslant -\mu_3 \sum_{s=1}^{N} z(u^s)\,\mathcal{E}_1 \overline{w}. \tag{1.55}$$

Finally, in place of

$$\sum_{s=1}^{N} z'(u^s) B^s \overline{w}$$

we write

$$\sum_{s=1}^{N} z'(u^s) A_{ij} u^s_{ij} \overline{w}$$

and estimate it:

$$\left|\sum_{s=1}^{N} z'(u^s) A^s_{ij} u^s_{ij} \overline{w}\right| \leqslant \frac{\varepsilon}{8} \sum_{s=1}^{N} A_{ij} u^s_{ki} u^s_{kj} + \frac{4}{\varepsilon} \frac{\left(\sum_{s=1}^{N} |z'(u^s)|\right)^2}{\sum_{s=1}^{N} z(u^s)} \operatorname{tr} \|A_{ij}\| \overline{w}. \tag{1.56}$$

Putting $\epsilon = \frac{1}{2}$ in (1.51), (1.52), (1.54) and (1.56), and taking (1.39) into account, we obtain from the equality of the form (1.12) and from the estimates obtained here and in the proof of Theorem 1.1 the following inequality at x_*:

$$\begin{aligned}
\sum_{s=1}^{N} z(u^s) A_{ij} \overline{w}_{ij} \geqslant -\sum_{s=1}^{N} z''(u^s) A_{ij} u^s_i u^s_j \overline{w} - \Bigg\{ (\mu_0 + 1 + 4\nu\nu_1^2 + 8\nu) \frac{\left(\sum_{s=1}^{N} |z'(u^s)|\right)^2}{\sum_{s=1}^{N} z(u^s)} \\
+ (\nu_1\mu_2 + \mu_4) \sum_{s=1}^{N} |z'(u^s)| + \Bigg[(4\mu_1 + 2\mu_3 + 2\nu_2\mu_2 + 16\nu\nu_2^2) + \frac{4\mu_0}{\nu\rho^2}\left(\frac{\zeta'}{\zeta}\right)^2 \\
+ \frac{4(4N+2)\mu}{\nu^\sigma \rho^2}\left(\frac{\zeta'}{\zeta}\right)^2 + \frac{2(4+2\sqrt{n})\mu}{\nu^\sigma \rho^2} \cdot \frac{|\zeta''| + |\zeta'|}{\zeta} \\
+ \frac{2(\nu_1\mu_2 + \mu_4)}{\sqrt{\nu}\rho} \frac{|\zeta'|}{\zeta} + \frac{4\nu\nu_1^2}{\nu\rho^2}\left(\frac{\zeta'}{\zeta}\right)^2 \Bigg] \sum_{s=1}^{N} z(u^s) \Bigg\} \mathcal{E}_1 \overline{w}.
\end{aligned} \tag{1.57}$$

We take $\zeta(\xi)$ to be the function (1.27) and $z(u)$ to be the function (1.29). Then, using (1.28) and (1.30) and the inequality $w(x_*) > \theta^2$, we get from (1.57) that

$$\sum_{s=1}^{N} z(u^s) A_{ij}\bar{w}_{ij} \geqslant \Bigg[\frac{1}{(\varkappa+1)^2 m^2} - \frac{(\mu_0+1+4\nu\nu_1^2+8\nu) N\left(\frac{\varkappa+1}{\varkappa}\right)^2}{\ln\varkappa\,(\varkappa+1)^2 m^2}$$

$$-\frac{(\nu_1\mu_2+\mu_4)N}{\varkappa m} - (4\mu_1+2\mu_3+2\nu_2\mu_2+16\nu\nu_2^2) N \ln(\varkappa+1) m \qquad (1.58)$$

$$-\frac{16(\mu_0+\nu\nu_1^2)}{\sigma^2\rho^2\theta^2} - \frac{8(8N+3\sqrt{n}+10)\mu}{\sigma^2\rho^2\theta^{2\sigma}} - \frac{4(\nu_1\mu_2+\mu_4)}{\sigma\rho\theta^\sigma}\Bigg] \mathcal{E}_1\bar{w}.$$

Let

$$\varkappa = \exp\{4(\mu_0+1+4\nu\nu_1^2+8\nu)N\}. \qquad (1.59)$$

Then the second term on the right-hand side of (1.58) is bounded below by

$$-\frac{1}{2}\frac{1}{(\varkappa+1)^2 m^2}.$$

Now we require that the quantities $m = 1 + m_1 - m_2$, μ_1, μ_2, μ_3, μ_4, ν, ν_1 and ν_2 satisfy the conditions

$$\nu_1\mu_2+\mu_4 \leqslant \frac{\varkappa}{4N(\varkappa+1)^2 m}, \qquad (1.60)$$

$$4\mu_1+2\mu_3+2\nu_2\mu_2+16\nu\nu_2^2 \leqslant \frac{1}{8N(\varkappa+1)^2 m^2 \ln(\varkappa+1) m}, \qquad (1.61)$$

and we take θ to be

$$\theta = \max\left\{\frac{16\sqrt{3(\mu_0+\nu\nu_1^2)}(\varkappa+1)m}{\sigma\rho}, \left[\frac{16\sqrt{3(8N+3\sqrt{n}+10)\mu}(\varkappa+1)m}{\sigma\rho}\right]^{1/\sigma}, \left[\frac{3\cdot 2^6(\nu_1\mu_2+\mu_4)(\varkappa+1)^2 m^2}{\sigma\rho}\right]^{1/\sigma}\right\}. \qquad (1.62)$$

Then from (1.57) we get the impossible inequality

$$\sum_{s=1}^{N} z(u^s) A_{ij}\bar{w}_{ij} \geqslant \frac{1}{16(\varkappa+1)^2 m^2}\mathcal{E}_1\bar{w}$$

at a point x_* where $\bar{w}(x)$ has a maximum. This means that it is impossible for the inequalities (1.13) to hold simultaneously. Therefore (1.16) holds for values of θ determined by (1.62). But then we can establish (1.44) as in the proof of Theorem 1.1 (cf. (1.35)). This proves Theorem 1.2.

We suppose now that the region Ω is the whole space R^n. In this case, every solution $u(x) \in C^3(R^n)$ of the system (0.1) is called an entire solution (as in [1] and [2]). It is quite simple to obtain the following results from Theorems 1.1 and 1.2.

Theorem 1.3. *For* $x \in R^n$, $\forall u$ *and* $|p| > L$, *where* $L = \text{const} > 0$, *let the coefficients of the elliptic system* (0.1) *satisfy the conditions* (1.1) *and* (1.2) *for* $\mu_1 = 0$ *and the condition* (1.37) *for* $\mu_2 = \mu_3 = \mu_4 = 0$. *Let* $u(x)$ *be an entire solution of the system* (0.1) *which satisfies the condition*

$$u = o(|x|), \quad |x| \to \infty. \tag{1.63}$$

Then at any point $x_0 \in R^n$

$$|u_x(x_0)| \leqslant 2L. \tag{1.64}$$

Proof. First of all, we indicate precisely how (1.63) is to be understood. We assume that there exists an increasing function $\phi(\rho)$ such that $|u^s(x)| \leq C\phi(|x|)$, where $C = \text{const} > 0$, $s = 1, \dots, N$, and $\phi(\rho)/\rho \to 0$ as $\rho \to \infty$. But then for any fixed point $x_0 \in R^n$ it follows from (1.63) that

$$\rho^{-1} \max_{x \in K_\rho(x_0)} |u(x)| \to 0 \quad \text{as} \quad \rho \to \infty. \tag{1.65}$$

Let

$$m_1^\rho = \min_{\substack{K_\rho(x_0) \\ s=1,\dots,N}} u^s(x), \qquad m_2^\rho = \max_{\substack{K_\rho(x_0) \\ s=1,\dots,N}} u^s(x),$$

$$m^\rho = m_2^\rho - m_1^\rho, \qquad \bar{m}^\rho = \max_{\substack{K_\rho(x_0) \\ s=1,\dots,N}} |u^s(x)|.$$

Since $m^\rho = m_2^\rho - m_1^\rho \leq |m_2^\rho| + |m_1^\rho| \leq 2\bar{m}^\rho$, it follows from (1.65) that

$$\frac{m^\rho}{\rho} \to 0 \quad \text{as} \quad \rho \to \infty. \tag{1.66}$$

Applying the result of Theorem 1.1 to the solution $u(x)$ in the sphere $K_\rho(x_0)$ (it is clear that all of the hypotheses of Theorem 1.1 are satisfied), we get (1.5) with θ given by

$$\theta \equiv \theta_\rho = \max\left\{\frac{\sqrt{2^9 \mu_0}\,(\varkappa + 1)}{\sigma} \cdot \frac{m^\rho}{\rho},\ \frac{16\sqrt{(8N + 3\sqrt{n} + 10)\,\mu\,(\varkappa + 1)}}{\sigma}\left(\frac{m^\rho}{\rho}\right)^{1/\sigma}\right\}. \tag{1.67}$$

From (1.66) it follows that $\theta_\rho \to 0$ as $\rho \to \infty$. Therefore, letting $\rho \to \infty$ in the inequality $|u_x(x_0)| \leq 2 \max(L, \theta_\rho)$, we get (1.64). This proves Theorem 1.3.

Theorem 1.4. *For* $x \in R^n$, $\forall u$ *and* $|p| > L$, *let the coefficients of the elliptic system* (1.38) *satisfy* (1.1) *and* (1.2) *for* $\mu_1 = 0$, (1.40) *for* $\mu_3 = 0$, (1.41) *for* $\mu_2 = \mu_3 = \mu_4 = 0$, *and* (1.42) *for* $\nu_2 = 0$. *Let* $u(x)$ *be an entire solution of the system* (1.38) *satisfying* (1.63). *Then* (1.64) *holds at every point.*

Proof. Applying the result of Theorem 1.2 to the solution $u(x)$ in the sphere $K_\rho(x_0)$, we get the bound $|u_x(x_0)| \leq 2 \max(L, \theta_\rho)$, where

$$\theta_\rho = \max\left\{\frac{16\sqrt{3(\mu_0 + \nu\nu_1^2)}\,(\varkappa + 1)}{\sigma} \cdot \frac{m^\rho}{\rho},\ \frac{16\sqrt{3\,(8N + 3\sqrt{n} + 10)\,\mu\,(\varkappa + 1)}}{\sigma}\left(\frac{m^\rho}{\rho}\right)^{1/\sigma}\right\}.$$

From (1.66) it follows that $\theta_\rho \to 0$ as $\rho \to \infty$. Therefore the estimate of $|u_x(x_0)|$ holds. This proves Theorem 1.4.

§2. Theorems of Liouville type for elliptic systems

In the first two theorems included here, we obtain results of the following kind: if the oscillation of an entire solution of an elliptic system is sufficiently small, the solution is constant.(2) Hence it follows that a system of the kind under investigation can have entire solutions (different from constants) only if they have sufficiently large oscillations.

Theorem 2.1. *Suppose that, for $x \in R^n$, $\forall u$ and for $|p| > l$, where l is any positive number, the coefficients of the system* (0.1) *satisfy the conditions*

$$\|A\| \leqslant \mu \frac{\mathcal{E}_1}{|p|^{2-r}}, \quad 0 \leqslant r < 2, \quad |p|\,|A^\tau_{|p|}| \leqslant \left(\frac{\mu_0}{Nn}\frac{A^\tau \mathcal{E}_1}{|p|^2}\right)^{1/2},$$
$$|\delta^s A^\tau| \leqslant \left(\frac{\mu_1}{n} A^\tau \mathcal{E}_1\right)^{1/2},$$
$$|B^s| \leqslant \frac{\mu_2}{N}\mathcal{E}_1, \quad |\tilde{\delta}^s B^s| \leqslant \frac{\mu_3}{2N}\mathcal{E}_1 |p|, \quad \frac{\partial^s B^s}{\partial u^s} \geqslant -\frac{\mu_3}{2N}\mathcal{E}_1, \quad |B^s| \leqslant \left(\frac{\mu_4}{Nn}\mathcal{E}_1 \lambda\right)^{1/2}, \tag{2.1}$$
$$s = 1, 2, \ldots, N.$$

where μ and μ_0 are positive constants which depend, in general, on l, and μ_1, μ_2, μ_3 and μ_4 are fixed nonnegative numbers. Let $u(x)$ be an entire solution of the system (0.1) *whose oscillation*

$$\operatorname*{osc}_{R^n} u(x) = \max_{s=1,2,\ldots,N} \operatorname*{osc}_{R^n} u^s(x)$$

satisfies the conditions

$$m \leqslant \frac{\varkappa}{4N(\varkappa+1)^2 \mu_2}, \quad m^2 \ln(\varkappa+1) m \leqslant \frac{1}{32N(\varkappa+1)^2(\mu_1+\mu_3+\mu_4)}, \tag{2.2}$$

where $m = 1 + \operatorname{osc}_{R^n} u(x)$ and $\kappa = e^{4(\mu_0+1)N}$. *Then $u(x)$ is constant.*

Proof. For a fixed $l > 0$, all of the hypotheses of Theorem 1.1$'$ are satisfied, with $\Omega = R^n$ and $L = l$. In particular, (1.5) follows from (2.2). Therefore at any point $x_0 \in R^n$ and for any $\rho > 0$ we have $|u_x(x_0)| \leq 2\max(l, \theta)$, where $\theta = \theta(\rho)$ approaches 0 as $\rho \to \infty$, as can easily be seen from (1.34). Therefore $|u_x(x_0)| \leq 2l$, whence it follows that $|u_x(x_0)| = 0$ at any point $x_0 \in R^n$ by virtue of the arbitrariness of $l > 0$. Thus for any $s = 1, \ldots, N$ we have that $|u^s_x| \equiv 0$ in R^n, from which the conclusion of Theorem 2.1 follows.

Theorem 2.2. *Suppose that, for $x \in R^n$, $\forall u$ and for $|p| > l$, where l is any positive number, the coefficients of the system* (1.38) *satisfy the conditions*

(2) For brevity, we say that a solution is constant if all of its components are constant.

$$\left.\begin{aligned}
&\|A\|\leqslant\mu\frac{\mathcal{E}_1}{|p|^{2-r}},\quad 0\leqslant r<2,\quad \operatorname{tr}\|A_{ij}\||p|^2\leqslant\nu\mathcal{E}_1,\\
&|p||A^{\tau}_{|p|}|\leqslant\sqrt{\frac{\mu_0}{Nn}\frac{A^{\tau}\mathcal{E}_1}{|p^2|}},\quad |\delta^s A^{\tau}|\leqslant\sqrt{\frac{\mu_1}{n}A^{\tau}\mathcal{E}_1},\\
&B^s(x,u,|p|)=B^s_1(x,u,|p|)+B^s_2(x,u,|p|),\quad \frac{\partial B^s}{\partial u^s}\geqslant-\frac{\mu_3}{2N}\mathcal{E}_1,\\
&|B^s_1|\leqslant\frac{\mu_2}{N}\mathcal{E}_1,\quad |\delta^s B^s_1|\leqslant\frac{\mu_3}{2N}\mathcal{E}_1|p|,\quad |p|(B^s_1)_{|p|}\leqslant\frac{\mu_4}{N}\mathcal{E}_1,\\
&|p|(B^s_2)_{|p|}\leqslant\nu_1|B^s_2|,\quad |\delta B^s_2|\leqslant\nu_2|B^s_2||p|,
\end{aligned}\right\}\qquad(2.3)$$

where the constants μ, $\mu_0>0$ *and* ν, $\nu_1\geq 0$ *depend, in general, on* l, *and the remaining constants* μ_1, μ_2, μ_3, μ_4 *and* ν_2 *are fixed nonnegative numbers. Let* $u(x)$ *be an entire solution of the system* (1.38) *whose oscillation* $\operatorname{osc}_{R^n}u(x)$ *satisfies the conditions*

$$\begin{aligned}
&m\leqslant\frac{\varkappa}{4N(\varkappa+1)^2(\nu_1\mu_2+\mu_4)},\\
&m^2\ln(\varkappa+1)m\leqslant\frac{1}{8N(\varkappa+1)^2(4\mu_1+2\mu_3+2\nu_2\mu_2+16\nu\nu_2^2)},
\end{aligned}\qquad(2.4)$$

where

$$m=1+\operatorname*{osc}_{R^n}u(x),\quad \varkappa=\exp\{4(\mu_0+1+4\nu\nu_1^2+8\nu_0)\mu\}.$$

Then $u(x)$ *is constant.*

Proof. For fixed $l>0$, all of the hypotheses of Theorem 1.2 are satisfied, with $\Omega=R^n$ and $L=l$. Therefore at any point $x_0\in R^n$ and for any $\rho>0$ we have $|u_x(x_0)|\leq 2\max(l,\theta)$, where $\theta=\theta(\rho)$ approaches 0 as $\rho\to\infty$, as can easily be seen from (1.44). Hence, as in the preceding case, it follows that $u(x)$ is constant.

Remark. It is clear that Theorem 2.1 is nontrivial only if μ_1, μ_2, μ_3 and μ_4 satisfy the inequalities

$$1<\frac{\varkappa}{4N(\varkappa+1)^2\mu_2},\quad \ln(\varkappa+1)<\frac{1}{32N(\varkappa+1)^2(\mu_1+\mu_2+\mu_3)}.$$

and Theorem 2.2 is nontrivial only if μ_1, μ_2, μ_3, μ_4 and ν_2 satisfy the inequalities

$$1<\frac{\varkappa}{4N(\varkappa+1)^2(\nu_1\mu_2+\mu_4)},\quad \ln(\varkappa+1)<\frac{1}{8N(\varkappa+1)^2(4\mu_1+2\mu_3+2\nu_2\mu_2+16\nu_0\nu_2^2)}.$$

Theorem 2.3. *Suppose that, for* $x\in R^n$, $\forall u$ *and for* $|p|>l$, *where* l *is an arbitrary positive number, the coefficients of a system of the form* (0.2) *satisfy the conditions*

$$\|A\|\leqslant\mu\frac{\mathcal{E}_1}{|p|^{2-r}},\quad 0\leqslant r<2,\quad \operatorname{tr}\|A_{ij}\||p|^2\leqslant\nu\mathcal{E}_1,\quad |p||A^{\tau}_{|p|}|\leqslant\left(\frac{\mu_0A^{\tau}\mathcal{E}_1}{Nn|p|^2}\right)^{1/2},\qquad(2.5)$$

$$\frac{\partial B^s}{\partial u^s} \geqslant 0, \quad |p|\,|B^s_{|p|}| \leqslant \nu_1 |B^s|, \tag{2.6}$$

where the constants μ, $\mu_0 > 0$ *and* ν, $\nu_1 \geq 0$ *depend, in general, on* l. *Let* $u(x)$ *be an entire solution of the system* (0.2) *for which*

$$u = o(|x|), \quad |x| \to \infty. \tag{2.7}$$

Then $u(x)$ *is constant.*

Proof. For a fixed $l > 0$, all of the hypotheses of Theorem 1.4 are satisfied, with $L = l$. Then from that theorem it follows that, at an arbitrary point $x_0 \in R^n$ we have $|u_x(x_0)| \leq l$, whence, by the arbitrariness of x_0 and l, it follows that $|u^s_x| \equiv c_s$, where $c_s = \text{const}$, $s = 1, \ldots, N$. Therefore $u(x)$ is constant. This proves Theorem 2.3.

If, in addition to restricting the growth of the solution as $|x| \to \infty$, we assume that $|u_x|$ is bounded in R^n, we can drop the restrictions (2.5) on the structure of $\|A_{ij}\|$. Suppose, for example, that $\max_{R^n}|u_x| \leq M$, $M = \text{const}$. Since the upper bounds of the left-hand sides in (2.5), taken over the set $|p| > l$, are then bounded above by a certain constant $c(l) > 0$, and the lower bounds of the right-hand sides in (2.5), taken over the same set $|p| > l$, are bounded below by some positive constant $d(l)$, it is easy to choose values of μ, ν and μ_0 for which (2.5) are satisfied. Then from Theorem 2.3 we deduce the following result.

Theorem 2.4. *Suppose that, for* $x \in R^n$, $\forall u$ *and for* $|p| > l$, *where* l *is an arbitrary positive number, the right-hand sides of the elliptic system* (0.2) *satisfy* (2.6), *and the matrix* $\|A_{ij}(|p|)\|$ *is arbitrary. Let* $u(x)$ *be an entire solution of the system* (0.2) *for which*

$$u = o(|x|), \quad |x| \to \infty, \quad \max_{R^n} |u_x| \leqslant M, \quad M = \text{const}. \tag{2.8}$$

Then $u(x)$ *is constant.*

Theorems 2.1 and 2.2 admit similar modifications.

In conclusion, we note that results which are analogous to Theorems 1.1, 1.3, and 2.1, and also (in the case $B^s_1 = 0$, $s = 1, \ldots, N$) to Theorems 2.3 and 2.4 can be proved similarly for quasilinear parabolic systems of the form

$$\left.\begin{array}{c} -u^s_t + A_{ij}(x, t, u, |u_x|)\, u^s_{ij} = B^s(x, t, u, u_x) \\ r = 1, 2, \ldots, N. \end{array}\right\}$$

BIBLIOGRAPHY

1. A. V. Ivanov, *Local estimates for the first derivatives of the solutions of second order quasilinear elliptic equations, and their application to Liouville type theorems*, Zap. Naučn. Sem. Leningrad. Otdel. Mat. Inst. Steklov. (LOMI) 30 (1972), 40–50 = J. Soviet Math. 4 (1975), 335–344. MR 48 #11754.

2. James B. Serrin, Jr., *Entire solutions of nonlinear Poisson equations*, Proc. London Math. Soc. (3) 24 (1972), 348–366. MR 44 #7146.

3. Raymond M. Redheffer, *On the inequality* $\Delta u \geq f(u, |\text{grad } u|)$, J. Math. Anal. Appl. 1 (1960), 277–299. MR 25 #3184.

4. A. V. Ivanov, *Local estimates of the maximum modulus of the first derivatives of the solutions of quasilinear nonuniformly elliptic and nonuniformly parabolic equations and of a system of general form*, Trudy Mat. Inst. Steklov. 110 (1970), 45–64 = Proc. Steklov Inst. Math. 110 (1970), 48–71 (1972). MR 45 #2318.

5. O. A. Ladyženskaja and N. N. Ural′ceva, *Linear and quasi-linear equations of elliptic type*, 2nd ed., "Nauka", Moscow, 1973; English transl. of 1st ed., Academic Press, New York, 1968. MR 39 #5941.

6. A. V. Ivanov, *The solvability of the Dirichlet problem for certain classes of second order elliptic systems*, Trudy Mat. Inst. Steklov. 125 (1973), 56–87 = Proc. Steklov Inst. Math. 125 (1973), 49–79 (1975). MR 50 #7809.

Translated by H. T. JONES

Trudy Mat. Inst. Steklov.
127 (1975)

Proc. Steklov Inst. Math.
127 (1975)

THE NUMBER OF SOLUTIONS OF A CERTAIN DIFFERENCE SCHEME

UDC 517.949.8

N. K. KORENEV

Abstract. It is proved that, for sufficiently small mesh length, a certain difference scheme approximating the stationary Navier-Stokes equation has the same number of solutions as the exact problem, provided the latter has a finite number of solutions and the Fréchet derivative of the exact operator has a bounded inverse at each of these solutions.

Bibliography: 4 titles.

In this article, we consider the relation between the number of solutions of a problem and the number of solutions of a certain difference scheme approximating the problem for the case of the stationary Navier-Stokes equation.

The application of abstract theorems in this situation is difficult, since the operator corresponding to the difference scheme and its derivative do not approximate the operator corresponding to the exact problem and its Fréchet derivative in the uniform operator topology even at isolated points.

Let us introduce the following notation. Ω is a bounded domain in R^3. The space $L_p(\Omega)$ is the space of functions the pth powers of which are integrable over Ω, and $\| \ \|_p$ is the norm in $L_p(\Omega)$. The space $\mathcal{H}(\Omega)$ is the closure in the Dirichlet integral metric of the set of smooth solenoidal vector-valued functions with compact support,

$$[\varphi, \psi] = \int_\Omega D_i\varphi_j D_i\psi_j dx,$$

$$|\varphi| = [\varphi, \varphi]^{1/2},$$

D_i being the operation of generalized differentiation with respect to the variable x_i. The set $\overline{\Omega}_h$ is the closure of the set of cubes $\omega^h(k) \equiv \omega^h(k_1, k_2, k_3) = \{x: k_i h < x_i < (k_i + 1)h,\ i = 1, 2, 3\}$ which are entirely contained in Ω, and $\partial\Omega_h$ is its boundary. The symbols $\overline{\Omega}_h$ and $\partial\Omega_h$ represent the respective sets of vertices of the cubes. The spaces L_p^h and $\mathcal{H}^h$ are the imitations of the spaces $L_p(\Omega)$ and $\mathcal{H}(\Omega)$ on the grid. The condition of being differentially solenoidal in $\mathcal{H}^h$ can be written in the form $\operatorname{div}_h \phi_h(k) = \phi_{hix_i}(k) = 0$. The scalar products and norms in these spaces will be provided with the index h.

AMS (MOS) subject classifications (1970). Primary 65N20; Secondary 35Q10, 76D05, 47H15.

Later, we will use the fact that for every $\phi \in \mathcal{H}(\Omega)$ one can find a sequence of functions $\phi_h \in \mathcal{H}^h$ such that

$$\| D_i\varphi - \tilde{\varphi}_{hx_i} \|_2 \underset{h\to 0}{\to} 0, \qquad i=1,\,2,\,3,$$

where the $\tilde{\phi}_{hx_i}$ are piecewise constant interpolants of the difference functions defined by the formula

$$\tilde{\varphi}_{hx_i}(x)\,|_{\omega^h(k)} = \varphi_{hx_i}(k)$$

(for details, see [1]).

The imbedding theorems can be applied to the functions defined on the grid. In particular, one can make use of the compactness of the imbedding of the grid analog of $\overset{\circ}{W}{}^1_p(\Omega)$ into the grid analog of $L_q(\Omega)$ with $q < np/(n-p)$ (for the proof of this, see [2] and [3]). Note in this regard that the functions equal to zero on $\partial\Omega_h$ can be continued by zero outside $\overline{\Omega}_h$. This turns out to be convenient when writing sum analogs of integral identitics.

1. The following auxiliary propositions are used later.

Lemma. *Let X be a Banach space, U the sphere $\{x \mid |x| < r\}$, and $\Phi\colon U \to X$ a mapping such that*

$$\Phi(x) = x + \Psi(x), \quad \Psi(0) = 0,$$
$$|\Psi(x) - \Psi(y)| \leqslant \alpha |x - y|, \quad 0 \leqslant \alpha < 1.$$

Then $\Phi(U)$ covers the sphere $|x| < r(1-\alpha)$, and on it Φ has an inverse which satisfies a Lipschitz condition with constant $1/(1-\alpha)$.

Remark 1. It is easily verified that the image of the sphere of radius $r(1-\alpha)$ under the mapping Φ^{-1} contains the sphere of radius $r(1-\alpha)/(1+\alpha)$. Indeed,

$$|x + \Psi(x)| \leqslant |x| + |\Psi(x)| = |x| + |\Psi(x) - \Psi(0)| \leqslant (1+\alpha)|x|,$$

from which the required assertion follows directly.

Remark 2. The assertion of the lemma remains valid if the mapping Ψ is Fréchet differentiable and the condition

$$|\Psi'|_{X\to X} \leqslant \alpha < 1, \quad x \in U,$$

is fulfilled.

The local inversion theorem can easily be proved using this lemma.

Theorem. *Let X and Y be Banach spaces, Φ a mapping from U to Y, where U is a neighborhood of $0 \in X$, and $\Phi(0_X) = 0_Y$.*

If Φ is Fréchet differentiable in U, if its derivative $\Phi'(x)$ depends continuously on x in the uniform operator topology, and if $\Phi'(0)$ is bounded and has a bounded inverse, then Φ is a homeomorphism of a sufficiently small neighborhood of the point $0 \in X$ onto some neighborhood of the point $0 \in Y$.

The proof is obvious, if we note that the mapping $I + (\Phi'(0)^{-1} \circ \Phi - I)$ satisfies the conditions of the preceding lemma in the sphere of radius r around 0, if r is chosen so that

$$|\Phi'(0)^{-1} \circ \Phi'(x) - I|_{X\to X} \leqslant \alpha < 1, \quad |x| < r$$

(for details, see [4]).

Remark. If for $\Phi'(0)$ there are known constants C_0 and C_1 such that

$$C_0|x| \leqslant |\Phi'(0)x| \leqslant C_1|x|,$$

then, taking into account the conclusion of the lemma, it is easy to obtain that $\Phi'(0)^{-1} \circ \Phi$ homeomorphically maps a neighborhood V, where $0_X \in V \supset B(0_X, r)$ and $V \supset B(0, r(1-\alpha)/(1+\alpha))$, onto a neighborhood $W \ni 0_Y$, where

$$B(0_Y, C_1 r(1-\alpha)) \supset W \supset B(0_Y, C_0 r(1-\alpha)).$$

2. Consider the problem

$$-\nu\Delta\mathbf{u} + u_i D_i \mathbf{u} = -\nabla P + \mathbf{f}(x), \tag{2.1}$$

$$\operatorname{div}\mathbf{u} = 0, \tag{2.2}$$

$$\mathbf{u}|_{\partial\Omega} = 0. \tag{2.3}$$

Assume that

$$\mathbf{f} \in \overset{\circ}{W}{}_2^{-1}(\Omega). \tag{2.4}$$

It is well known (see [1]) that finding a solution of system (2.1)–(2.3) in the space $\mathcal{H}(\Omega)$ reduces to solving the equation

$$\nu I\mathbf{u} - A(\mathbf{u}) = \mathbf{F}, \tag{2.5}$$

or, equivalently, to finding a $\mathbf{u}$ which satisfies the identity

$$[\nu I\mathbf{u} - A(\mathbf{u}) - \mathbf{F}, \varphi] = \int_\Omega (\nu D_i\mathbf{u}D_i\varphi - u_i\mathbf{u}D_i\varphi - \mathbf{f}\varphi)\,dx = 0 \tag{2.6}$$

for any function $\phi \in \mathcal{H}(\Omega)$. Here A is a nonlinear completely continuous operator from $\mathcal{H}(\Omega)$ into $\mathcal{H}(\Omega)$.

Lemma 2.1. *The operator A is continuously Fréchet differentiable in the uniform operator topology.*

Proof. Consider the Gâteaux differential of A at the point $\mathbf{u}$ in the direction $\mathbf{w}$:

$$[A'(\mathbf{u})\mathbf{w}, \varphi] = \frac{d}{dt}[A(\mathbf{u}+t\mathbf{w}), \varphi]|_{t=0} = \int_\Omega (u_i\mathbf{w} + w_i\mathbf{u}, D_i\varphi)\,dx. \tag{2.7}$$

Using this expression, it is easy to prove that the Gâteaux differential satisfies a Lipschitz condition. Indeed, let $\mathbf{u}', \mathbf{u}'' \in \mathcal{H}(\Omega)$, and $\delta\mathbf{u} = \mathbf{u}' - \mathbf{u}''$. Also, let $\delta A'$ denote $A'(\mathbf{u}') - A'(\mathbf{u}'')$. Then

$$[\delta A'\mathbf{w}, \varphi] = \int_\Omega (\delta u_i \mathbf{w} + w_i \delta\mathbf{u}, D_i\varphi)\, dx.$$

Bounding the right side of this equality by the Hölder inequality, we obtain

$$|[\delta A'\mathbf{w}, \varphi]| \leqslant C \|\delta\mathbf{u}\|_4 \|\mathbf{w}\|_4 |\varphi|. \tag{2.8}$$

In (2.8), set $\phi = \delta A'\mathbf{w}$. Then

$$|\delta A'\mathbf{w}| \leqslant C \|\delta\mathbf{u}\|_4 \|\mathbf{w}\|_4. \tag{2.9}$$

Taking the linearity of $\delta A'$ and the bound $\|\mathbf{w}\|_4 \leq C_\Omega|\mathbf{w}|$ into account, we obtain

$$|\delta A'| \leqslant C_1 \|\delta\mathbf{u}\|_4 \leqslant C_2 |\delta\mathbf{u}|. \tag{2.10}$$

Using a well-known theorem according to which Fréchet differentiability and the coincidence of the Fréchet derivative with the Gâteaux derivative follow from the continuity of the Gâteaux derivative, we can assert that the expression on the right side of (2.7) defines the Fréchet derivative of the operator A at the point $\mathbf{u}$. Continuity follows from (2.10).

Let us point out one more result, the difference analog of which will be used later.

Lemma 2.2. *Assume that the Fréchet derivative of the operator A has been calculated at the points $\mathbf{v}^{(s)}$, $s = 1, 2, \ldots$, contained in some sphere. Let the derivatives at these points be applied to the sequence $\mathbf{w}^{(s)}$ which converges weakly to 0 in $\mathcal{H}(\Omega)$. Then $A'(\mathbf{v}^{(s)})\mathbf{w}^{(s)}$ converges strongly to 0 in $\mathcal{H}(\Omega)$.*

The proof follows directly from the inequality

$$|A'(\mathbf{v}^{(s)})\mathbf{w}^{(s)}| \leqslant C \|\mathbf{v}^{(s)}\|_4 \|\mathbf{w}^{(s)}\|_4$$

and the complete continuity of the imbedding of $\mathcal{H}(\Omega)$ in $L_p(\Omega)$, $p < 6$.

Assume that equation (2.5) has exactly m solutions $\mathbf{u}^{(1)}, \ldots, \mathbf{u}^{(m)}$, and that the inequalities

$$\lambda|\mathbf{u}| \leqslant |\nu I\mathbf{u} - A'(\mathbf{u}^{(i)})\mathbf{u}| \leqslant \Lambda|\mathbf{u}|, \quad i = 1, \ldots, m, \tag{2.11}$$

are fulfilled at all $\mathbf{u} \in \mathcal{H}(\Omega)$.

Lemma 2.3. *The mapping $\Phi = \nu I - A$ is a homeomorphism of the neighborhoods $V^{(i)} \ni \mathbf{u}^{(i)}$, $i = 1, \ldots, m$, onto $W \ni \mathbf{F}$.*

Proof. Noting that from the continuous differentiability of A follows the continuous differentiability of Φ and, by virtue of (2.11), the invertibility of $\Phi'(\mathbf{u}^{(i)})$, $i = 1, \ldots, m$, we find that the hypotheses of the local inversion theorem are fulfilled.

3. We approximate equation (2.5) in the space $\mathcal{H}^h$ of differentially solenoidal functions equal to zero on $\partial\Omega_h$. We seek solutions $\mathbf{v}_h$ as elements of $\mathcal{H}^h$ which satisfy the identity

$$[\nu I\mathbf{v}_h - A_h(\mathbf{v}_h) - \mathbf{F}_h,\ \varphi_h]_h$$
$$= h^3 \sum_{\bar{\Omega}_h} \left\{ \left[\nu \mathbf{v}_{hx_i} - \frac{1}{2}(\overset{+i}{v}_h \overset{+i}{\mathbf{v}}_h + \overset{+i}{\mathbf{v}}_h \mathbf{v}_h) \right] \tilde{\varphi}_{hx_i} - \mathbf{F}_{hx_i} \tilde{\varphi}_{hx_i} \right\} = 0 \tag{3.1}$$

with an arbitrary $\phi_h \in \mathcal{H}^h$. $\mathbf{F}_h$ is a differentially solenoidal approximation of the element $\mathbf{F}$ which converges to $\mathbf{F}$ strongly in the norm of $\mathcal{H}(\Omega)$.

We obtain the following equation in $\mathcal{H}^h$:

$$\nu I\mathbf{v}_h - A_h(\mathbf{v}_h) = \mathbf{F}_h. \tag{3.2}$$

For solutions of this equation, we have an a priori bound which is obtained in the following manner. Set $\phi_h = \mathbf{v}_h$ in (3.1):

$$\nu[\mathbf{v}_h,\ \mathbf{v}_h]_h = [\mathbf{F}_h,\ \mathbf{v}_h]_h. \tag{3.3}$$

Note that the nonlinear term vanishes. From (3.3) we have

$$|\mathbf{v}_h|_h \leqslant \frac{1}{\nu} |\mathbf{F}_h|_h. \tag{3.4}$$

Bound (3.4) permits one to prove that the equation can be solved for every h. We will not go into this in detail. We only remark that, for every h, the mapping $\nu I - A_h$ is continuous from $\mathcal{H}^h$ to $\mathcal{H}^h$ and is coercive in the sense that

$$\frac{[\nu I\mathbf{v}_h - A_h(\mathbf{v}_h),\ \mathbf{v}_h]_h}{|\mathbf{v}_h|_h} \underset{|\mathbf{v}_h|_h \to +\infty}{\longrightarrow} +\infty.$$

This permits one to reduce the question of the solvability of (3.2) to the theorem of Brouwer about the existence of fixed points of continuous finite-dimensional mappings. The final answer is that equation (3.2) has at least one solution in the sphere of radius $(1/\nu)|\mathbf{F}_h|_h$.

Lemma 3.1. *Let* $\mathbf{u}_h^{(i)}$, $i = 1, \ldots, m$, *be* $\mathcal{H}^h$*-approximations of the exact solutions* $\mathbf{u}^{(i)}$ *which converge strongly in* $\mathcal{H}(\Omega)$ *to these exact solutions as* $h \to 0$. *Then an* $h_0 > 0$ *can be found such that for* $h < h_0$ *the inequalities* $\lambda|\mathbf{v}_h|_h \le |\Phi'_h(\mathbf{u}_h^{(i)})\mathbf{v}_h|_h \le \Lambda|\mathbf{v}_h|_h$, $i = 1, \ldots, m$, *are fulfilled for any* $\mathbf{v}_h \in \mathcal{H}^h$, *where* $\Phi_h = \nu I - A_h$. *The constants* λ *and* Λ *can be chosen to be the same as in* (2.11).

Proof. The right side of the inequality is obvious. We will prove the left side by contradiction. Assume that for some i a sequence $h \to 0$ and a sequence $\mathbf{v}_h$ corresponding to the sequence of h's have been found such that $|\mathbf{v}_h|_h = 1$ and $|\mathbf{g}_h|_h \to 0$ as $h \to 0$, where $\mathbf{g}_h = \Phi'_h(\mathbf{u}_h^{(i)})\mathbf{v}_h$. Assuming that $\mathbf{v}_h$ converges weakly to $\mathbf{v}^0$ in $\mathcal{H}(\Omega)$, let us consider the expression

$$[\Phi'_h(\mathbf{u}_h^{(i)})\,\mathbf{v}_h,\ \varphi_h]_h = [\mathbf{g}_h,\ \varphi_h]_h$$

with a sequence ϕ_h which converges strongly in $\mathcal{H}(\Omega)$ to $\phi \in \mathcal{H}(\Omega)$. Transferring to the piecewise constant interpolants and then taking the limit as $h \to 0$ gives

$$[\nu I \mathbf{v}^0 - A'(\mathbf{u}^{(i)}) \mathbf{v}^0, \varphi] = 0.$$

From this, taking into account the arbitrariness in the choice of ϕ and inequality (2.11), we conclude that $\mathbf{v}^0 = 0$. Let us prove that $\mathbf{v}_h$ converges strongly to 0. As a consequence of the equality $\nu \mathbf{v}_h = A_h'(\mathbf{u}_h^{(i)})\mathbf{v}_h + \mathbf{g}_h$ and the condition on $\mathbf{g}_h$, in order to do this it is sufficient to prove that $A_h'(\mathbf{u}_h^{(i)})\mathbf{v}_h$ converges strongly to 0 in $\mathcal{H}(\Omega)$. This follows immediately from the grid analog of the bound used in the proof of Lemma 2.2; that is, we have arrived at a contradiction to the normalization condition, which proves the lemma.

Lemma 3.2. *The Fréchet derivative of the operator* A_h *satisfies a Lipschitz condition for all* h; *that is, if* $\mathbf{v}_h', \mathbf{v}_h'' \in \mathcal{H}^h$, $\delta \mathbf{v}_h = \mathbf{v}_h' - \mathbf{v}_h''$, *and* $\delta A_h' = A_h'(\mathbf{v}_h') - A_h'(\mathbf{v}_h'')$, *then* $|\delta A_h'|_{\mathcal{H} \to \mathcal{H}} \le C \|\delta \mathbf{v}_h\|_{4,h} \le C_1 |\delta \mathbf{v}_h|_h$.

Proof. One need only repeat the proof of the analogous fact for the exact operator word for word, reformulating it for differences.

Consider the values of the mapping Φ_h at the points $\mathbf{u}_h^{(i)}$. We have

$$[\Phi_h(\mathbf{u}_h^{(i)}), \varphi_h]_h = [\nu I \mathbf{u}_h^{(i)} - A_h(\mathbf{u}_h^{(i)}), \varphi_h]_h = [\mathbf{F}_h^{(i)}, \varphi_h]_h, \tag{3.5}$$

where the symbol $\mathbf{F}_h^{(i)}$ denotes the value of the mapping in $\mathcal{H}^h$.

Lemma 3.3. *For sufficiently small* h, Φ_h *is a homeomorphism of the neighborhoods* $V_h^{(i)} \ni \mathbf{u}_h^{(i)}$ *onto the neighborhoods* $W_h^{(i)} \ni \mathbf{F}_h^{(i)}$, $i = 1, \ldots, m$, *where the neighborhoods* $V_h^{(i)}$ *and* $W_h^{(i)}$ *contain a sphere whose radius does not depend on* h.

Proof. It is sufficient to verify that the mapping

$$\Phi_h'(\mathbf{u}_h^{(i)})^{-1} \circ \Phi_h'(\mathbf{v}_h) - I,$$
$$i = 1, 2, \ldots, m,$$

is a contraction whenever

$$|\mathbf{u}_h^{(i)} - \mathbf{v}_h|_h < \rho \neq \rho(h),$$
$$i = 1, 2, \ldots, m.$$

In order to do this, transform expression (3.5) in the following manner:

$$|\Phi_h'(\mathbf{u}_h^{(i)})^{-1} \circ \Phi_h'(\mathbf{v}_h) - I| = |\Phi_h'(\mathbf{u}_h^{(i)})^{-1} \circ \{\Phi_h'(\mathbf{v}_h) \pm \Phi_h'(\mathbf{u}_h^{(i)})\} - I|$$
$$\le |\Phi_h'(\mathbf{u}_h^{(i)})| \, |\Phi_h'(\mathbf{v}_h) - \Phi_h'(\mathbf{u}_h^{(i)})|.$$

All of the norms in the above expressions are operator norms. By Lemma 3.1, the inverse of $\Phi_h'(\mathbf{u}_h^{(i)})$ is uniformly bounded for sufficiently small h, and therefore the transformations we have made are meaningful. It remains to prove that the norm of the last expression is small. In order to do this, we make use of the form of Φ_h'.

We have

$$\Phi_h'(\mathbf{v}_h) - \Phi_h'(\mathbf{u}_h^{(i)}) = (\nu I - A_h'(\mathbf{v}_h)) - (\nu I - A_h'(\mathbf{u}_h^{(i)})) = A_h'(\mathbf{u}_h^{(i)}) - A_h'(\mathbf{v}_h).$$

By Lemma 3.2, the norm of the last expression is bounded by $C|\mathbf{u}_h^{(i)} - \mathbf{v}_h|_h$ with a constant which does not depend on h. The lemma is proved.

Lemma 3.4. *Let* $\mathbf{F}_h$ *be an approximation in* $\mathcal{H}^h$ *of the element* $\mathbf{F}$ *which converges strongly to* $\mathbf{F}$ *in* $\mathcal{H}(\Omega)$. *Let* $\mathbf{F}_h^{(i)}$ *be the value of the operator* Φ_h *at the points* $\mathbf{u}_h^{(i)}$. *Then* $|\mathbf{F}_h - \mathbf{F}_h^{(i)}|_h \to 0$ *as* $h \to 0$.

Proof. Consider

$$|\mathbf{F}_h - \mathbf{F}_h^{(i)}|_h^2 = |\mathbf{F}_h|_h^2 - 2[\mathbf{F}_h, \mathbf{F}_h^{(i)}]_h + |\mathbf{F}_h^{(i)}|_h^2.$$

From the construction of $\mathbf{F}_h$, we have $|\mathbf{F}_h|_h^2 \to |\mathbf{F}|^2$ as $h \to 0$.

For the second term we obtain

$$[\mathbf{F}_h, \mathbf{F}_h^{(i)}]_h \underset{h\to 0}{\to} [\mathbf{F}, \nu I\mathbf{u}^{(i)} - A(\mathbf{u}^{(i)})] = |\mathbf{F}|^2.$$

For the last term we have

$$|\mathbf{F}_h^{(i)}|_h^2 = [\mathbf{F}_h^{(i)}, \nu I\mathbf{u}_h^{(i)} - A_h(\mathbf{u}_h^{(i)})]_h \underset{h\to 0}{\to} |\mathbf{F}|^2.$$

The lemma is proved.

Lemma 3.5. *For sufficiently small* h, *equation* (3.2) *has at least* m *solutions.*

Proof. By virtue of the fact that the elements $\mathbf{F}_h^{(i)}$, $i = 1, \dots, m$, and $\mathbf{F}_h$ have a common limit in $\mathcal{H}(\Omega)$, and by virtue of the conclusion of Lemma 3.3, one can assert that, for sufficiently small h, $W_h \equiv \bigcap_i W_h^{(i)} \neq \emptyset$. But then again, by decreasing h, one can get $\mathbf{F}_h$ into W_h. Consider the preimages of the neighborhood W_h. Use the former notation $V_h^{(i)}$ for them. By construction, each $V_h^{(i)}$ contains the element $\mathbf{u}_h^{(i)}$. In addition, since $\nu I - A_h$ is a homeomorphism, and since W_h contains the element $\mathbf{F}_h$, the preimage of $\mathbf{F}_h$, which we denote by $\mathbf{v}_h^{(i)}$, is in $V_h^{(i)}$. Thus we have obtained that there exist $\mathbf{v}_h^{(i)}$ such that

$$\nu I\mathbf{v}_h^{(i)} - A_h(\mathbf{v}_h^{(i)}) = \mathbf{F}_h.$$

Lemma 3.6. *Starting from some* h, *equation* (3.2) *has no more than* m *solutions.*

Proof. Assume there has been found a sequence of mesh lengths $h \to 0$ and a corresponding sequence of solutions $\mathbf{v}_h$, none of which coincide with any of the $\mathbf{v}_h^{(i)}$. Taking (3.4) into account, we can assume that the sequence $\mathbf{v}_h$ converges weakly in $\mathcal{H}(\Omega)$. Moreover, by virtue of the assumption that equation (2.5) has m solutions, the weak limit coincides with one of the exact solutions.

Consider the identity

$$\nu[\mathbf{v}_h, \mathbf{v}_h]_h = [\mathbf{F}_h, \mathbf{v}_h]_h,$$

which is valid for all $\mathbf{v}_h$. Taking the limit as $h \to 0$, we obtain

$$\lim_{h\to 0} \nu[\mathbf{v}_h, \mathbf{v}_h]_h = [\mathbf{F}, \mathbf{u}^{(i)}],$$

where $\mathbf{u}^{(i)}$ is the weak limit of the sequence v_h. Since $[\mathbf{F}, \mathbf{u}^{(i)}] = \nu[\mathbf{u}^{(i)}, \mathbf{u}^{(i)}]$, we obtain that v_h converges to $\mathbf{u}^{(i)}$ in norm, and hence that v_h converges strongly to $\mathbf{u}^{(i)}$ in $\mathcal{H}(\Omega)$. Making use of Lemma 3.5, we obtain that, for sufficiently small h, v_h coincides with one of the $\mathrm{v}_h^{(i)}$. The lemma is proved.

Taking into account the conclusions of Lemmas 3.4–3.6, one can state the following theorem.

Theorem 3.1. *If equation* (2.5) *has exactly* m *solutions, and assumption* (2.11) *is fulfilled, then, starting from some* h, *equation* (3.2) *also has* m *solutions, and they converge strongly in* $\mathcal{H}(\Omega)$ *to the exact solutions.*

Proof. It is necessary to clarify only the assertion about the strong convergence. In order to do this, consider the two equalities

$$\nu[\mathbf{u}_h^{(i)}, \mathbf{u}_h^{(i)}]_h = [\mathbf{F}_h^{(i)}, \mathbf{u}_h^{(i)}]_h,$$
$$\nu[\mathbf{v}_h^{(i)}, \mathbf{v}_h^{(i)}]_h = [\mathbf{F}_h, \mathbf{v}_k^{(i)}]_h.$$

Subtracting one from the other, we obtain

$$\nu(|\mathbf{u}_h^{(i)}|_h^2 - |\mathbf{v}_h^{(i)}|^2) = [\mathbf{F}_h^{(i)}, \mathbf{u}_h^{(i)}]_h - [\mathbf{F}_h, \mathbf{v}_h^{(i)}]_h.$$

Taking the limit as $h \to 0$, we obtain that the norms $|\mathrm{v}_h^{(i)}|_h$ converge to the corresponding $|\mathbf{u}^{(i)}|$. This concludes the proof of the theorem.

In conclusion, the author expresses his gratitude to O. A. Ladyženskaja for her guidance.

BIBLIOGRAPHY

1. O. A. Ladyženskaja, *Mathematical problems in the dynamics of a viscous incompressible fluid*, 2nd rev. aug. ed., "Nauka", Moscow, 1970; English transl. of 1st ed., *The mathematical theory of viscous incompressible flow*, Gordon and Breach, New York, 1961; rev. 1969. MR 42 #6442; 27 #5034b; 40 #7610.

2. S. L. Sobolev, *Sur l'évaluation de quelques sommes pour une fonction définie sur un réseau*, Izv. Akad. Nauk SSSR Ser. Mat. 4 (1940), 5–16. (Russian; French summary) MR 1, 298.

3. O. A. Ladyženskaja, *The mixed problem for a hyperbolic equation*, GITTL, Moscow, 1953. (Russian) MR 17, 160.

4. Jacob T. Schwartz, *Nonlinear functional analysis*, Lecture Notes, Courant Inst. Math. Sci., New York, 1966.

Translated by J. LAVERY

Trudy Mat. Inst. Steklov.
127 (1975)

Proc. Steklov Inst. Math.
127 (1975)

ON A CONDITION FOR THE CONVERGENCE OF "OVERDETERMINED" APPROXIMATIONS

UDC 517.949.8

N. K. KORENEV

Abstract. We prove that for a wide class of domains with nonsmooth boundaries, a function defined in such a domain and zero on its boundary can be approximated by "mesh" functions that are zero on the boundary of the mesh domain approximating the original domain from within.

Bibliography: 2 titles.

In this note we consider the problem of the difference approximation of functions $u(x) \in \overset{\circ}{W}{}_p^1(\Omega)$. It turns out that the simplest Steklov averages for a function $u(x)$ which is extended by zero outside the domain are nonzero on the boundary of the mesh domain contained in the original domain. In investigations of the character of the convergence of the solutions of difference schemes corresponding to problems for second order partial differential equations with homogeneous boundary conditions, it turns out to be convenient to "overdetermine" the approximating functions by putting them equal to zero at the boundary points of the mesh. Some research workers have proved that these "overdetermined" mesh approximations and their difference quotients converge in $L_p(\Omega)$ to the function itself and its derivative, respectively, if the boundary of the domain is smooth. We prove below that this is valid for a very wide class of domains.

Let Ω be a bounded domain in R^n with the boundary $\partial\Omega$, let $\overset{\circ}{W}{}_p^1(\Omega)$ denote the space of functions for which the pth powers $(p \geq 1)$ of the first generalized derivatives are summable over Ω, and which are zero on the boundary of the domain, and let D_i denote the operation of generalized differentiation with respect to x_i.

Let k be a vector with integer components. For $h > 0$ we construct a partition of R^n into cubes $\omega^h(k) = \omega^h(k_1, \ldots, k_n) = \{x\colon k_i h < x_i < (k_i + 1)h,\ i = 1, \ldots, n\}$. Let $\overline{\Omega}_h$ denote the closure of the union of all the cubes $\omega^h(k)$ completely belonging to Ω, and let $\partial\Omega_h$ denote the boundary of $\overline{\Omega}_h$. The symbols $\overline{\Omega}_h$, $\partial\Omega_h$, and $\Omega_h = \overline{\Omega}_h \setminus \partial\Omega_h$ also denote the corresponding sets of vertices.

We assume that the function $\phi(x) \in \overset{\circ}{W}{}_p^1(\Omega)$ is extended by zero outside Ω, and that $\phi \in \overset{\circ}{W}{}_p^1(R^n)$. We define the approximation of ϕ by the formula

AMS (MOS) subject classifications (1970). Primary 65N10.

$$\varphi_h(k)=\frac{1}{h^n}\int\limits_{\omega^h(k)}\varphi(x)\,dx. \tag{1}$$

Here $\phi_h(k)$ is defined at all vertices of the partition of R^n. $\phi_{hx_i}(k)$ is a right difference quotient with respect to x_i; and $\tilde{\phi}_h(x)$ and $\tilde{\phi}_{hx_i}(x)$ are the piecewise constant completions of $\phi_h(k)$ and $\phi_{hx_i}(k)$, respectively, for which the value of the completions on $\omega^h(k)$ is assumed to be the value of the functions at the vertex k.

We state some auxiliary lemmas.

Lemma 1. *Let* $\phi(x)\in \overset{\circ}{W}{}^1_p(\Omega)$, *and let* $\phi_h(k)$ *be its mesh approximation as defined by* (1). *Then*

$$h^n\sum_{k\in R^n}|\varphi_{hx_i}(k)|^p\leqslant\|D_i\varphi\|^p_{p,\,\Omega},$$

where $\|D_i\phi\|_{p,\Omega}$ *is the* $L_p(\Omega)$*-norm of the derivative* $D_i\phi$.

Proof. We have

$$\varphi_{hx_i}(k)=\frac{1}{h^{n+1}}\int\limits_{\omega^h(k)}\left(\int\limits_{x_i}^{x_i+h}D_i\varphi(x_1,\ldots,x_{i-1},\xi_i,x_{i+1},\ldots,x_n)\,d\xi_i\right)dx,$$

from which we note that the value of the difference quotient is represented by the superposition of two averages. The first of these is given by

$$(D_i\varphi)_h(k_1,\ldots,k_{i-1},x_i,k_{i+1},\ldots,k_n)=\frac{1}{h^n}\int\limits_{\omega^h(k_1,\ldots,k_{i-1},x_i,k_{i+1},\ldots,k_n)}D_i\varphi(\xi)\,d\xi,$$

where $\omega^h(k_1,\ldots,k_{i-1},x_i,k_{i+1},\ldots,k_n)=\{\xi\colon k_lh<\xi_l<(k_l+1)h,\ l=1,2,\ldots,i-1,i+1,\ldots,n,\ x_i<\xi_i<x_i+h\}$.

The second average has the form

$$\varphi_{hx_i}(k)=\frac{1}{h}\int\limits_{k_ih}^{(k_i+1)h}(D_i\varphi)_h\,dx_i;$$

that is, the first average is a combination of a discrete average and a Steklov average, while the second is purely discrete.

The assertion of Lemma 1 follows from the following inequalities, which can be verified easily:

$$h\sum_{k_i}|\varphi_{hx_i}(k)|^p\leqslant\int\limits_{-\infty}^{+\infty}|(D_i\varphi)_h|^p\,dx_i, \tag{2}$$

$$h^{n-1}\sum_{(k_1,\ldots,k_{i-1},k_{i+1},\ldots,k_n)}\int\limits_{-\infty}^{+\infty}|(D_i\varphi)_h|^p\,dx_i\leqslant\int\limits_{R^n}|D_i\varphi|^p\,dx. \tag{3}$$

We do not prove these inequalities; the derivation of inequalities of the type (2), (3) for Steklov averages is given in [1].

The next lemma follows from Lemma 1.

Lemma 2. *Under the conditions of Lemma* 1, $\|D_i\phi - \tilde{\phi}_{hx_i}\|_{p,\Omega} \to 0$ *as* $h \to 0$.

The proof is carried out by using Lemma 1 in exactly the same way as the proof in [1] of the analogous result for Steklov averages.

We also need the following assertion.

Lemma 3. *Let* $B(\xi, r)$ *denote the ball of radius* r *with center* ξ. *For every function* $u(x) \in W_1^1(B)$,

$$\int_B |u|\,dx \leqslant \beta \frac{r^n}{\operatorname{meas} \mathcal{E}_0} \operatorname{meas}^{1/n} B \int_B |\nabla u|\,dx,$$

where $\beta = \beta(n)$ *and* $\mathcal{E}_0$ *is the set of those* $x \in B$ *at which* $u(x) = 0$.

Proof. See [2], Chapter II, §3, Lemma 3.4.

Corollary. *If* $u \in W_p^1(B)$, *then*

$$\left(\int_B |u|\,dx\right)^p \leqslant C \left|\frac{r^n}{\operatorname{meas}\mathcal{E}_0}\right|^p \operatorname{meas}^{\frac{p}{n}+p-1} B \int_B |\nabla u|^p\,dx. \tag{4}$$

The inequality (4) follows from the inequality in Lemma 3 if we raise the latter to the power p and use Hölder's inequality to estimate the integral on the right-hand side.

We define $\dot{\phi}_h(k)$ by

$$\dot{\varphi}_h(k) = \begin{cases} \varphi_h(k), & k \in \Omega_h, \\ 0, & k \in R^n \setminus \Omega_h. \end{cases} \tag{5}$$

Lemma 4. *Let* $\phi(x) \in \overset{\circ}{W}{}_p^1(\Omega)$, $p \geq 1$, *and let* $\dot{\phi}_h(k)$ *be the approximation defined by* (5). *If*

$$\begin{gathered} \operatorname{meas} \Omega_r \xrightarrow[r\to 0]{} 0, \\ \Omega_r = \bigcup_{\xi \in \partial\Omega} B(\xi, r), \end{gathered} \tag{6}$$

and the following inequality holds uniformly with respect to r:

$$\frac{\operatorname{meas} B(\xi, r)}{\operatorname{meas}\{(R^n \setminus \Omega) \cap B(\xi, r)\}} \leqslant C, \quad \forall \xi \in \partial\Omega, \tag{7}$$

then

$$\lim_{h\to 0} \|D_i\varphi - \tilde{\dot{\varphi}}_{hx_i}\|_{p,\Omega} = 0, \quad i = 1, \ldots, n. \tag{8}$$

Proof. By the inequality

$$\|D_i\varphi - \tilde{\dot{\varphi}}_{hx_i}\|_{p,\Omega} \leqslant \|D_i\varphi - \tilde{\varphi}_{hx_i}\|_{p,\Omega} + \|\tilde{\varphi}_{hx_i} - \tilde{\dot{\varphi}}_{hx_i}\|_{p,\Omega} \tag{9}$$

and Lemma 2 we conclude that to prove (8) it is sufficient to establish conditions under which the last term in (9) has the limit zero. We estimate it from above:

$$\|\tilde{\varphi}_{hx_i} - \tilde{\dot{\varphi}}_{hx_i}\|_{p,\Omega}^p \leqslant \|\tilde{\varphi}_{hx_i} - \tilde{\dot{\varphi}}_{hx_i}\|_{p,R^n}^p = h^n \sum_{k \in R^n} |\varphi_{hx_i}(k) - \dot{\varphi}_{hx_i}(k)|^p. \tag{10}$$

We write each term in (10) as

$$\varphi_{hx_i}(k)-\dot{\varphi}_{hx_i}(k)=\frac{1}{h}\{[\varphi_h(\ldots, k_i+1, \ldots)-\varphi_h(\ldots, k_i, \ldots)] \\ -[\dot{\varphi}_h(\ldots, k_i+1, \ldots)-\dot{\varphi}_h(\ldots, k_i, \ldots)]\}. \tag{11}$$

Note that the expression (11) is zero when either

$$(k_1, \ldots, k_i, \ldots, k_n),\ (k_1, \ldots, k_i+1, \ldots, k_n) \in \Omega_h, \tag{12}$$

or

$$(k_1, \ldots, k_i, \ldots, k_n),\ (k_1, \ldots, k_i+1, \ldots, k_n) \in R^n \setminus \bar{\Omega}_h, \\ \omega^h(\ldots, k_i, \ldots) \cap \Omega = \omega^h(\ldots, k_i+1, \ldots) \cap \Omega = \varnothing. \tag{13}$$

The expression (11) is nonzero when either

$$\{(\ldots, k_i, \ldots),\ (\ldots, k_i+1, \ldots)\} \cap \partial\Omega_h \neq \varnothing, \tag{14}$$

or

$$(\ldots, k_i, \ldots),\ (\ldots, k_i+1, \ldots) \in R^n \setminus \bar{\Omega}_h, \tag{15}$$

and at least one of the cubes associated with it has a nonempty intersection with Ω.

It is clear from the above conclusions that the summation on the right-hand side of (10) is over those k for which the pairs $(\ldots, k_i, \ldots)$, $(\ldots, k_i+1, \ldots)$ satisfy (14) and (15). All the remaining terms are zero.

With due regard for (11), the left-hand side of (10) is estimated as follows:

$$\|\tilde{\varphi}_{hx_i}-\tilde{\dot{\varphi}}_{hx_i}\|_{p,\Omega}^p \leqslant h^n \sum_k \left[\frac{2}{h^{n+1}} \int\limits_{\omega^h(\ldots, k_i, \ldots) \cup \omega^h(\ldots, k_i+1, \ldots)} |\varphi(x)|\, dx\right]^p, \tag{16}$$

where the summation is over those k satisfying (14) or (15).

Then we note that we can add to the cubes $\omega^h(\ldots, k_i, \ldots)$ and $\omega^h(\ldots, k_i+1, \ldots)$, whose vertices satisfy (14) or (15), the cube ω_e^h that has at least one vertex in common with one of the cubes of the pair, and

$$\omega_e^h \cap (R^n \setminus \Omega) \neq \varnothing. \tag{17}$$

Here ω_e^h can be identified with the cube of the original pair that satisfies (17). We obtain the existence of ω_e^h for each pair from the following argument.

If, for example, for a pair of the type (14) it is impossible to find an ω_e^h satisfying (17), then the vertices $(\ldots, k_i, \ldots)$ and $(\ldots, k_i+1, \ldots)$ are interior; that is, they are of the type (12). Similarly for a pair of the type (15).

We write

$$\chi_h(\bar{k}) = \omega^h(\ldots, k_i, \ldots) \cup \omega^h(\ldots, k_i+1, \ldots) \cup \omega_e^h. \tag{18}$$

The triad $\chi_h(\bar{k})$ of cubes satisfies the conditions

$$\chi_h(\bar{k}) \cap \Omega \neq \varnothing,\quad \chi_h(\bar{k}) \cap (R^n \setminus \Omega) \neq \varnothing, \tag{19}$$

from which it follows immediately that

$$\exists \xi_k \in \chi_h(\bar{k}), \quad \xi_{\bar{k}} \in \partial\Omega. \tag{20}$$

Note that the point $\xi_{\bar{k}}$ associated with triads that differ according to the composition of the cubes can be taken to be the same point if the intersection of these triads is nonempty.

It is easy to see that the number of times that the point $\xi_{\bar{k}}$ occurs in the triads $\chi_h(k)$ that differ in composition is finite. It is estimated in an elementary way as the number of pairs of cubes that can be associated with a cube containing $\xi_{\bar{k}}$ and that have at least one vertex in common with it and with one another.

We construct a covering of the triad by balls $B(\xi_{\bar{k}}, r)$ such that

$$rh^{-1} \leqslant C, \quad \forall h, \tag{21}$$

$$B(\xi_{\bar{k}}, r) \supset \chi_h(k), \quad \xi_{\bar{k}} \in \chi_h(k). \tag{22}$$

Now we consider all these balls. Any point belonging to their union is contained in only a finite number of the balls.

For, the number of triads χ_h that fall in a fixed ball is finite, and by (21) it is independent of h. Hence the number of balls that have a nonempty intersection with a fixed ball $B(\xi_{\bar{k}}, r)$ is finite and independent of h.

We denote the multiplicity of the covering by Π. Now we estimate the integrals on the right-hand side of (16):

$$\int_{\omega^h \cup \omega^h} |\varphi(x)|\, dx \leqslant \int_{\chi_h(\bar{k})} |\varphi(x)|\, dx \leqslant \int_{B(\xi_{\bar{k}}, r)} |\varphi(x)|\, dx. \tag{23}$$

We use (4) to estimate the integral on the right-hand side of (23); we obtain

$$\left(\int_{B(\xi_{\bar{k}}, r)} |\varphi|\, dx\right)^p \leqslant C \left|\frac{r^n}{\operatorname{meas} \mathscr{E}_0}\right|^p \operatorname{meas}^{\frac{p}{n}|p\ 1} B \int_{B(\xi_{\bar{k}}, r)} |\nabla\varphi|^p\, dx. \tag{24}$$

By (20) we have

$$B(\xi_{\bar{k}}, r) \supseteq \mathscr{E}_0 \supseteq B(\xi_{\bar{k}}, r) \cap (R^n \diagdown \Omega).$$

With due regard for the condition (7) we obtain that the coefficient on the right-hand side of (24) is not greater than

$$C \operatorname{meas}^{\frac{p}{n}+p-1} B. \tag{25}$$

We recall that C denotes different constants independent of h. From condition (21) we conclude that (25) has the form CH^{np-n+p}.

Then, taking (24) and (26) into account, we estimate the integral on the left-hand side of (16) as follows:

$$\|\tilde{\varphi}_{hx_i} - \tilde{\varphi}_{hx_i}\|_{p, R^n}^p \leqslant C \sum_{\bar{k}} \int_{B(\xi_{\bar{k}}, r)} |\nabla\varphi|^p\, dx.$$

Bearing in mind that the multiplicity of the repetition of each point in $\bigcup_{\bar{k}} B(\xi_{\bar{k}}, r)$

does not exceed Π, we obtain

$$\|\tilde{\varphi}_{hx_i} - \tilde{\tilde{\varphi}}_{hx_i}\|^p_{p,R^n} \leqslant C\Pi \int\limits_{\Omega_r} |\nabla\varphi|^p\,dx.$$

Condition (6), the result that $\nabla\phi$ belongs to $L_p(R^n)$, and the absolute continuity of the Lebesgue integral give the conclusion of Lemma 3.

In conclusion the author thanks Professor O. A. Ladyženskaja for her supervision.

BIBLIOGRAPHY

1. I. P. Natanson, *Theory of functions of a real variable*, 2nd rev. ed., GITTL, Moscow, 1957; English transl., Vols. 1, 2,* Ungar, New York, 1955, 1961. MR **16**, 120; **16**, 884; **26** #6309.

2. O. A. Ladyženskaja and N. N. Ural'ceva, *Linear and quasi-linear equations of elliptic type*, "Nauka", Moscow, 1964; English transl., Academic Press, New York, 1968. MR **35** #1955; **39** #5941.

Translated by L. W. LONGDON

**Editor's note.* Volume 1 (Chapters I–IX) of the translation is based on the first (1950) edition.

Trudy Mat. Inst. Steklov.
127 (1975)

Proc. Steklov Inst. Math.
127 (1975)

ON SOME MODEL NONSTATIONARY SYSTEMS IN THE THEORY OF NON-NEWTONIAN FLUIDS

UDC 517.994

A. P. OSKOLKOV

Abstract. A study is made of a class of quasilinear nonstationary systems of equations, generalizing the system of Navier-Stokes equations and describing flows of certain types of non-Newtonian fluids. Uniqueness theorems are proved for various generalized solutions of initial-boundary value problems and also the Cauchy problem. For the problems studied a priori estimates are established for the generalized solutions of energy type. Certain linearized problems are solved in closed form.

Bibliography: 18 titles.

It is a known fact that the motion of a continuous incompressible medium may be described by the system of equations

$$\frac{d\mathbf{v}}{dt} = \operatorname{div} T + \mathbf{f}, \quad \operatorname{div} \mathbf{v} = 0, \tag{0.1}$$

where $\mathbf{v}(x, t)$ is the velocity, $T = (T_{ik})$ is the symmetric stress tensor, $\mathbf{f}(x, t)$ is the vector of the external forces, and $d/dt \equiv \partial/\partial t + \mathbf{v} \cdot \operatorname{grad}$ is the Stokes derivative (or substantial derivative). Introduction of the stress tensor into the equation of motion serves the purpose of taking into account reactions occurring in the medium during the course of its motion. By establishing a relationship between the stress tensor T and the other kinematic and thermodynamic variables, we in effect determine and classify differing types of media in motion, for example, a fluid, an elastic medium, etc. A relationship of this kind between the stress tensor T and the other variables is referred to as the constitutive equation [1]. The simplest example of a constitutive equation is that for an ideal fluid, namely, $T = -pE$ (where $p(x, t)$ is the pressure and E is the unit matrix).

A widely used system of axioms describing the motion of a fluid, one which is sufficient for almost all present-day applications in hydrodynamics, is the Stokes system of axioms [1]. A medium whose constitutive equation obeys the Stokes axioms is customarily referred to as a Stokesian fluid. For an incompressible Stokesian fluid the constitutive equation has the form [1], [2]

AMS (MOS) subject classifications (1970). Primary 35A05, 35B15, 35D99, 35Q10, 76D05; Secondary 76B15.

$$T=f(D)\equiv -pE+\beta D+\gamma D^2, \tag{0.2}$$

where $D=(D_{ik}(\mathbf{v}))$, $D_{ik}(\mathbf{v})=\frac{1}{2}(\partial v_i/\partial x_k+\partial v_k/\partial x_i)$ is the symmetric velocity deformation tensor, and β and γ are scalar functions of the principal invariants I, II, and III of the tensor D, which have, in the case of an incompressible fluid, the forms $\mathrm{I}=\operatorname{div}\mathbf{v}=0$, $\mathrm{II}=\Sigma_{i,k}D_{ik}^2\equiv\hat{\mathbf{v}}^2$, and $\mathrm{III}=\det D$.

If in (0.2) we have $\beta=2\nu\equiv\mathrm{const}>0$ and $\gamma\equiv 0$, we obtain the Newtonian law

$$T=-pE+2\nu D, \tag{0.3}$$

which displays a linear dependence between the deformation velocity and the stresses in the fluid. Substituting (0.3) into (0.1), we obtain the well-known Navier-Stokes equations

$$\frac{\partial\mathbf{v}}{\partial t}-\nu\Delta\mathbf{v}+v_k\frac{\partial\mathbf{v}}{\partial x_k}+\operatorname{grad}p=\mathbf{f},\quad \operatorname{div}\mathbf{v}=0. \tag{0.4}$$

The constant ν here is spoken of as the kinematic viscosity coefficient.

If in (0.2) we have $\beta=\beta(\hat{\mathbf{v}}^2)$ and $\gamma\equiv 0$, then, substituting a stress tensor of this kind into the equation of motion (0.1), we obtain the so-called Navier-Stokes equation with a nonlinear viscosity [1], [2]:

$$\frac{\partial v_i}{\partial t}+v_k\frac{\partial v_i}{\partial x_k}-\frac{\partial}{\partial x_k}(\beta(\hat{\mathbf{v}}^2)D_{ik})+\frac{\partial p}{\partial x_i}=f_i,\quad \operatorname{div}\mathbf{v}=0. \tag{0.5}$$

Initial-boundary value problems for the Navier-Stokes equations have been studied by many authors [1], [2]. The most complete and definitive results in this direction were obtained by O. A. Ladyženskaja [2]–[4]. Moreover, she made a detailed study of the global solvability of initial-boundary value problems for a broad class of Navier-Stokes equations with a nonlinear viscosity [2].

Up to the present time the description of viscous fluid motions has, in practice, been accomplished through use of the Navier-Stokes equations (0.4). However, there are many fluids which do not obey the classical Newtonian law (0.3) through whose use the Navier-Stokes equations were obtained from the equations of motion (0.1). Prominent in this regard are weak-concentration aqueous polymer solutions, suspensions, bitumens, phenols, and other fluids, possessing hereditary, visco-elastic, and nonelastic properties [5]–[8], including lag and relaxation effects. The most general constitutive equation for the description of the motion of isotropic fluids has the form [6], [8]

$$F\left(t,\ T_{ij},\ \frac{dT_{ij}}{dt},\ \frac{d^2T_{ij}}{dt^2},\ \ldots,\ D_{ij},\ \frac{dD_{ij}}{dt},\ \frac{d^2D_{ij}}{dt^2},\ \ldots\right)=0 \tag{0.6}$$

and relates the stress and deformation tensors and their various (substantial) derivatives with respect to the time, wherein explicit dependence of F on t indicates the influence of the history of the deformation. There is a broad class of fluids which have nonhereditary motions and whose deformations relax visco-elastically, i.e. the response of such a fluid to an applied external force is not instantaneous but appears after a certain lag; the constitutive relation for such a fluid can be written [6], [8] in the form

$$T = -pE + 2\nu D + \sum_{l=1}^{N} 2\varkappa_l \frac{d^l D}{dt^l}, \quad N = 1, 2, \ldots, \frac{d}{dt} \equiv \frac{\partial}{\partial t} + \mathbf{v} \cdot \mathrm{grad}. \tag{0.7}$$

Here $\nu \equiv \mathrm{const} > 0$ is the kinematic viscosity coefficient, and the nonnegative constants κ_l, $l = 1, \ldots, N$, $\kappa_N > 0$, are the viscosity coefficients for shearing action of various orders. Here the numbers $\tau_l = \kappa_l \nu^{-1}$, $l = 1, \ldots, N$, are referred to as deformation relaxation times or retardation times of the various orders. Following [6], we shall refer in the sequel to fluids with constitutive relations of the form (0.7) as non-Newtonian fluids.

The simplest model of a non-Newtonian fluid is a linearly retarded elastic fluid, referred sometimes as a Voigt or Kelvin body [6], [8]. Such a fluid is characterized by the constitutive relation

$$T = -pE + 2\nu D + 2\varkappa \frac{\partial D}{\partial t}, \quad \nu, \varkappa \equiv \mathrm{const} > 0, \tag{0.8}$$

its motion being described by the system of equations

$$\frac{\partial \mathbf{v}}{\partial t} - \nu \Delta \mathbf{v} + v_k \frac{\partial \mathbf{v}}{\partial x_k} - \varkappa \frac{\partial \Delta \mathbf{v}}{\partial t} + \mathrm{grad}\, p = f, \quad \mathrm{div}\, \mathbf{v} = 0. \tag{0.9}$$

Constant stresses $T_{ij} \equiv \mathrm{const}$ in a Voigt body produce a retarded deformation, which approaches the elastic (instantaneous) deformation asymptotically with a decreasing rate [8]:

$$\frac{\partial D_{ij}}{\partial t} = \frac{\sigma_{ij}}{2\nu\tau} e^{-t/\tau}, \quad t \to \infty, \quad \tau = \varkappa \nu^{-1}, \tag{0.10}$$

where $\sigma_{ij} = T_{ij} + pE_{ij}$ is the stress deviator.

As an immediate generalization of a Voigt body we have the case of a linearly retarded elastic fluid with the constitutive relation

$$T = -pE + 2\nu D + \sum_{l=1}^{N} 2\varkappa_l \frac{\partial^l D}{\partial t^l}, \quad \nu, \varkappa_l \equiv \mathrm{const} > 0, \tag{0.11}$$

for which the equations of motion are of the form

$$\frac{\partial \mathbf{v}}{\partial t} - \nu \Delta \mathbf{v} + v_k \frac{\partial \mathbf{v}}{\partial x_k} - \sum_{l=1}^{N} \varkappa_l \frac{\partial^l \Delta \mathbf{v}}{\partial t^l} + \mathrm{grad}\, p = f, \quad \mathrm{div}\, \mathbf{v} = 0, \quad N = 2, 3, \ldots. \tag{0.12}$$

Initial-boundary value problems for the equations (0.9) are formulated in exactly the same way as for the Navier-Stokes equations. The same methods used for studying the Navier-Stokes equations apply to these equations also; detailed studies of this kind have been given in [9] and [10]. Moreover, it proves to be the case that the system (0.9) regularizes the system of Navier-Stokes equations (to which it reduces for $\kappa = 0$) in the sense that for initial-boundary value problems for the equations (0.9) there exist classes of functions for which these problems have a globally unique solution. Similar results hold for the equations (0.12) also.

A model of a non-Newtonian fluid, next in complexity, is that of a nonlinearly retarded elastic fluid with the constitutive equation

$$T = -pE + 2\nu D + 2\varkappa \frac{dD}{dt}, \quad \nu,\ \varkappa \equiv \text{const} > 0. \tag{0.13}$$

Motion of such a fluid is described by the system of equations

$$\begin{gathered}
\frac{\partial v_i}{\partial t} - \nu\Delta v_i + v_k\frac{\partial v_i}{\partial x_k} - \varkappa\frac{\partial \Delta v_i}{\partial t} - \varkappa v_k\frac{\partial \Delta v_i}{\partial x_k} \\
- \varkappa\frac{\partial v_j}{\partial x_k}\left(\frac{\partial^2 v_i}{\partial x_j \partial x_k} + \frac{\partial^2 v_k}{\partial x_i \partial x_j}\right) + \frac{\partial p}{\partial x_i} = f_i, \\
\operatorname{div}\mathbf{v} = 0.
\end{gathered} \tag{0.14}$$

To fluids with the constitutive equation (0.13) there belong, for example, weakly concentrated aqueous solutions of polymers, flows of which are well described in terms of retarded properties [11], [12].

Let us study the motion of a fluid in a bounded region $\Omega \subset E_n$, $n = 2, 3$, with a smooth boundary $\partial\Omega$; let $Q_T = \Omega \times (0, T)$, $0 < T < \infty$, and $\partial Q_T = \partial\Omega \times (0, T)$; and let us assume that at all times the fluid adheres to the walls of the fixed container Ω. The system (0.14) is then to be solved in Q_T subject to the initial-boundary conditions

$$\mathbf{v}|_{t=0} = \mathbf{v}_0(x), \quad x \in \Omega; \quad \mathbf{v}|_{\partial Q_T} = 0. \tag{0.15}$$

The initial-boundary value problem (0.14), (0.15) has at most one sufficiently smooth solution, for example, a classical solution [9]; however, a study of the solvability of this problem is difficult owing to the presence in the system (0.13) of the strongly nonlinear terms

$$-\varkappa\frac{\partial v_j}{\partial x_k}\left(\frac{\partial^2 v_i}{\partial x_j \partial x_k} + \frac{\partial^2 v_k}{\partial x_i \partial x_j}\right),$$

containing, besides the principal terms, the second derivatives of v with respect to x. It is therefore natural, in a first approximation, to assume that the relative velocities and deformations are small; neglecting these terms, we then consider the much simpler quasilinear system

$$\frac{\partial \mathbf{v}}{\partial t} - \nu\Delta\mathbf{v} + v_k\frac{\partial \mathbf{v}}{\partial x_k} - \varkappa\left(\frac{\partial \Delta\mathbf{v}}{\partial t} + v_k\frac{\partial \Delta\mathbf{v}}{\partial x_k}\right) + \operatorname{grad} p = f, \quad \operatorname{div}\mathbf{v} = 0, \tag{0.16}$$

which, as before, is to be solved in Q_T subject to the initial-boundary conditions (0.15). The initial-boundary value problem (0.16), (0.15), and also the stationary boundary value problem corresponding to it, namely,

$$-\nu\Delta\mathbf{v} + v_k\frac{\partial \mathbf{v}}{\partial x_k} - \varkappa v_k\frac{\partial \Delta\mathbf{v}}{\partial x_k} + \operatorname{grad} p = f, \quad \operatorname{div}\mathbf{v} = 0, \quad x \in \Omega; \quad \mathbf{v}|_{\partial\Omega} = 0 \tag{0.17}$$

were examined in detail in [9], [13], and [14].

If Ω is a two-dimensional region, then, for the flows described by (0.16), we can introduce a stream function $\psi(x_1, x_2, t)$ through the relations $v_1 = \partial\psi/\partial x_2$ and $v_2 = -\partial\psi/\partial x_1$. This stream function then satisfies the equation

$$\frac{\partial}{\partial t}(\Delta\psi - \varkappa\Delta^2\psi) - \nu\Delta^2\psi - \left\{\frac{\partial\psi}{\partial x_1}\frac{\partial}{\partial x_2}(\Delta\psi - \varkappa\Delta^2\psi) - \frac{\partial\psi}{\partial x_2}\frac{\partial}{\partial x_1}(\Delta\psi - \varkappa\Delta^2\psi)\right\}$$
$$- \varkappa\left\{\frac{\partial^2\psi}{\partial x_1\partial x_2}\left(\frac{\partial^2}{\partial x_1^2} - \frac{\partial^2}{\partial x_2^2}\right)\Delta\psi - \frac{\partial^2\Delta\psi}{\partial x_1\partial x_2}\left(\frac{\partial^2}{\partial x_1^2} - \frac{\partial^2}{\partial x_2^2}\right)\psi\right\}$$
$$= -\frac{\partial f_2}{\partial x_1} + \frac{\partial f_1}{\partial x_2} \equiv g(x_1, x_2, t), \qquad (0.18)$$

which is to be solved in Q_T subject to the boundary conditions

$$\psi_{\partial Q_T} = 0, \quad \left.\frac{\partial\psi}{\partial n}\right|_{\partial Q_T} = 0 \qquad (0.19)$$

and the initial Cauchy condition

$$\psi|_{t=0} = \psi_0(x_1, x_2), \quad \psi_{0x_1} = -v_{20}(x), \quad \psi_{0x_2} = v_{10}(x), \quad x \in \Omega. \qquad (0.20)$$

The equation (0.18) contains the strongly nonlinear terms

$$-\varkappa\left\{\frac{\partial^2\psi}{\partial x_1\partial x_2}\left(\frac{\partial^2}{\partial x_1^2} - \frac{\partial^2}{\partial x_2^2}\right)\Delta\psi - \frac{\partial^2\Delta\psi}{\partial x_1\partial x_2}\left(\frac{\partial^2}{\partial x_1^2} - \frac{\partial^2}{\partial x_2^2}\right)\psi\right\},$$

in which, as in the principal linear terms also, fourth derivatives of ψ with respect to x appear. Neglecting these terms in a first approximation, we obtain the quasilinear equation

$$\frac{\partial}{\partial t}(\Delta\psi - \varkappa\Delta^2\psi) - \nu\Delta^2\psi - \left\{\frac{\partial\psi}{\partial x_1}\frac{\partial}{\partial x_2}(\Delta\psi - \varkappa\Delta^2\psi)\right.$$
$$\left. - \frac{\partial\psi}{\partial x_2}\frac{\partial}{\partial x_1}(\Delta\psi - \varkappa\Delta^2\psi)\right\} = g(x, t), \qquad (0.21)$$

which, as before, is to be solved in Q_T subject to the initial-boundary conditions (0.19) and (0.20). The initial-boundary value problem (0.21), (0.19), (0.20), and the stationary boundary value problem corresponding to it, namely,

$$\nu\Delta^2\psi + \frac{\partial\psi}{\partial x_1}\frac{\partial}{\partial x_2}(\Delta\psi - \varkappa\Delta^2\psi) - \frac{\partial\psi}{\partial x_2}\frac{\partial}{\partial x_1}(\Delta\psi - \varkappa\Delta^2\psi) = g(x), \quad x \in \Omega;$$
$$\psi|_{\partial\Omega} = \left.\frac{\partial\psi}{\partial n}\right|_{\partial\Omega} = 0 \qquad (0.22)$$

were studied in [15].

We return now to non-Newtonian fluids in which the constitutive equation has the form (0.7) with $N = 2, 3, \ldots$. We substitute the stress tensor (0.7) into the equation of motion (0.1) and, taking into account the study of the Navier-Stokes equations given in [1] and the study of non-Newtonian fluids described above with $N = 1$, we neglect in a first approximation the strongly nonlinear terms containing all possible products of the derivatives of the velocities v with respect to x and t. We then obtain the following system of equations [9]:

$$L(\mathbf{v}) \equiv \frac{\partial \mathbf{v}}{\partial t} - \nu\Delta\mathbf{v} + v_k \frac{\partial \mathbf{v}}{\partial x_k} - \sum_{l=1}^{N} \varkappa_l \left(\frac{\partial^l \Delta\mathbf{v}}{\partial t^l} + l v_k \frac{\partial^l \Delta\mathbf{v}}{\partial t^{l-1}\partial x_k}\right)$$

$$+ \operatorname{grad} p = \mathbf{f}, \quad \operatorname{div} \mathbf{v} = 0, \quad N = 2, 3, \ldots. \tag{0.23}$$

We consider two problems for the system (0.23): the Cauchy problem consisting in determining in $E_n \times (0, T)$ solutions of the system (0.23), which satisfy the initial Cauchy conditions

$$\left.\frac{\partial^m \mathbf{v}}{\partial t^m}\right|_{t=0} = \mathbf{v}_{0m}(x), \quad x \in E_n, \quad m = 0, 1, \ldots, N-1, \tag{0.24}$$

and the initial-boundary problem consisting in determining in $Q_T = \Omega \times (0, T)$ (Ω is a bounded region in E_n, $n = 2, 3$) solutions of the system (0.23) which satisfy the initial-boundary conditions

$$\left.\frac{\partial^m \mathbf{v}}{\partial t^m}\right|_{t=0} = \mathbf{v}_{0m}(x), \quad x \in \Omega, \quad m = 0, 1, \ldots, N-1; \quad \mathbf{v}\big|_{\partial Q_T} = 0, \tag{0.25}$$

$$(\Delta\mathbf{v})_n\big|_{\partial Q_T} = 0. \tag{0.26}$$

If Ω is a two-dimensional region, we can also introduce for the flows described by the system (0.23) a stream function ψ through the relations $v_1 = \partial\psi/\partial x_2$ and $v_2 = -\partial\psi/\partial x_1$, which will satisfy the equation

$$\frac{\partial}{\partial t}\left(\Delta\psi - \sum_{l=1}^{N} \varkappa_l \frac{\partial^{l-1}\Delta^2\psi}{\partial t^{l-1}}\right) - \nu\Delta^2\psi - \left\{\frac{\partial\psi}{\partial x_1}\frac{\partial}{\partial x_2}\left(\Delta\psi - \sum_{l-1}^{N} l\varkappa_l \frac{\partial^{l-1}\Delta^2\psi}{\partial t^{l-1}}\right)\right.$$

$$\left. - \frac{\partial\psi}{\partial x_2}\frac{\partial}{\partial x_1}\left(\Delta\psi - \sum_{l=1}^{N} l\varkappa_l \frac{\partial^{l-1}\Delta^2\psi}{\partial t^{l-1}}\right)\right\} - \sum_{l=1}^{N} l\varkappa_l \left[\frac{\partial^2\psi}{\partial x_1\partial x_2}\left(\frac{\partial^2}{\partial x_1^2} - \frac{\partial^2}{\partial x_2^2}\right)\frac{\partial^{l-1}\Delta\psi}{\partial t^{l-1}}\right.$$

$$\left. - \frac{\partial^2}{\partial x_1\partial x_2}\frac{\partial^{l-1}\Delta\psi}{\partial t^{l-1}}\left(\frac{\partial^2}{\partial x_1^2} - \frac{\partial^2}{\partial x_2^2}\right)\psi\right] = \frac{\partial f_1}{\partial x_2} - \frac{\partial f_2}{\partial x_1} \equiv g, \quad N = 2, 3, \ldots. \tag{0.27}$$

If in a first approximation we neglect, as we did in the case $N = 1$, the strongly nonlinear terms containing fourth derivatives with respect to x of $\partial^{l-1}\psi/\partial t^{l-1}$, $l = 1, \ldots, N$, we obtain the quasilinear equation

$$l(\psi) \equiv \frac{\partial}{\partial t}\left(\Delta\psi - \sum_{l=1}^{N} \varkappa_l \frac{\partial^{l-1}\Delta^2\psi}{\partial t^{l-1}}\right) - \nu\Delta^2\psi - \left\{\frac{\partial\psi}{\partial x_1}\frac{\partial}{\partial x_2}\left(\Delta\psi - \sum_{l=1}^{N} l\varkappa_l \frac{\partial^{l-1}\Delta^2\psi}{\partial t^{l-1}}\right)\right.$$

$$\left. - \frac{\partial\psi}{\partial x_2}\frac{\partial}{\partial x_1}\left(\Delta\psi - \sum_{l=1}^{N} l\varkappa_l \frac{\partial^{l-1}\Delta^2\psi}{\partial t^{l-1}}\right)\right\} = g(x, t), \quad N = 2, 3, \ldots. \tag{0.28}$$

For the equation (0.28) we also consider two problems: the Cauchy problem, consisting in determining in $R_T = E_2 \times (0, T)$, $0 < T < \infty$, solutions of the equation (0.28) satisfying the initial conditions

$$\left.\frac{\partial^m \psi}{\partial t^m}\right|_{t=0} = \psi_{0m}(x), \quad \psi_{0mx_1} = -v_{20m}(x), \quad \psi_{0mx_2} = v_{10m}(x), \quad x \in \Omega,$$
$$m = 0, 1, \ldots, N-1, \tag{0.29}$$

and the initial-boundary value problem, consisting in determining in $Q_T = \Omega \times (0, T)$ (Ω is a bounded domain in E_2 with a smooth boundary) solutions of the equation (0.28) which satisfy the initial conditions (0.29), the boundary conditions (0.19), and the boundary condition

$$\left.\frac{\partial \Delta\psi}{\partial n}\right|_{\partial Q_T} = 0. \tag{0.30}$$

Our main effort in this paper amounts to proving uniqueness theorems for various generalized solutions of the initial-boundary value problems (0.23), (0.25) and (0.28), (0.19), (0.29) (see §1), and also in proving various global a priori estimates for generalized solutions of the initial-boundary value problems (0.23), (0.25), (0.26), and (0.28), (0.19), (0.30), (0.29) (see §§2 and 3); we also prove various a priori estimates "in the small" for generalized solutions of the Cauchy problems (0.23), (0.24) and (0.28), (0.29) (see §4). In addition, in §5 we examine the question of the existence of small-amplitude gravitational waves on the surface of an unbounded non-Newtonian fluid, and we also solve the problem concerning the motion of a non-Newtonian fluid in a halfspace due to the oscillating motion of the plane bounding the halfspace.

In solving the problems considered in the present paper, we use the method (which we also employed in [9], [10] and [13]–[15]) developed by O. A. Ladyženskaja in treating the hydrodynamics of viscous and of ideal fluids [2], [3]. The author takes this opportunity to thank O. A. Ladyženskaja for her interest in his work and for valuable suggestions.

§1. Uniqueness theorems for boundary value problems

In this section we prove uniqueness theorems for various generalized solutions of the initial-boundary value problems studied in this paper.

Let Ω be a two-dimensional or three-dimensional bounded domain. By a generalized solution of the initial-boundary value problem (0.23), (0.25) we shall mean a function $\mathbf{v}(x, t) \in \overset{\circ}{J}(Q_T)$ which, together with the derivatives $\mathbf{v}_x$ and $\partial^{l-1}\Delta\mathbf{v}/\partial t^{l-1}$, $l = 1, \ldots, N-1$, is bounded in Q_T, has generalized derivatives $\partial\mathbf{v}/\partial t, \partial\mathbf{v}_x/\partial t, \ldots, \partial^N\mathbf{v}_x/\partial t^N \in L_2(Q_T)$, and satisfies the initial-boundary conditions (0.25) and the integral identity

$$\iint\limits_{Q_t} \left[\mathbf{v}_t\mathbf{\Phi} + \nu\mathbf{v}_x\mathbf{\Phi}_x + \sum_{l=1}^{N} \varkappa_l \frac{\partial^l \mathbf{v}_x}{\partial t^l}\mathbf{\Phi}_x + v_k\frac{\partial \mathbf{v}}{\partial x_k}\mathbf{\Phi} + v_k \sum_{l=1}^{N} l\varkappa_l \frac{\partial^{l-1}\Delta\mathbf{v}}{\partial t^{l-1}}\mathbf{\Phi}_{x_k} \right] dQ$$
$$= \iint\limits_{Q_t} \mathbf{f\Phi}\, dQ, \quad 0 < t \leqslant T, \tag{1.1}$$

for arbitrary $\mathbf{\Phi}(x, t) \in \overset{\circ}{J}(Q_T) \cap \overset{\circ}{W}{}_2^{2,0}(Q_T)$. We then have

Theorem 1.1. *The initial-boundary value problem* (0.23), (0.25) *has at most one generalized solution.*

In fact, if the problem (0.23), (0.25) has two generalized solutions $\mathbf{v}_1$ and $\mathbf{v}_2$, then their difference $\omega = \mathbf{v}_1 - \mathbf{v}_2$ satisfies the integral identity

$$\iint_{Q_t}\Big[\omega_t\Phi + \nu\omega_x\Phi_x + \sum_{l=1}^{N}\varkappa_l\frac{\partial^l\omega_x}{\partial t^l}\Phi_x + v_{1k}\frac{\partial\omega}{\partial x_k}\Phi + \omega_k\frac{\partial \mathbf{v}_2}{\partial x_k}\Phi$$

$$+ v_{1k}\sum_{l=1}^{N} l\varkappa_l\frac{\partial^{l-1}\Delta\omega}{\partial t^{l-1}}\Phi_{x_k} + \omega_k\sum_{l=1}^{N} l\varkappa_l\frac{\partial^{l-1}\Delta\mathbf{v}_2}{\partial t^{l-1}}\Phi_{x_k}\Big]dQ = 0, \quad 0 < t \leqslant T, \tag{1.2}$$

and, in addition,

$$\frac{\partial^m\omega}{\partial t^m}\bigg|_{t=0} = 0, \quad x \in \Omega, \quad m = 0, 1, \ldots, N-1; \quad \omega|_{\partial Q_T} = 0. \tag{1.3}$$

In (1.2) we put

$$\Phi = \sum_{l=1}^{N} l\varkappa_l\frac{\partial^{l-1}\omega}{\partial t^{l-1}}.$$

Integrating by parts and using the homogeneous boundary conditions (0.25) and (1.3), by virtue of which $\Phi|_{\partial Q_T} = 0$, we obtain

$$\iint_{Q_t} v_{1k}\sum_{l=1}^{N} l\varkappa_l\frac{\partial^{l-1}\Delta\omega}{\partial t^{l-1}}\cdot\frac{\partial}{\partial x_k}\sum_{l=1}^{N} l\varkappa_l\frac{\partial^{l-1}\omega}{\partial t^{l-1}}dQ \equiv \iint_{Q_t} v_{1k}\Delta\Phi\frac{\partial\Phi}{\partial x_k}dQ$$

$$= \iint_{Q_t} v_{1k}\frac{\partial\Phi_j}{\partial x_k}\frac{\partial^2\Phi_j}{\partial x_i^2}dQ = \iint_{\partial Q_t} v_{1k}\frac{\partial\Phi_j}{\partial x_k}\frac{\partial\Phi_j}{\partial x_i}\cos(\mathbf{n}, x_i)\,dS$$

$$- \iint_{Q_t}\frac{\partial\Phi_j}{\partial x_i}\Big(v_{1k}\frac{\partial\Phi_j}{\partial x_k}\Big)_{x_i}dQ = -\iint_{Q_t}\Big(\frac{\partial v_{1k}}{\partial x_i}\frac{\partial\Phi_j}{\partial x_k} + v_{1k}\frac{\partial^2\Phi_j}{\partial x_i\partial x_k}\Big)\frac{\partial\Phi_j}{\partial x_i}dQ$$

$$= -\iint_{Q_t}\frac{\partial v_{1k}}{\partial x_i}\frac{\partial\Phi_j}{\partial x_i}\frac{\partial\Phi_j}{\partial x_k}dQ \leqslant C_1(\max_{Q_T}|\mathbf{v}_{1x}|)\iint_{Q_t}|\Phi_x|^2dQ$$

$$\leqslant C_2(N, C_1)\iint_{Q_t}\sum_{l=1}^{N}\Big|\frac{\partial^{l-1}\omega_x}{\partial t^{l-1}}\Big|^2dQ, \quad 0 < t \leqslant T. \tag{1.4}$$

Further, using Hölder's inequality, we obtain, for the $\Phi(x, t)$ chosen,

$$\iint_{Q_t}\Big[\Big(v_{1k}\frac{\partial\omega}{\partial x_k} + \omega_k\frac{\partial\mathbf{v}_2}{\partial x_k}\Big)\Phi + \omega_k\sum_{l=1}^{N} l\varkappa_l\frac{\partial^{l-1}\Delta\mathbf{v}_2}{\partial t^{l-1}}\Phi_{x_k}\Big]dQ$$

$$\leqslant C_8\Big(\max_{Q_T}|\mathbf{v}_1|, \max_{Q_T}|\mathbf{v}_{2x}|, \sum_{l=1}^{N}\max_{Q_T}\Big|\frac{\partial^{l-1}\Delta\mathbf{v}_2}{\partial t^{l-1}}\Big|\Big)\iint_{Q_t}\sum_{l=1}^{N}\Big(\Big|\frac{\partial^{l-1}\omega}{\partial t^{l-1}}\Big|^2$$

$$+ \Big|\frac{\partial^{l-1}\omega_x}{\partial t^{l-1}}\Big|^2\Big)dQ, \quad 0 < t \leqslant T. \tag{1.5}$$

From the Newton-Leibniz formula and the homogeneous initial conditions (1.3) it follows that

$$\iint\limits_{Q_t} \left| \frac{\partial^{l-1}\omega}{\partial t^{l-1}} \right|^2 dQ \leqslant T^{2(N-l-1)} \iint\limits_{Q_t} \left| \frac{\partial^{N-1}\omega}{\partial t^{N-1}} \right|^2 dQ, \quad l=1, \ldots, N; \quad 0<t\leqslant T. \tag{1.6}$$

If, in addition, we apply an inequality which results from Friedrichs' inequality, namely,

$$\iint\limits_{Q_t} \left| \frac{\partial^{l-1}\omega}{\partial t^{l-1}} \right|^2 dQ \leqslant C_\Omega \iint\limits_{Q_t} \left| \frac{\partial^{l-1}\omega_x}{\partial t^{l-1}} \right|^2 dQ, \quad l=1, \ldots, N; \quad 0<t\leqslant T, \tag{1.7}$$

we obtain from (1.2) the inequality

$$\int\limits_{\Omega_t} \left| \frac{\partial^{N-1}\omega_x}{\partial t^{N-1}} \right|^2 dx \leqslant C_4(C_2, C_3, T, C_\Omega) \iint\limits_{Q_t} \left| \frac{\partial^{N-1}\omega_x}{\partial t^{N-1}} \right|^2 dQ, \quad 0<t\leqslant T, \tag{1.8}$$

and then it follows from (1.6), (1.7) and the homogeneous initial conditions (1.3) that $\omega(x, t) \equiv 0$ in Q_T.

Let Ω be a two-dimensional bounded domain. By a strong solution of the problem (0.23), (0.25) we shall mean a function $\mathrm{v}(x, t) \in \mathring{J}(Q_T)$ which, together with the derivative v_x, is bounded in Q_T, has generalized derivatives v_t, $\partial^l \mathrm{v}_x/\partial t^l$, $\partial^{l-1}\Delta \mathrm{v}/\partial t^{l-1} \in L_2(Q_T)$, $l = 1, \ldots, N$, has derivatives $\partial^{l-1}\Delta \mathrm{v}_x/\partial t^{l-1} \in L_{2,1}(Q_T)$, $l = 1, \ldots, N$, and satisfies the initial-boundary conditions (0.25) and the integral identity

$$\iint\limits_{Q_t} \Bigg[\mathrm{v}_t\Phi + \nu \mathrm{v}_x\Phi_x + \sum_{l=1}^{N} \varkappa_l \frac{\partial^l \mathrm{v}_x}{\partial t^l}\Phi_x + v_k \frac{\partial \mathrm{v}}{\partial x_k}\Phi - v_k \sum_{l=1}^{N} l\varkappa_l \frac{\partial^{l-1}\Delta \mathrm{v}_{x_k}}{\partial t^{l-1}}\Phi \Bigg] dQ = \iint\limits_{Q_t} \mathbf{f}\Phi dQ, \quad 0<t\leqslant T, \tag{1.9}$$

for arbitrary $\Phi(x, t) \in \mathring{J}(Q_T) \cap \mathring{W}_2^{2,0}(Q_T)$. We show now that the following theorem is valid.

Theorem 1.2. *In the two-dimensional bounded domain* Ω *the initial-boundary problem* (0.23), (0.25) *has at most one strong solution.*

In fact, the difference $\omega(x, t)$ of two possible strong solutions of the problem (0.23), (0.25) satisfies the integral identity

$$\iint\limits_{Q_t} \Bigg[\omega_t\Phi + \nu\omega_x\Phi_x + \sum_{l=1}^{N} \varkappa_l \frac{\partial^l \omega_x}{\partial t^l}\Phi_x + v_{1k}\frac{\partial \omega}{\partial x_k}\Phi + \omega_k \frac{\partial \mathrm{v}_2}{\partial x_k}\Phi - v_{1k}\sum_{l=1}^{N} l\varkappa_l \frac{\partial^l \Delta\omega}{\partial t^{l-1}\partial x_k}\Phi - \omega_k \sum_{l=1}^{N} l\varkappa_l \frac{\partial^l \Delta \mathrm{v}_2}{\partial t^{l-1}\partial x_k}\Phi \Bigg] dQ = 0, \quad 0<t\leqslant T,$$

$$\forall \Phi(x, t) \in \mathring{J}(Q_T) \cap \mathring{W}_2^{2,0}(Q_T). \tag{1.10}$$

and the homogeneous initial-boundary conditions (1.3). As was done in the proof of Theorem 1.1, we put

$$\Phi=\sum_{l=1}^{N} l\varkappa_l \frac{\partial^{l-1}\omega}{\partial t^{l-1}}.$$

Then, first of all, proceeding as we did in obtaining (1.4), we have

$$\iint\limits_{Q_t} v_{1k}\sum_{l=1}^{N} l\varkappa_l \frac{\partial^l \Delta\omega}{\partial t^{l-1}\partial x_k}\sum_{l=1}^{N} l\varkappa_l \frac{\partial^{l-1}\omega}{\partial t^{l-1}}\,dQ=-\iint\limits_{Q_t} v_{1k}\sum_{l=1}^{N} l\varkappa_l \frac{\partial^{l-1}\Delta\omega}{\partial t^{l-1}}\sum_{l=1}^{N} l\varkappa_l \frac{\partial^l \omega}{\partial t^{l-1}\partial x_k}\,dQ$$

$$\leqslant C_2\iint\limits_{Q_t}\sum_{l=1}^{N}\left|\frac{\partial^{l-1}\omega_x}{\partial t^{l-1}}\right|^2 dQ,\quad 0<t\leqslant T. \tag{1.11}$$

Further, using Hölder's inequality and the well-known inequality (see [2], Chapter 1)

$$\|u\|^4_{L_4(\Omega)}\leqslant 2\|u\|^2_{L_2(\Omega)}\cdot\|u_x\|^2_{L_2(\Omega)}. \tag{1.12}$$

valid for an arbitrary function $u(x)\in \overset{\circ}{W}{}^1_2(\Omega)$ and an arbitrary two-dimensional domain Ω, and applying Friedrichs' inequality, we obtain

$$\iint\limits_{Q_t}\left[v_{1k}\frac{\partial\omega}{\partial x_k}+\omega_k\frac{\partial \mathbf{v}_2}{\partial x_k}-\omega_k\sum_{l=1}^{N} l\varkappa_l \frac{\partial^l\Delta\mathbf{v}_2}{\partial t^{l-1}\partial x_k}\right]\sum_{l=1}^{N} l\varkappa_l\frac{\partial^{l-1}\omega}{\partial t^{l-1}}\,dQ$$

$$\leqslant C_5(\max_{Q_T}|\mathbf{v}|,\ \max_{Q_T}|\mathbf{v}_x|,\ C_\Omega)\iint\limits_{Q_t}\sum_{l=1}^{N}\left|\frac{\partial^{l-1}\omega_x}{\partial t^{l-1}}\right|^2 dQ$$

$$+C_6(N)\sum_{l,\,m=1}^{N}\int_0^t\left\{\left(\int\limits_\Omega\left|\frac{\partial^{m-1}\Delta\mathbf{v}_{2x}}{\partial t^{m-1}}\right|^2dx\right)^{1/2}\left(\int\limits_\Omega \omega^4 dx\right)^{1/4}\left(\int\limits_\Omega\left|\frac{\partial^{l-1}\omega}{\partial t^{l-1}}\right|^4dx\right)^{1/4}\right\}dt$$

$$\leqslant C_5\iint\limits_{Q_t}\sum_{l=1}^{N}\left|\frac{\partial^{l-1}\omega_x}{\partial t^{l-1}}\right|^2 dQ+C_7(C_6,\ C_\Omega)\int_0^t\left\{\sum_{m=1}^{N}\left(\int\limits_\Omega\left|\frac{\partial^{m-1}\Delta\mathbf{v}_{2x}}{\partial t^{m-1}}\right|^2dx\right)^{1/2}\right.$$

$$\left.\times\sum_{l=1}^{N}\int\limits_\Omega\left|\frac{\partial^{l-1}\omega_x}{\partial t^{l-1}}\right|^2dx\right\}dt\leqslant C_8(C_5,\ C_7)\left[t+\sum_{m=1}^{N}\int_0^t\left(\int\limits_\Omega\left|\frac{\partial^{m-1}\Delta\mathbf{v}_{2x}}{\partial t^{m-1}}\right|^2dx\right)^{1/2}dt\right]$$

$$\times\max_{0\leqslant\tau\leqslant t}\sum_{l=1}^{N}\int\limits_{\Omega_\tau}\left|\frac{\partial^{l-1}\omega_x}{\partial t^{l-1}}\right|^2dx,\quad 0<t\leqslant T. \tag{1.13}$$

With the aid of (1.11), (1.13), (1.6), (1.7) and the inequality

$$\max_{0\leqslant\tau\leqslant t}\int\limits_{\Omega_\tau}\left|\frac{\partial^{l-1}\omega}{\partial t^{l-1}}\right|^2dx\leqslant C_9(T,\ N)\max_{0\leqslant\tau\leqslant t}\int\limits_{\Omega_\tau}\left|\frac{\partial^{N-1}\omega}{\partial t^{N-1}}\right|^2dx,\quad l=1,\ \ldots,\ N;\ 0<t\leqslant T, \tag{1.14}$$

which follows from the Newton-Leibniz formula and the homogeneous initial conditions (1.3), we obtain from (1.10), using the operation of maximization with respect to t, the inequality

$$\max_{0\leqslant\tau\leqslant t}\int\limits_{\Omega_\tau}\left|\frac{\partial^{N-1}\omega_x}{\partial t^{N-1}}\right|^2dx\leqslant C_{10}(C_2,\ C_8,\ C_9,\ C_\Omega)$$

$$\times\left[t+\sum_{m=1}^{N}\int_0^t\left(\int\limits_{\Omega_\tau}\left|\frac{\partial^{m-1}\Delta\mathbf{v}_{2x}}{\partial t^{m-1}}\right|^2dx\right)^{1/2}d\tau\right]\max_{0\leqslant\tau\leqslant t}\int\limits_{\Omega_\tau}\left|\frac{\partial^{N-1}\omega_x}{\partial t^{N-1}}\right|^2dx,\quad 0<t\leqslant T. \tag{1.15}$$

If we now choose t_1 to satisfy the condition

$$C_{10}\left[t_1+\sum_{m=1}^{N}\int_0^{t_1}\left(\int_{\Omega_\tau}\left|\frac{\partial^{m-1}\Delta \mathrm{v}_{2x}}{\partial t^{m-1}}\right|^2 dx\right)^{1/2} d\tau\right]<1, \tag{1.16}$$

we find from (1.15) and (1.14) that $\omega(x, t) \equiv 0$ in Q_t with $t < t_1$. Following this we may show that $\omega(x, t)$ is equal to zero throughout Q_T by proceeding stepwise in t (cf. [2], Chapter 6).

As before, let us assume that Ω is a two-dimensional bounded domain. By a weak solution of the initial-boundary value problem (0.23), (0.25) we shall mean a function $\mathrm{v}(x, t) \in \mathring{J}(Q_T)$ which, together with the derivative v_x, is bounded in Q_T, has generalized derivatives v_t, $\partial^l \mathrm{v}_x/\partial t^l \in L_2(Q_T)$, $l = 1, \ldots, N$, and the derivatives $\partial^{l-1}\Delta \mathrm{v}/\partial t^{l-1} \in L_2(Q_T) \cap L_{4,1}(Q_T)$, $l = 1, \ldots, N$, satisfies the initial-boundary conditions (0.25), and also satisfies the integral identity (1.1) for arbitrary $\Phi(x, t) \in \mathring{J}(Q_T) \cap \mathring{W}_2^{2,0}(Q_T)$. We then have

Theorem 1.3. *In the two-dimensional bounded domain* Ω *the initial-boundary value problem* (0.23), (0.25) *has at most one weak solution.*

In fact, the difference $\omega(x, t)$ of two possible weak solutions of the problem (0.23), (0.25) satisfies the integral identity (1.2) and the initial-boundary conditions (1.3). As before, let us put

$$\Phi(x, t)=\sum_{l=1}^{N} l\varkappa_l \frac{\partial^{l-1}\omega}{\partial t^{l-1}}.$$

Then for the fundamental nonlinear term

$$\iint_{Q_t} v_{1k}\sum_{l=1}^{N} l_l \frac{\partial^{l-1}\Delta\omega}{\partial t^{l-1}}\Phi_{x_k} dQ$$

we will have the estimate (1.4), and, instead of (1.5), we obtain, using the Hölder and Friedrichs inequalities and applying (1.12), the following estimate:

$$\iint_{Q_t}\left\{\left[v_{1k}\frac{\partial\omega}{\partial x_k}+\omega_k\frac{\partial \mathrm{v}_2}{\partial x_k}\right]\sum_{l=1}^{N} l\varkappa_l\frac{\partial^{l-1}\omega}{\partial t^{l-1}}+\omega_k\sum_{l=1}^{N} l\varkappa_l\frac{\partial^{l-1}\Delta\omega}{\partial t^{l-1}}\sum_{m=1}^{N} m\varkappa_m\frac{\partial^m\omega}{\partial t^{m-1}\partial x_k}\right\}dQ$$
$$\leqslant C_5\iint_{Q_t}\sum_{l=1}^{N}\left|\frac{\partial^{l-1}\omega_x}{\partial t^{l-1}}\right|^2 dQ+C_{11}(N)\sum_{l,\,m=1}^{N}\int_0^t\left\{\left(\int_\Omega\left|\frac{\partial^{m-1}\Delta \mathrm{v}_2}{\partial t^{m-1}}\right|^4 dx\right)^{1/4}\right.$$
$$\left.\times\left(\int_\Omega \omega^4 dx\right)^{1/4}\left(\int_\Omega\left|\frac{\partial^{l-1}\omega_x}{\partial t^{l-1}}\right|^2 dx\right)^{1/2}\right\}dt\leqslant C_5\iint_{Q_t}\sum_{l=1}^{N}\left|\frac{\partial^{l-1}\omega_x}{\partial t^{l-1}}\right|^2 dQ$$
$$+C_{12}(C_{11}, C_\Omega)\int_0^t\left\{\sum_{m=1}^{N}\left(\int_\Omega\left|\frac{\partial^{m-1}\Delta \mathrm{v}_2}{\partial t^{m-1}}\right|^4 dx\right)^{1/4}\sum_{l=1}^{N}\int_\Omega\left|\frac{\partial^{l-1}\omega_x}{\partial t^{l-1}}\right|^2 dx\right\}dt\leqslant$$

$$\leqslant C_{13}(C_5, C_{12})\left[t+\sum_{m=1}^{N}\int_0^t\left(\int_\Omega\left|\frac{\partial^{m-1}\Delta \mathbf{v}_2}{\partial t^{m-1}}\right|^4 dx\right)^{1/4} dt\right]$$
$$\times \max_{0\leqslant\tau\leqslant t}\sum_{l=1}^{N}\int_{\Omega_\tau}\left|\frac{\partial^{l-1}\boldsymbol{\omega}_x}{\partial t^{l-1}}\right|^2 dx,\ 0<t\leqslant T. \tag{1.17}$$

Using (1.4), (1.17), (1.6), (1.7), and (1.14), we find, from the identity (1.2) for the chosen $\Phi(x, t)$, upon using the operation of maximization with respect to t, the inequality

$$\max_{0\leqslant\tau\leqslant t}\int_{\Omega_\tau}\left|\frac{\partial^{N-1}\boldsymbol{\omega}_x}{\partial t^{N-1}}\right|^2 dx\leqslant C_{14}(C_2, C_{13}, C_9, T, N, C_\Omega)\left[t\right.$$
$$\left.+\sum_{m=1}^{N}\int_0^t\left(\int_{\Omega_\tau}\left|\frac{\partial^{m-1}\Delta \mathbf{v}_2}{\partial t^{m-1}}\right|^4 dx\right)^{1/4} d\tau\right]\cdot\max_{0\leqslant\tau\leqslant t}\int_{\Omega_\tau}\left|\frac{\partial^{N-1}\boldsymbol{\omega}_x}{\partial t^{N-1}}\right|^2 dx,\ 0<t\leqslant T, \tag{1.18}$$

from which we conclude, as we did from the inequality (1.15) in proving Theorem 1.2, proceeding stepwise in t for $t\in[0, T]$, that $\omega(x, t)=0$ in Q_T. It is obvious that uniqueness theorems analogous to the Theorems 1.1 and 1.3 also hold for the solutions of the initial-boundary value problem (0.26), (0.25).

We return now to the two-dimensional initial-boundary value problem (0.28), (0.19), (0.29). By a generalized solution of this problem we shall mean a function $\psi(x, t)\in L_2(\Omega_t)$ for which the following exist and belong to $L_2(\Omega_t)$, $0\leq t\leq T$: all derivatives of ψ appearing in the equation (0.28) and all the derivatives $\partial^{l-1}\Delta^2\psi_x/\partial t^{l-1}\in L_{4,1}(Q_T)$, $l=1, \ldots, N$; the function ψ is also to satisfy the equation (0.28) (in the sense of equality in $L_1(\Omega_t)$, $0<t\leq T$) as well as the boundary conditions (0.19) and the initial conditions (0.29). We then have

Theorem 1.4. *In the two-dimensional bounded domain Ω with a boundary belonging to the class C^2 the initial-boundary value problem* (0.28), (0.19), (0.29) *has at most one generalized solution.*

Suppose, in fact, that ψ_1 and ψ_2 are two possible generalized solutions of this problem, and that $\omega=\psi_1-\psi_2$. Then ω satisfies (in the sense of equality in $L_1(\Omega_t)$, $0<t\leq T$) the equation

$$\frac{\partial}{\partial t}\left(\Delta\omega-\sum_{l=1}^{N}\varkappa_l\frac{\partial^{l-1}\Delta^2\omega}{\partial t^{l-1}}\right)-\nu\Delta^2\omega-\left\{\frac{\partial\psi_1}{\partial x_1}\frac{\partial}{\partial x_2}\left(\Delta\omega-\sum_{l=1}^{N}l\varkappa_l\frac{\partial^{l-1}\Delta^2\omega}{\partial t^{l-1}}\right)\right.$$
$$\left.-\frac{\partial\psi_1}{\partial x_2}\frac{\partial}{\partial x_1}\left(\Delta\omega-\sum_{l=1}^{N}l\varkappa_l\frac{\partial^{l-1}\Delta^2\omega}{\partial t^{l-1}}\right)\right\}-\left\{\frac{\partial\omega}{\partial x_1}\frac{\partial}{\partial x_2}\left(\Delta\psi_2-\sum_{l=1}^{N}l\varkappa_l\frac{\partial^{l-1}\Delta^2\psi_2}{\partial t^{l-1}}\right)\right.$$
$$\left.-\frac{\partial\omega}{\partial x_2}\frac{\partial}{\partial x_1}\left(\Delta\psi_2-\sum_{l=1}^{N}l\varkappa_l\frac{\partial^{l-1}\Delta^2\psi_2}{\partial t^{l-1}}\right)\right\}=0,\ (x, t)\in Q_T, \tag{1.19}$$

and the homogeneous initial-boundary conditions

$$\left.\frac{\partial^m \omega}{\partial t^m}\right|_{t=0}=0,\ x\in\Omega,\ m=0,\ 1,\ \ldots,\ N-1;\ \omega|_{\partial Q_T}=\left.\frac{\partial\omega}{\partial n}\right|_{\partial Q_T}=0. \tag{1.20}$$

From the boundary conditions (1.20) (or, equivalently, (0.19)) it follows that

$$\left.\frac{\partial\omega}{\partial x_1}\right|_{\partial Q_T}=\left.\frac{\partial\omega}{\partial x_2}\right|_{\partial Q_T}=0. \tag{1.21}$$

Further, it is readily seen (see [2], Chapter 1, or [15]) that $\omega(x,\ t)$ obeys the inequalities

$$\int_{\Omega_t}|\nabla\omega|^4\,dx\leqslant C_\Omega^2\left(\int_{\Omega_t}|\Delta\omega|^2\,dx\right)^2,\ \ 0\leqslant t\leqslant T, \tag{1.22}$$

$$\int_{\Omega_t}|\Delta\omega|^2\,dx\leqslant C(\Omega)\int_{\Omega_t}|\Delta^2\omega|^2\,dx,\ \ 0\leqslant t\leqslant T, \tag{1.23}$$

and also the following inequalities analogous to (1.14):

$$\max_{0\leqslant\tau\leqslant t}\int_{\Omega_\tau}\left|\frac{\partial^{l-1}\Delta^2\omega}{\partial t^{l-1}}\right|^2 dx\leqslant C_{15}(T,\ N)\max_{0\leqslant\tau\leqslant t}\int_{\Omega_\tau}\left|\frac{\partial^{N-1}\Delta^2\omega}{\partial t^{N-1}}\right|^2 dx, \tag{1.24}$$

$$l=1,\ \ldots,\ N;\ 0<t\leqslant T.$$

We multiply the equation (1.19) by

$$\Delta\omega-\sum_{l=1}^N l\varkappa_l\frac{\partial^{l-1}\Delta^2\omega}{\partial t^{l-1}}$$

and integrate the result over Q_t, $0<t\leq T$. Then, using the fact that for arbitrary $\Phi(x,\ t)\in W_2^{1,0}(Q_T)$ and arbitrary $\psi_1(x,\ t)\in W_2^{2,0}(Q_T)$, satisfying the boundary conditions (0.19), and hence also the boundary conditions (1.21), the identity

$$\iint_{Q_t}\left(\frac{\partial\psi_1}{\partial x_1}\frac{\partial\Phi}{\partial x_2}\ \ \frac{\partial\psi_1}{\partial x_2}\frac{\partial\Phi}{\partial x_1}\right)\Phi dQ=0 \tag{1.25}$$

holds, we obtain the equation

$$\iint_{Q_t}\frac{\partial}{\partial t}\left(\Delta\omega-\sum_{l=1}^N\varkappa_l\frac{\partial^{l-1}\Delta^2\omega}{\partial t^{l-1}}\right)\cdot\left(\Delta\omega-\sum_{l=1}^N l\varkappa_l\frac{\partial^{l-1}\Delta^2\omega}{\partial t^{l-1}}\right)dQ-\nu\iint_{Q_t}\Delta^2\omega\left(\Delta\omega\right.$$

$$\left.-\sum_{l=1}^N l\varkappa_l\frac{\partial^{l-1}\Delta^2\omega}{\partial t^{l-1}}\right)dQ-\iint_{Q_t}\left\{\frac{\partial\omega}{\partial x_1}\frac{\partial}{\partial x_2}\left(\Delta\psi_2-\sum_{l=1}^N l\varkappa_l\frac{\partial^{l-1}\Delta^2\psi_2}{\partial t^{l-1}}\right)\right.$$

$$\left.-\frac{\partial\omega}{\partial x^2}\frac{\partial}{\partial x_1}\left(\Delta\psi_2-\sum_{l=1}^N l\varkappa_l\frac{\partial^{l-1}\Delta^2\psi_2}{\partial t^{l-1}}\right)\right\}$$

$$\times\left(\Delta\omega-\sum_{l=1}^N l\varkappa_l\frac{\partial^{l-1}\Delta^2\omega}{\partial t^{l-1}}\right)dQ=0,\ 0<t\leqslant T. \tag{1.26}$$

From (1.26), applying the Hölder and Cauchy inequalities, and using (1.22)–(1.24) and the operation of maximization with respect to t (cf. the method of obtaining (1.18)), we obtain

$$\max_{0\leqslant\tau\leqslant t}\int_{\Omega_\tau}\left|\frac{\partial^{N-1}\Delta^2\omega}{\partial t^{N-1}}\right|^2 dx\leqslant C_{16}\left[t+\sum_{m=1}^{N}\left(\int_0^t\left|\frac{\partial^{m-1}\Delta^2\psi_{2x}}{\partial t^{m-1}}\right|^4 dx\right)^{1/4} d\tau\right]$$
$$\times\max_{0\leqslant\tau\leqslant t}\int_{\Omega_\tau}\left|\frac{\partial^{N-1}\Delta^2\omega}{\partial t^{N-1}}\right|^2 dx, \qquad (1.27)$$
$$0<t\leqslant T,$$

from which, as in the proof of Theorems 1.2 and 1.3, we obtain, proceeding by steps in t for $t\in[0, T]$, upon using (1.24) and (1.23), the result $\omega(x, t)\equiv 0$ in Q_T.

§2. A priori estimates for solutions of the initial-boundary value problem (0.23), (0.25), (0.26)

Let Ω be a two-dimensional or three-dimensional bounded domain with boundary $\partial\Omega$ of class C^2. By a strong generalized solution of the initial-boundary value problem (0.23), (0.25), (0.26) we mean a function $\mathbf{v}(x, t)\in\mathring{J}(Q_T)$ which has generalized derivatives

$$\mathbf{v}_x,\ \mathbf{v}_t,\ \mathbf{v}_{xx},\ \frac{\partial^l\mathbf{v}_x}{\partial t^l},\ \frac{\partial^l\Delta\mathbf{v}}{\partial t^l},\quad l=1,\ \ldots,\ N-1,$$

belonging to $L_2(\Omega_t)$, $0\leq t\leq T$, and derivatives $\partial^l\Delta\mathbf{v}_x/\partial t^l$, $l=1, \ldots, N-1$, and $\partial^N\Delta\mathbf{v}/\partial t^N$, belonging to $L_2(Q_T)$, and satisfies the initial-boundary conditions (0.25), (0.26) and the integral identity

$$L_t(\mathbf{v},\ \mathbf{\Phi})\equiv\iint_{Qt}\left\{\left(\mathbf{v}_t-\nu\Delta\mathbf{v}-\sum_{l=1}^{N-1}\varkappa_l\frac{\partial^l\Delta\mathbf{v}}{\partial t^l}\right)\mathbf{\Phi}+\varkappa_N\frac{\partial^N\mathbf{v}_x}{\partial t^N}\mathbf{\Phi}_x\right.$$
$$\left.-v_k\left(\mathbf{v}-\sum_{l=1}^{N}l\varkappa_l\frac{\partial^{l-1}\Delta\mathbf{v}}{lt^{l-1}}\right)\mathbf{\Phi}_{xk}\right\}dQ=\iint_{Qt}\mathbf{f\Phi}dQ,\ 0<t\leqslant T, \qquad (2.1)$$

for arbitrary $\mathbf{\Phi}(x, t)\in\mathring{J}(Q_T)\cap W_2^{1,0}(Q_T)$. Our main effort in this section is to derive an a priori global estimate for a strong generalized solution of the problem (0.23), (0.25), (0.26).

Theorem 2.1. *Let* $\mathbf{f}(x, t)\in L_2(Q_T)$, $\mathbf{v}_{0m}(x)\in W_2^3(\Omega)\cap\mathring{W}_2^2(\Omega)\cap\mathring{J}(\Omega)$, $(\Delta\mathbf{v}_{0m})_n|_{\partial\Omega}=0$, $m=0, 1, \ldots, N-1$. *Then for an arbitrary strong generalized solution of the problem* (0.23), (0.25), (0.26) *we have the estimate*

$$I_T(\mathbf{v})\equiv\iint_{Q_T}\left|\frac{\partial^N\mathbf{v}_x}{\partial t^N}\right|^2 dQ+\sum_{l=0}^{N-1}\left\{\max_{0\leqslant t\leqslant T}\int_{\Omega_t}\left(\left|\frac{\partial^l\mathbf{v}_x}{\partial t^l}\right|^2+\left|\frac{\partial^l\Delta\mathbf{v}}{\partial t^l}\right|^2\right)dx\right.$$
$$\left.+\max_{Q_T}\left|\frac{\partial^l\mathbf{v}}{\partial t^l}\right|\right\}\leqslant C_1, \qquad (2.2)$$

where the constant C_1 *depends only on* $\|\mathbf{f}\|_{L_2(Q_T)}$, *the norms* $\|\mathbf{v}_{0m}\|_{W_2^3(\Omega)}$,

$m = 0, 1, \ldots, N-1$; κ_l, $l = 1, \ldots, N$ ($\kappa_N > 0$); *and* N *and* T; *moreover,* $C_1 \to \infty$ *if* $\kappa_N \to 0$.

To prove (2.2) we start by putting

$$\Phi = \mathbf{v} - \sum_{l=1}^{N} l\varkappa_l \frac{\partial^{l-1}\Delta\mathbf{v}}{\partial t^{l-1}}$$

in the integral identity (2.1). Then, using the fact that the integral containing the nonlinear terms is equal to zero, i.e.

$$\iint_{Q_t} v_k \frac{\partial}{\partial x_k}\left(\mathbf{v} - \sum_{l=1}^{N} l\varkappa_l \frac{\partial^{l-1}\Delta\mathbf{v}}{\partial t^{l-1}}\right)^2 dQ = 0, \quad 0 < t \leqslant T, \tag{2.3}$$

we obtain the equation

$$\iint_{Q_t}\left\{\frac{\partial}{\partial t}\left(\mathbf{v} - \sum_{l=1}^{N} \varkappa_l \frac{\partial^{l-1}\Delta\mathbf{v}}{\partial t^{l-1}}\right) - \nu\Delta\mathbf{v} - \mathbf{f}\right\}\left(\mathbf{v} - \sum_{l=1}^{N} l\varkappa_l \frac{\partial^{l-1}\Delta\mathbf{v}}{\partial t^{l-1}}\right) dQ = 0,$$
$$0 < t \leqslant T. \tag{2.4}$$

We extract from the left side of this equation the fundamental positive integrals

$$J_t \equiv \frac{1}{2}\sum_{l=1}^{N} l\varkappa_l^2 \int_{\Omega_t}\left|\frac{\partial^{l-1}\Delta\mathbf{v}}{\partial t^{l-1}}\right|^2 dx + \sum_{m=2}^{N} m\varkappa_{m-1}\varkappa_m \iint_{Q_t}\left|\frac{\partial^{m-1}\Delta\mathbf{v}}{\partial t^{m-1}}\right|^2 dQ + \nu\varkappa_1 \iint_{Q_t} |\Delta\mathbf{v}|^2 dQ$$
$$+ 2\nu\varkappa_2 \int_{\Omega_t} |\Delta\mathbf{v}|^2 dx + \frac{1}{2}\int_{\Omega_t} \mathbf{v}^2 dx + \nu \iint_{Q_t} \mathbf{v}_x^2 dQ, \quad 0 < t \leqslant T, \tag{2.5}$$

and we then estimate the remaining terms with the aid of the Hölder and Cauchy inequalities and an inequality of the type (1.6), obtainable from the Newton-Leibniz formula,

$$\iint_{Q_t}\left|\frac{\partial^{l-1}\omega}{\partial t^{l-1}}\right|^2 dQ \leqslant 2T\int_{\Omega_0}\left|\frac{\partial^{l-1}\omega}{\partial t^{l-1}}(x, 0)\right|^2 dx + 2T^2 \iint_{Q_t}\left|\frac{\partial^{l}\omega}{\partial t^{l}}\right|^2 dQ, \; 0 < t \leqslant T, \tag{2.6}$$

in which we substitute for ω the values $\Delta\mathbf{v}$ ($l = 1, \ldots, N-1$) and $\mathbf{v}$ (for $l = 2$). Collecting all these terms, we obtain the following inequality from (2.4):

$$J_t \leqslant C_2 \sum_{l=1}^{N} \iint_{Q_t}\left|\frac{\partial^{l-1}\Delta\mathbf{v}}{\partial t^{l-1}}\right|^2 dQ + C_3 \sum_{m=0}^{N-1} \|\mathbf{v}_{0m}\|^2_{W_2^3(\Omega)} + \|\mathbf{f}\|^2_{L_2(Q_t)}, \quad 0 < t \leqslant T, \tag{2.7}$$

where the constants C_2 and C_3 depend only on κ_l, $l = 1, \ldots, N$, and T and N. From (2.7), with the aid of Gronwall's lemma (see [2]) and Sobolev's imbedding theorems, we obtain

$$\sum_{l=0}^{N-1}\left\{\max_{0 \leqslant t \leqslant T}\int_{\Omega_t}\left(\left|\frac{\partial^l\Delta\mathbf{v}}{\partial t^l}\right|^2 + \left|\frac{\partial^l\mathbf{v}_x}{\partial t^l}\right|\right) dx + \max_{Q_T}\left|\frac{\partial^l\mathbf{v}}{\partial t^l}\right|\right\} \leqslant C_4, \tag{2.8}$$

where the constant C_4 depends only on $\|\mathbf{f}\|_{L_2(Q_T)}$, $\|\mathbf{v}_{0m}\|_{W_2^3(\Omega)}$, $m = 0, 1, \ldots, N - 1$; κ_l, $l = 1, \ldots, N$; T and N.

In the integral identity (2.1) we now put $\Phi = \partial^N \mathbf{v}/\partial t^N$. Integrating by parts, we obtain the equation

$$\varkappa_N \iint_{Q_T} \left| \frac{\partial^N \mathbf{v}_x}{\partial t^N} \right|^2 dQ = -\iint_{Q_T} [\mathbf{v}_t - \mathbf{f}] \frac{\partial^N \mathbf{v}}{\partial t^N} dQ - \iint_{Q_T} \left(\nu \mathbf{v}_x + \sum_{l=1}^{N-1} \varkappa_l \frac{\partial^l \mathbf{v}_x}{\partial t^l} \right) \frac{\partial^N \mathbf{v}_x}{\partial t^N} dQ$$
$$+ \iint_{Q_T} v_k \left(\mathbf{v} - \sum_{l=1}^{N} l \varkappa_l \frac{\partial^{l-1} \Delta \mathbf{v}}{\partial t^{l-1}} \right) \frac{\partial^N}{\partial t^N} \frac{\partial \mathbf{v}}{\partial x_k} dQ, \tag{2.9}$$

from which, upon applying Hölder's and Friedrichs' inequalities to estimate the integrals on the right-hand side and using the estimate (2.8), we obtain

$$\iint_{Q_T} \left| \frac{\partial^N \mathbf{v}_x}{\partial t^N} \right|^2 dQ \leqslant C_5 \left(\frac{1}{\varkappa_N},\ C_4,\ \|\mathbf{f}\|_{L_2(Q_T)} \right). \tag{2.10}$$

Adding (2.8) and (2.10), we obtain (2.2) with $C_1 - C_4 + C_5$.

Besides the system (0.23), we consider a quasilinear system with a nonlinearity much weaker than that in (0.23), namely,

$$\frac{\partial \mathbf{v}}{\partial t} - \nu \Delta \mathbf{v} + a_k(x, t) \frac{\partial \mathbf{v}}{\partial x_k} - \sum_{l=1}^{N} \varkappa_l \left(\frac{\partial^l \Delta \mathbf{v}}{\partial t^l} + l v_k \frac{\partial^l \Delta \mathbf{v}}{\partial t^{l-1} \partial x_k} \right) + \operatorname{grad} p = \mathbf{f},$$
$$\operatorname{div} \mathbf{v} = 0, \quad N = 2, 3, \ldots, \tag{2.11}$$

where $\mathbf{a}(x, t)$ is a known vector belonging to $\mathring{J}(Q_T) \cap C(Q_T)$. As before, we solve the system (2.11) in Q_T, subject to the initial-boundary conditions (0.25), (0.26), and we define a strong generalized solution of the initial-boundary value problem (2.11), (0.25), (0.26) to be a function $\mathbf{v} \in \mathring{J}(Q_T) \cap C(Q_T)$ which has generalized derivatives $\mathbf{v}_x$, $\mathbf{v}_t$, $\mathbf{v}_{xx}$, $\partial^l \mathbf{v}_x/\partial t^l$, $\partial^l \Delta \mathbf{v}/\partial t^l$, $l = 1, \ldots, N - 1$, belonging to $L_2(\Omega_t)$, $0 \le t \le T$, $\partial^l \Delta \mathbf{v}_x/\partial t^l \in L_2(Q_T)$, $l = 1, \ldots, N - 1$, and the derivative $\partial^N \mathbf{v}_x/\partial t^N \in L_2(Q_T)$; in addition, this function is to satisfy the initial-boundary conditions (0.25), (0.26) and also the integral identity

$$\iint_{Q_t} \left\{ \left(\mathbf{v}_t - \nu \Delta \mathbf{v} - \sum_{l=1}^{N-1} \varkappa_l \frac{\partial^l \Delta \mathbf{v}}{\partial t^l} \right) \Phi + \varkappa_N \frac{\partial^N \mathbf{v}_x}{\partial t^N} \Phi_x + a_k \mathbf{v}_{k_x} \Phi \right.$$
$$\left. + v_k \sum_{l=1}^{N} l \varkappa_l \frac{\partial^{l-1} \Delta \mathbf{v}}{\partial t^{l-1}} \Phi_{x_k} \right\} dQ = \iint_{Q_t} \mathbf{f} \Phi \, dQ, \quad 0 < t \leqslant T, \tag{2.12}$$

for arbitrary $\Phi(x, t) \in \mathring{J}(Q_T) \cap W_2^{1,0}(Q_T)$. For strong generalized solutions of the problem (2.11), (0.25), (0.26) we have the following theorem, analogous to Theorem 2.1.

Theorem 2.2. *Let* $\mathbf{f}(x, t) \in L_2(Q_T)$, $\mathbf{v}_{0m}(x) \in W_2^3(\Omega) \cap \mathring{W}_2^2(\Omega) \cap \mathring{J}(\Omega)$, and $(\Delta \mathbf{v}_{0m})_n|_{\partial\Omega} = 0$, $m = 0, 1, \ldots, N - 1$. *Then the following estimate holds for an arbitrary strong generalized solution of the initial-boundary value problem* (2.11), (0.25), (0.26):

$$I_T(\mathbf{v}) \leqslant C_6, \tag{2.13}$$

where the constant C_6 *depends only on the same quantities as* C_1 *and, in addition, on* $\max_{Q_T}|\mathrm{a}(x, t)|$.

We note that to obtain an analog of (2.8) it is necessary to put

$$\Phi = \sum_{l=1}^{N} l\varkappa_l \frac{\partial^{l-1}\Delta \mathbf{v}}{\partial t^{l-1}}.$$

in the integral identity (2.12).

It is of interest to examine the question of how solutions of the initial-boundary value problem (2.11), (0.25), (0.26) behave as the shear viscosity coefficients κ_l, $l = 1, \ldots, N$,(1) tend towards zero. For $N = 1$ this problem was studied in [8]. We consider here the case $N = 2$, $\kappa_1 = 0$, and prove the following theorem.

Theorem 2.3. *In the initial-boundary value problem* (2.11), (0.25), (0.26) *let* $N = 2$, $\kappa_1 = 0$ *and* $\kappa_2 \equiv \kappa > 0$. *Let* $\mathrm{f}(x, t) \in \overset{\circ}{W}{}_2^{1,1}(Q_T) \cap L_2(\Omega_t)$, $0 \leq t \leq T$; $\mathrm{v}_{0m}(x) \in W_2^3(\Omega) \cap \overset{\circ}{W}{}_2^2(\Omega) \cap \overset{\circ}{J}(\Omega)$, $(\Delta \mathrm{v}_{0m})_n|_{\partial\Omega} = 0$, $m = 0, 1$; *and* $\mathrm{a}(x, t) \in \overset{\circ}{J}(Q_T) \cap \overset{\circ}{C}{}^{1,1}(Q_T)$. *Then for an arbitrary strong generalized solution* $\mathrm{v}^{\kappa}(x, t)$ *of the initial-boundary value problem* (2.11), (0.25), (0.26) *we have the following estimate for each* $\kappa > 0$:

$$\varkappa\left\{\iint_{Q_T}\left|\frac{\partial^2 \mathbf{v}_x^\varkappa}{\partial t^2}\right|^2 dQ + \max_{0\leqslant t\leqslant T}\int_{\Omega_t}\left|\frac{\partial \Delta \mathbf{v}^\varkappa}{\partial t}\right|^2 dx\right\} + \iint_{Q_T}\left|\frac{\partial \mathbf{v}_x^\varkappa}{\partial t}\right|^2 dQ$$

$$+ \max_{0\leqslant t\leqslant T}\int_{\Omega_t}\left(\left|\frac{\partial \mathbf{v}^\varkappa}{\partial t}\right|^2 + |\Delta \mathbf{v}^\varkappa|^2\right) dx \leqslant C_7, \tag{2.14}$$

where the constant C_7 *depends only on* $\|\mathrm{f}\|_{W_2^{1,1}(Q_T)}$, $\max_{0\le t\le T}\|\mathrm{f}\|_{L_2(\Omega_t)}$, $\|\mathrm{a}\|_{C^{1,1}(Q_T)}$, $\|\mathrm{v}_{0m}\|_{W_2^3(\Omega)}$, $m = 0, 1$; *and on* T *and* ν *of the domain* Ω, *and is independent of* $\kappa > 0$. *As* $\kappa \to 0$, $\mathrm{v}^{\kappa}(x, t)$ *tends towards a unique solution, belonging to* $\overset{\circ}{W}{}_2^{2,1}(Q_T) \cap \overset{\circ}{J}(Q_T)$, *of the following initial-boundary value problem for the system of Navier-Stokes equations, linearized in the Oseen sense* (*see* [2]):

$$\frac{\partial \mathbf{v}}{\partial t} - \nu\Delta\mathbf{v} + a_k(x, t)\frac{\partial \mathbf{v}}{\partial x_k} + \operatorname{grad} p = \mathbf{f}, \ \operatorname{div}\mathbf{v} = 0, \ (x, t) \in Q_T;$$

$$\mathbf{v}|_{t=0} = \mathbf{v}_0(x), \ \mathbf{v}|_{\partial Q_T} = 0. \tag{2.15}$$

We begin by proving the last assertion of the theorem, assuming (2.14) to have already been proved. It is obvious that a strong generalized solution of the problem (2.11), (0.25), (0.26) satisfies, for $N = 2$, $\kappa_1 = 0$ and $\kappa_2 = \kappa$, the integral identity

---

(1) For the initial-boundary value problem (0.23), (0.25), (0.26) this question has as yet not been resolved.

$$\iint\limits_{Q_t}\Big[(\mathbf{v}_t^\varkappa-\nu\Delta\mathbf{v}^\varkappa)\,\Phi+\varkappa\frac{\partial^2\mathbf{v}_x^\varkappa}{\partial t^2}\,\Phi_x+a_k\mathbf{v}_{xk}^\varkappa\Phi+2\varkappa\mathbf{v}_k^\varkappa\frac{\partial\Delta\mathbf{v}^\varkappa}{\partial t}\,\Phi_{xk}-\mathbf{f}\Phi\Big]dQ=0,$$

$$0<t\leqslant T, \qquad (2.16)$$

for arbitrary $\Phi\in\mathring{J}(Q_T)\cap W_2^{1,0}(Q_T)$. By virtue of (2.14) we can select from the set $\{\mathbf{v}^\kappa\}$ of strong generalized solutions of the problem (2.11), (0.25), (0.26) at least one subsequence $\{\mathbf{v}^{\kappa_i}\}$ which as $\kappa_i\to 0$ converges strongly in $L_2(Q_T)$ to a limit function $\mathbf{v}(x,t)\in L_2(Q_T)$, and for which $\mathbf{v}_x^{\kappa_i}$, $\mathbf{v}_t^{\kappa_i}$, $\Delta\mathbf{v}^{\kappa_i}$ and $\partial\mathbf{v}_x^{\kappa_i}/\partial t$ converge weakly in $L_2(Q_T)$ to $\mathbf{v}_x$, $\mathbf{v}_t$, $\Delta\mathbf{v}$ and $\partial\mathbf{v}_x/\partial t$, respectively. In addition, by virtue of (2.14),

$$\varkappa_i\iint\limits_{Q_T}\Big\{\frac{\partial\Delta\mathbf{v}^{\varkappa_i}}{\partial t}\,\Phi+\frac{\partial^2\mathbf{v}_x^{\varkappa_i}}{\partial t^2}\,\Phi_x+2v_k^{\varkappa_i}\frac{\partial\Delta\mathbf{v}^{\varkappa_i}}{\partial t}\,\Phi_{xk}\Big\}dQ\to 0,\ \ \varkappa_i\to 0. \qquad (2.17)$$

If, after this, we pass to the limit as $\kappa_i\to 0$ in (2.16), we find that the limit function $\mathbf{v}(x,t)\in\mathring{W}_2^{2,1}(Q_T)\cap\mathring{J}(Q_T)$ satisfies the integral identity

$$\iint\limits_{Q_t}(\mathbf{v}_t-\nu\Delta\mathbf{v}+a_k\mathbf{v}_{xk})\,\Phi dQ=\iint\limits_{Q_t}\mathbf{f}\Phi dQ,\ 0<t\leqslant T, \qquad (2.18)$$

from which, as a result of the arbitrariness in the choice of the function Φ, we find that $\mathbf{v}(x,t)$ satisfies the Oseen system (2.15). Satisfaction of the initial Cauchy condition $\mathbf{v}|_{t=0}=\mathbf{v}_0(x)$ is proved by known methods (see [2], Chapter 6).

We now prove (2.14). To do this we begin by putting $\Phi=\partial\Delta\mathbf{v}^\kappa/\partial t$ in the integral identity (2.12), defining a strong generalized solution of the problem (2.11), (0.25), (0.26) with $N=2$, $\kappa_1=0$ and $\kappa_2=\kappa$. Then, using the fact that the integral with the nonlinear term is equal to zero, i.e.

$$\iint\limits_{Q_t}v_k^\varkappa\frac{\partial\Delta\mathbf{v}^\varkappa}{\partial t}\Big(\frac{\partial\Delta\mathbf{v}^\varkappa}{\partial t}\Big)_{xk}dQ=0,\ 0<t\leqslant T, \qquad (2.19)$$

we obtain the equation

$$J_\varkappa\equiv\iint\limits_{Q_t}\Big|\frac{\partial\mathbf{v}_x^\varkappa}{\partial t}\Big|^2dQ+\frac{\nu}{2}\int\limits_{\Omega_t}|\Delta\mathbf{v}^\varkappa|^2dx+\frac{\varkappa}{2}\int\limits_{\Omega_t}\Big|\frac{\partial\Delta\mathbf{v}^\varkappa}{\partial t}\Big|^2dx=\frac{\nu}{2}\int\limits_{\Omega_0}|\Delta\mathbf{v}_{00}|^2dx$$

$$+\frac{\varkappa}{2}\int\limits_{\Omega_0}|\Delta\mathbf{v}_{01}|^2dx-\iint\limits_{Q_t}f\frac{\partial\Delta\mathbf{v}^\varkappa}{\partial t}dQ+\iint\limits_{Q_t}a_kv_{xk}^\varkappa\frac{\partial\Delta\mathbf{v}^\varkappa}{\partial t}dQ,\ 0<t\leqslant T. \qquad (2.20)$$

Integrating by parts in the third and fourth integrals of (2.20) and applying the Hölder and Cauchy inequalities, we have

$$-\iint\limits_{Q_t}f\frac{\partial\Delta\mathbf{v}^\varkappa}{\partial t}dQ=\iint\limits_{Q_t}f_x\frac{\partial\mathbf{v}_x^\varkappa}{\partial t}dQ\leqslant\delta\iint\limits_{Q_t}\Big|\frac{\partial\mathbf{v}_x^\varkappa}{\partial t}\Big|^2dQ$$

$$+C_\delta\iint\limits_{Q_t}|\mathbf{f}_x|^2dQ,\quad\forall\delta>0, \qquad (2.21)$$

$$\iint\limits_{Q_t} a_k v^{\varkappa}_{x_k} \frac{\partial \Delta \mathbf{v}^{\varkappa}}{\partial t}\, dQ = -\iint\limits_{Q_t} \frac{\partial^2 \mathbf{v}^{\varkappa}}{\partial t \partial x_i}\left(a_k \frac{\partial^2 \mathbf{v}^{\varkappa}}{\partial x_i \partial x_k} + \frac{\partial a_k}{\partial x_i}\frac{\partial \mathbf{v}^{\varkappa}}{\partial x_k}\right) dQ \leqslant \delta \iint\limits_{Q_t} \left|\frac{\partial \mathbf{v}^{\varkappa}_x}{\partial t}\right|^2 dQ$$
$$+ C_8\left(\frac{1}{\delta},\ \max_{Q_T}|\mathbf{a}|,\ |\mathbf{a}_x|\right)\iint\limits_{Q_t}(|\mathbf{v}^{\varkappa}_{xx}|^2 + |\mathbf{v}^{\varkappa}_x|^2)\, dQ,\quad \forall \delta > 0. \tag{2.22}$$

Then, from (2.20), choosing $\delta > 0$ in (2.21) and (2.22) sufficiently small, we obtain

$$J_{\varkappa} \leqslant C_9(\|\mathbf{v}_{0m}\|_{W^2_2(\Omega)}) + C_{10}\iint\limits_{Q_t}|\mathbf{f}_x|^2 dQ + C_{11}(C_8)\iint\limits_{Q_t}(|\mathbf{v}^{\varkappa}_{xx}|^2$$
$$+ |\mathbf{v}^{\varkappa}_x|^2)\, dQ,\quad 0 < t \leqslant T, \tag{2.23}$$

from which, applying Sobolev's imbedding theorems, the second fundamental inequality for elliptic operators, and the Gronwall lemma, we obtain (cf. [2], Chapter 6)

$$\iint\limits_{Q_T}\left|\frac{\partial \mathbf{v}^{\varkappa}_x}{\partial t}\right|^2 dQ + \max_{0\leqslant t\leqslant T}\int\limits_{\Omega_t}\left[\varkappa\left|\frac{\partial \Delta \mathbf{v}^{\varkappa}}{\partial t}\right|^2 + \nu|\Delta \mathbf{v}^{\varkappa}|^2\right] dx \leqslant C_{12}; \tag{2.24}$$

here C_{12} depends only on $\|\mathbf{f}\|_{L_2(Q_T)}$, $\|\mathbf{v}_{0m}\|_{W^2_2(\Omega)}$, $\max_{Q_T}(|\mathbf{a}|, |\mathbf{a}_x|)$, T and ν, and is independent of $\kappa > 0$.

In (2.12) we now put $\Phi = \partial^2 \mathbf{v}^{\kappa}/\partial t^2$. We obtain

$$\frac{1}{2}\int\limits_{\Omega_t}\left|\frac{\partial \mathbf{v}^{\varkappa}}{\partial t}\right|^2 dx + \varkappa\iint\limits_{Q_t}\left|\frac{\partial^2 \mathbf{v}^{\varkappa}_x}{\partial t^2}\right|^2 dQ = \frac{1}{2}\int\limits_{\Omega_0}|\mathbf{v}_{01}|^2 dx + \iint\limits_{Q_t} f\frac{\partial^2 \mathbf{v}^{\varkappa}}{\partial t^2}\, dQ$$
$$+ \iint\limits_{Q_t} a_k \mathbf{v}^{\varkappa}_{x_k}\frac{\partial^2 \mathbf{v}^{\varkappa}}{\partial t^2}\, dQ - 2\varkappa\iint\limits_{Q_t}\mathbf{v}^{\varkappa}_k\frac{\partial \Delta \mathbf{v}^{\varkappa}}{\partial t}\left(\frac{\partial^2 \mathbf{v}^{\varkappa}}{\partial t^2}\right)_{x_k} dQ$$
$$+ \nu\iint\limits_{Q_t}\Delta \mathbf{v}^{\varkappa}\frac{\partial^2 \mathbf{v}^{\varkappa}}{\partial t^2}\, dQ,\quad 0 < t \leqslant T. \tag{2.25}$$

To estimate the integrals on the right side of (2.25) we integrate by parts with respect to x and t, apply the Hölder and Cauchy inequalities, and use (2.24). In addition, we make use of the fact that, by virtue of (2.24) and the Sobolev imbedding theorems,

$$\max_{Q_T}|\mathbf{v}^{\varkappa}| \leqslant C_{13}(C_{12},\ \nu,\ \Omega). \tag{2.26}$$

We then have

$$\iint\limits_{Q_t}\mathbf{f}\frac{\partial^2 \mathbf{v}^{\varkappa}}{\partial t^2}\, dQ = \int\limits_{\Omega_t}\mathbf{f}\frac{\partial \mathbf{v}^{\varkappa}}{\partial t}\, dx\Bigg|_{t=0}^{t=t} - \iint\limits_{Q_t}\frac{\partial \mathbf{f}}{\partial t}\frac{\partial \mathbf{v}^{\varkappa}}{\partial t}\, dQ \leqslant \delta\iint\limits_{Q_t}\left|\frac{\partial \mathbf{v}^{\varkappa}}{\partial t}\right|^2 dQ$$
$$+ \delta\int\limits_{\Omega_t}\left|\frac{\partial \mathbf{v}^{\varkappa}}{\partial t}\right|^2 dx + C_{\delta}\int\limits_{\Omega_t}|\mathbf{f}|^2 dx + \frac{1}{2}\int\limits_{\Omega_0}(\mathbf{f}^2 + \mathbf{v}^2_{01})\, dx,\quad \forall \delta > 0; \tag{2.27}$$

$$\iint\limits_{Q_t} a_k \mathbf{v}^{\varkappa}_{x_k}\frac{\partial^2 \mathbf{v}^{\varkappa}}{\partial t^2}\, dQ = \int\limits_{\Omega_t} a_k \mathbf{v}^{\varkappa}_{x_k}\frac{\partial \mathbf{v}^{\varkappa}}{dt}\, dx\Bigg|_{t=0}^{t=t} - \iint\limits_{Q_t}\frac{\partial \mathbf{v}^{\varkappa}}{\partial t}\left(\frac{\partial a_k}{\partial t}\mathbf{v}^{\varkappa}_{x_k} + a_k\frac{\partial^2 \mathbf{v}^{\varkappa}}{\partial t \partial x_k}\right) dQ$$
$$\leqslant \delta\int\limits_{\Omega_t}\left|\frac{\partial \mathbf{v}^{\varkappa}}{\partial t}\right|^2 dx + C_{14}\left(\frac{1}{\delta},\ C_{12},\ \max_{Q_t}|\mathbf{a}|,\ |\mathbf{a}_t|,\ \|\mathbf{v}_{00}\|_{W^1_2(\Omega)},\right.$$
$$\left.\|\mathbf{v}_{01}\|_{L_2(\Omega)}\right),\quad \forall \delta > 0; \tag{2.28}$$

$$-2\varkappa \iint\limits_{Q_t} v_k^{\varkappa} \frac{\partial \Delta \mathbf{v}^{\varkappa}}{\partial t} \frac{\partial^3 \mathbf{v}^{\varkappa}}{\partial t^2 \partial x_k} dQ \leqslant C_{15}(C_{13}) \left[\varkappa \iint\limits_{Q_t} \left| \frac{\partial \Delta \mathbf{v}^{\varkappa}}{\partial t} \right|^2 dQ \right]^{1/2} \times \left[\varkappa \iint\limits_{Q_t} \left| \frac{\partial^3 \mathbf{v}^{\varkappa}}{\partial t^2 \partial x} \right|^2 dQ \right]^{1/2} \leqslant \varkappa\delta \iint\limits_{Q_t} \left| \frac{\partial^2 \mathbf{v}_x^{\varkappa}}{\partial t^2} \right|^2 dQ + C_{16}\left(\frac{1}{\delta}, C_{15}, C_{12}, T\right), \ \forall \delta > 0; \tag{2.29}$$

$$\nu \iint\limits_{Q_t} \Delta \mathbf{v}^{\varkappa} \frac{\partial^2 \mathbf{v}^{\varkappa}}{\partial t^2} dQ = \nu \int\limits_{\Omega_t} \Delta \mathbf{v}^{\varkappa} \frac{\partial \mathbf{v}^{\varkappa}}{\partial t} dx \Big|_{t=0}^{t=t} - \nu \iint\limits_{Q_t} \frac{\partial \Delta \mathbf{v}^{\varkappa}}{\partial t} \frac{\partial \mathbf{v}^{\varkappa}}{\partial t} dQ = \nu \int\limits_{\Omega_t} \Delta \mathbf{v}^{\varkappa} \frac{\partial \mathbf{v}^{\varkappa}}{\partial t} dx \Big|_{t=0}^{t=t} + \nu \iint\limits_{Q_t} \left| \frac{\partial \mathbf{v}_x^{\varkappa}}{\partial t} \right|^2 dQ \leqslant \delta \int\limits_{\Omega_t} \left| \frac{\partial \mathbf{v}^{\varkappa}}{\partial t} \right|^2 dx + C_{17}\left(\frac{1}{\delta}, C_{12}, \|\mathbf{v}_{00}\|_{W_2^2(\Omega)}, \|\mathbf{v}_{01}\|_{L_2(\Omega)}, \nu\right), \ \forall \delta > 0. \tag{2.30}$$

From (2.27)–(2.30), in which $\delta > 0$ is chosen sufficiently small, and the Gronwall lemma, we obtain from (2.25) the estimate

$$\max_{0 \leqslant t \leqslant T} \int\limits_{\Omega_t} \left| \frac{\partial \mathbf{v}^{\varkappa}}{\partial t} \right|^2 dx + \varkappa \iint\limits_{Q_T} \left| \frac{\partial^2 \mathbf{v}_x^{\varkappa}}{\partial t^2} \right|^2 dQ \leqslant C_{18}, \tag{2.31}$$

in which the constant C_{18} depends only on $\|\mathbf{f}\|_{W_2^{1,1}(Q_T)}$, $\max_{0 \le t \le T} \|\mathbf{f}\|_{L_2(\Omega_t)}$, $\|\mathbf{v}_{0m}\|_{W_2^2(\Omega)}$, $m = 0, 1$, $\|\mathbf{a}\|_{C^{1,1}(Q_T)}$, T, ν, and the domain Ω, and is independent of $\kappa > 0$.

Adding (2.24) and (2.31), we obtain the estimate (2.14) with $C_7 = C_{12} + C_{18}$. This completes the proof of Theorem 2.3.

§3. A priori estimates for solutions of the initial-boundary value problem (0.28), (0.19), (0.30), (0.29)

Let Ω be a two-dimensional bounded domain with boundary $\partial\Omega$ of class C^4. We define a strong generalized solution of the initial-boundary value problem (0.28), (0.19), (0.30), (0.29) to be a function $\psi(x, t)$, $(x, t) \in Q_T = \Omega \times (0, t)$, which has generalized derivatives $\Delta\psi_x$, $\Delta\psi_t$, $\Delta^2\psi$, $\partial^l \Delta\psi_x/\partial t^l$, $\partial^l \Delta^2\psi/\partial t^l$, $l = 1, 2, \ldots, N-1$, belonging to $L_2(\Omega_t)$, $0 \le t \le T$, and derivatives

$$\frac{\partial^l \Delta^2 \psi_x}{\partial t^l}, \quad l = 1, \ldots, N-1, \quad \frac{\partial^N \Delta\psi_x}{\partial t^N}, \quad \frac{\partial^N \Delta^2 \psi}{\partial t^N} \in L_2(Q_T),$$

and satisfies the initial-boundary conditions (0.19), (0.30), (0.29) and the integral identity

$$l_t(\psi, \Phi) \equiv \iint\limits_{Q_t} \left\{ \left(\Delta\psi_t - \nu\Delta^2\psi - \sum_{l=1}^{N-1} \varkappa_l \frac{\partial^l \Delta^2 \psi}{\partial t^l} \right) \Phi + \varkappa_N \frac{\partial^N \Delta\psi_x}{\partial t^N} \Phi_x - \left(\Delta\psi - \sum_{l=1}^{N} l\varkappa_l \frac{\partial^{l-1} \Delta^2 \psi}{\partial t^{l-1}} \right) \left(\frac{\partial\psi}{\partial x_2} \frac{\partial\Phi}{\partial x_1} - \frac{\partial\psi}{\partial x_1} \frac{\partial\Phi}{\partial x_2} \right) \right\} dQ = \iint\limits_{Q_t} g\Phi \, dQ, \ 0 < t \leqslant T, \tag{3.1}$$

for arbitrary $\Phi(x, t) \in W_2^{1,0}(Q_T)$. We then have

Theorem 3.1. *Let* $g(x, t) \in L_2(Q_T)$, $\psi_{0m}(x) \in W_2^5(\Omega)$ *and* $\psi_{0m\,\partial\Omega} = \partial\psi_{0m}/\partial n|_{\partial\Omega} = \partial\Delta\psi_{0m}/\partial n|_{\partial\Omega} = 0$, $m = 0, 1, \ldots, N-1$. *Then for an arbitrary strong generalized solution of the problem* (0.28), (0.19), (0.30), (0.29) *we have the estimate*

$$J_T(\psi) \equiv \iint_{Q_T} \left| \frac{\partial^N \Delta\psi_x}{\partial t^N} \right| dQ + \sum_{l=0}^{N-1} \left[\max_{0 \leqslant t \leqslant T} \int_{\Omega_t} \left(\left| \frac{\partial^l \Delta\psi_x}{\partial t^l} \right|^2 + \left| \frac{\partial^l \Delta^2\psi}{\partial t^l} \right|^2 \right) dx + \max_{Q_T} \left| \frac{\partial^l \Delta\psi}{\partial t^l} \right| \right] \leqslant C_1, \qquad (3.2)$$

where the constant C_1 *depends only on* $\|g\|_{L_2(Q_T)}$, *the norms* $\|\psi_{0m}\|_{W_2^5(\Omega)}$, $m = 0, 1, \ldots, N-1$; $\varkappa_l$, $l = 1, \ldots, N$ $(\varkappa_N > 0)$, *and* N *and* T; *moreover*, $C_1 \to \infty$ *if* $\varkappa_N \to 0$.

To prove (3.2) we start by putting

$$\Phi = \Delta\psi - \sum_{l=1}^{N} l\varkappa_l \frac{\partial^{l-1}\Delta^2\psi}{\partial t^{l-1}}$$

in the integral identity (3.1). Then, using the fact that the integral containing the nonlinear terms, namely

$$\iint_{Q_t} \left(\Delta\psi - \sum_{l=1}^{N} l\varkappa_l \frac{\partial^{l-1}\Delta^2\psi}{\partial t^{l-1}} \right) \frac{\partial}{\partial(x_1, x_2)} \left(\psi, \Delta\psi - \sum_{l=1}^{N} l\varkappa_l \frac{\partial^{l-1}\Delta^2\psi}{\partial t^{l-1}} \right) dQ = 0, \quad 0 < t \leqslant T, \qquad (3.3)$$

is equal to zero, we obtain the equation

$$\iint_{Q_t} \left(\Delta\psi_t - \nu\Delta^2\psi - \sum_{l=1}^{N} \varkappa_l \frac{\partial^l \Delta^2\psi}{\partial t^l} - g \right) \left(\Delta\psi - \sum_{l=1}^{N} l\varkappa_l \frac{\partial^{l-1}\Delta^2\psi}{\partial t^{l-1}} \right) dQ = 0, \quad 0 < t \leqslant T. \qquad (3.4)$$

In (3.4) we isolate the fundamental positive integrals:

$$I_t \equiv \frac{1}{2} \sum_{l=1}^{N} l\varkappa_l^2 \int_{\Omega_t} \left| \frac{\partial^{l-1}\Delta^2\psi}{\partial t^{l-1}} \right|^2 dx + \sum_{m=2}^{N} m\varkappa_m \varkappa_{m-1} \iint_{Q_t} \left| \frac{\partial^{m-1}\Delta^2\psi}{\partial t^{m-1}} \right|^2 dQ + \nu\varkappa_1 \iint_{Q_t} |\Delta^2\psi|^2 dQ + 2\nu\varkappa_2 \int_{\Omega_t} |\Delta^2\psi|^2 dx + \frac{1}{2} \int_{\Omega_t} |\Delta\psi|^2 dx + \nu \iint_{Q_t} |\Delta\psi_x|^2 dQ, \quad 0 < t \leqslant T, \qquad (3.5)$$

and we estimate the remaining terms with the aid of the Hölder and Cauchy inequalities and the inequality (2.6) with $\omega = \Delta^2\psi$ (for $l = 1, \ldots, N-1$) and $\omega = \Delta\psi$ (for $l = 2$). From (3.4) we then obtain

$$I_t \leqslant C_2(T, N, \varkappa_l) \sum_{l=1}^{N} \iint_{Q_t} \left| \frac{\partial^{l-1}\Delta^2\psi}{\partial t^{l-1}} \right|^2 dQ + C_3(T, N, \varkappa_l) \sum_{m=0}^{N-1} \|\psi_{0m}\|^2_{W_2^5(\Omega)} + \|g\|^2_{L_2(Q_t)}, \quad 0 < t \leqslant T. \qquad (3.6)$$

Next we obtain the following inequality from (3.6) with the aid of Gronwall's lemma ([2], Chapter 6) and the Sobolev imbedding theorems:

$$\sum_{l=0}^{N-1}\left[\max_{0\leqslant t\leqslant T}\int_{\Omega_t}\left(\left|\frac{\partial^l\Delta^2\psi}{\partial t^l}\right|^2+\left|\frac{\partial^l\Delta\psi_x}{\partial t^l}\right|^2\right)dx+\max_{Q_T}\left(\left|\frac{\partial^l\Delta\psi}{\partial t^l}\right|+\left|\frac{\partial^l\psi_x}{\partial t^l}\right|\right)\right]\leqslant C_4, \tag{3.7}$$

where the constant C_4 depends only on $\|g\|_{L_2(Q_T)}$, the norm $\|\psi_{0m}\|_{W_2^5(\Omega)}$, $m = 0, 1, \ldots, N-1$, and κ_l, $l = 1, \ldots, N$.

In (3.1) we now put $\Phi = \partial^N\Delta\psi/\partial t^N$. Integrating by parts, we obtain

$$\begin{aligned}\varkappa_N\iint_{Q_T}\left|\frac{\partial^N\Delta\psi_x}{\partial t^N}\right|^2 dQ &= -\iint_{Q_T}[\Delta\psi_t-g]\frac{\partial^N\Delta\psi}{\partial t^N}\,dQ\\ &-\iint_{Q_T}\left(\nu\Delta\psi_x+\sum_{l=1}^{N-1}\varkappa_l\frac{\partial^l\Delta\psi_x}{\partial t^l}\right)\frac{\partial^N\Delta\psi_x}{\partial t^N}\,dQ\\ &+\iint_{Q_T}\left(\Delta\psi-\sum_{l=1}^{N}l\varkappa_l\frac{\partial^{l-1}\Delta^2\psi}{\partial t^{l-1}}\right)\left(\frac{\partial\psi}{\partial x_2}\frac{\partial^N\Delta\psi_{x_1}}{\partial t^N}-\frac{\partial\psi}{\partial x_1}\frac{\partial^N\Delta\psi_{x_2}}{\partial t^N}\right)dQ,\end{aligned} \tag{3.8}$$

from which, upon applying Hölder's inequality and Cauchy's inequality with ϵ, and using (3.7) to estimate the integrals on the right-hand side, we obtain

$$\iint_{Q_T}\left|\frac{\partial^N\Delta\psi_x}{\partial t^N}\right|^2 dQ\leqslant C_5\left(\frac{1}{\varkappa_N},\ C_4,\ \|g\|_{L_2(Q_T)}\right). \tag{3.9}$$

Adding (3.7) and (3.9), we obtain (3.2) with $C_1 = C_4 + C_5$.

§4. A priori estimates for solutions of the Cauchy problems (0.23), (0.24) and (0.28), (0.29)

Let $R_T = E_n \times (0, T)$, $T > 0$, $n = 2, 3$. We define a generalized solution of the Cauchy problem (0.23), (0.24) of order $j \geq 1$ to be a function $\mathrm{v}(x, t) \in J(R_T)$ which, together with the derivatives $D_x^\alpha D_t^l\mathrm{v}$, $\alpha = 0, 1, \ldots, j$; $l = 0, 1, \ldots, N-1$, is bounded in R_T, has generalized derivatives $D_x^{j+1}D_t^l\mathrm{v}$ and $D_x^\alpha D_t^l\Delta\mathrm{v}$ belonging to $L_2(E_n)$, $0 \leq t \leq T$, $\alpha = 0, 1, \ldots, j$, $l = 0, 1, \ldots, N-1$, and $D_x^{\alpha+1}D_t^N\mathrm{v} \in L_2(R_T)$, $\alpha = 0, 1, \ldots, j$, satisfies the initial conditions (0.24) and, together with $p(x, t) \in W_2^{j+1,0}(R_T)$, the system (0.23) almost everywhere in R_T. We prove the following assertion.

Theorem 4.1. *Let* $\mathrm{f}(x, t) \in W_2^{j,0}(E_n)$, $0 \leq t \leq T$, *and* $\int_0 \|\mathrm{f}\|^2_{W_2^{j,0}(E_n)}\,dt \equiv A_j < \infty$, *and let* $\mathrm{v}_{0m}(x) \in W_2^{j+3}(E_n) \cap J(E_n)$, $m = 0, 1, \ldots, N-1$. *The following estimate is then valid for a generalized solution of the Cauchy problem* (0.23), (0.24) *of order* $j \geq 1$ *for all* $t^* < t_1$ *(the relation* (4.8) *gives an upper bound for* t_1*):*

$$\begin{aligned}J_{t^*}(\mathrm{v})\equiv\iint_{R_{t^*}}\sum_{\alpha=0}^{j}|D_x^{\alpha+1}D_t^N\mathrm{v}|^2\,dQ+\sum_{l=0}^{N-1}\sum_{\alpha=0}^{j}\Big\{\max_{0\leqslant t\leqslant t^*}\int_{E_n}[|D_x^\alpha D_t^l\Delta\mathrm{v}|^2\\ +|D_x^{\alpha+1}D_t^l\mathrm{v}|^2]\,dx+\max_{R_{t^*}}|D_x^\alpha D_t^l\mathrm{v}|\Big\}\leqslant C_1,\end{aligned} \tag{4.1}$$

where the constant C_1 depends only on A_j, the norms $\|\mathbf{v}_{0m}\|_{W_2^{j+3}(E_n)}$, $m = 0, 1, \ldots, N-1$, κ_l, $l = 1, \ldots, N$; T and $t^* < t_1$; moreover, $C_1 \to \infty$ if $t^* \to t_1$ or $\kappa_N \to 0$. A generalized solution of the Cauchy problem (0.23), (0.24) of order j for $j \geq 2$ is unique (for all $T > 0$).

To prove (4.1) we multiply both sides of the equation (0.23) scalarly in $W_2^j(E_n)$, $0 < t \leq T$, by

$$\sum_{l=1}^{N} l\kappa_l \frac{\partial^{l-1}\Delta\mathbf{v}}{\partial t^{l-1}} \equiv \mathbf{w}.$$

Using the fact that

$$(\operatorname{grad} p,\ \mathbf{w})_{W_2^j(E_n)} = 0, \quad t > 0, \tag{4.2}$$

$$\int_{E_n} v_k D_x^\alpha \mathbf{w} D_x^\alpha \mathbf{w}_{x_k} dx = 0, \quad \alpha = 0, 1, \ldots, j; \quad t > 0, \tag{4.3}$$

we obtain (cf. [3])

$$\iint_{R_t} \Big\{ \sum_{\alpha=0}^{j} \Big[D_x^\alpha \mathbf{v}_t D_x^\alpha \mathbf{w} - \sum_{l=1}^{N} \kappa_l D_x^\alpha D_t^l \Delta\mathbf{v} \cdot D_x^\alpha \mathbf{w} - \nu D_x^\alpha \Delta\mathbf{v} \cdot D_x^\alpha \mathbf{w}$$
$$+ D_x^\alpha (v_k \mathbf{v}_{x_k}) D_x^\alpha \mathbf{w} \Big] - \sum_{\sigma=1}^{j} (D_x v_k D_x^{\alpha-1} \mathbf{w}_{x_k} + \ldots + D_x^\sigma v_k \cdot \mathbf{w}_{x_k}) D_x^\alpha \mathbf{w} \Big\} dQ$$
$$= \iint_{R_t} \sum_{\alpha=0}^{j} D_x^\alpha \mathbf{f} \cdot D_x^\alpha \mathbf{w} dQ, \quad 0 < t \leqslant T. \tag{4.4}$$

From (4.4), upon integrating by parts, applying the Hölder and Cauchy inequalities, and inequalities of the type (2.6), and using the Sobolev imbedding theorems

$$\|u\|_{L_4(\Omega)} \leqslant C_2 \|u\|_{W_2^1(\Omega)}, \quad \max_{\Omega} |u| \leqslant C_3 \|u\|_{W_2^2(\Omega)}, \tag{4.5}$$

valid even for $\Omega = E_n$, $n = 2, 3$ (see [16] and [3]), to estimate the nonlinear terms, we obtain for

$$y(t) \equiv \sum_{l=0}^{N-1} \sum_{\alpha=0}^{j} \iint_{R_t} (|D_x^\alpha D_t^l \Delta\mathbf{v}|^2 + |D_x^\alpha D_t^l \mathbf{v}|^2)\, dQ, \quad t \geqslant 0, \tag{4.6}$$

the differential inequality (cf. [3])

$$y'(t) \leqslant C_4 + C_5 y^{3/2}(t), \quad t \geqslant 0; \quad y(0) = 0. \tag{4.7}$$

Here the constant C_4 depends only on A_j, the norms $\|\mathbf{v}_{0m}\|_{W_2^{j+3}(E_n)}$; $m = 0, 1, \ldots, N-1$, the coefficients ν and κ_l, $l = 1, \ldots, N$, and absolute constants C_2 and C_3, while the constant C_5 depends only on ν, κ_l ($l = 1, \ldots, N$), T, C_2 and C_3.

Integrating (4.7), we obtain for

$$t^* < t_1 = (\sqrt{C_4 C_5} + C_5 y(0))^{-1} \tag{4.8}$$

the estimate

$$\sum_{\alpha=0}^{j} \sum_{l=0}^{N-1} \max_{0 \leqslant t \leqslant t^*} \int_{E_n} (|D_x^\alpha D_t^l \Delta \mathbf{v}|^2 + |D_x^\alpha D_t^l \mathbf{v}|^2)\, dx \leqslant C_6, \quad \forall t^* < t_1, \tag{4.9}$$

where the constant C_6 depends only on C_4, C_5, and $t^* < t_1$. We then have, from (4.5), the obvious inequality

$$\int_{E_n} |D_x^{\alpha+1} D_t^l \mathbf{v}|^2\, dx \leqslant 2 \int_{E_n} (|D_x^\alpha D_t^l \Delta \mathbf{v}|^2 + |D_x^\alpha D_t^l \mathbf{v}|^2)\, dx,$$
$$l = 0, 1, \ldots, N-1; \quad \alpha = 0, 1, \ldots, j, \tag{4.10}$$

and the estimate (4.9), the following type of estimate:

$$\sum_{\alpha=0}^{j} \sum_{l=0}^{N-1} \left\{ \max_{0 \leqslant t \leqslant t^*} \int_{E_n} |D_x^{\alpha+1} D_t^l \mathbf{v}|^2\, dx + \max_{R_{t^*}} |D_x^\alpha D_t^l \mathbf{v}| \right\}$$
$$\leqslant C_7(C_3, C_6, t^*), \quad \forall t^* < t_1. \tag{4.11}$$

We now multiply both sides of (0.23) scalarly in $W_2^j(E_n)$, $0 \leq t < t_1$, by $D_t^N \mathbf{v}$ and then integrate the resulting equation with respect to t from 0 to an arbitrary $t^* < t_1$. Integrating by parts, applying Hölder's inequality and Cauchy's inequality with ϵ, and then using (4.9) and (4.11), we obtain

$$\sum_{\alpha=0}^{j} \iint_{R_{t^*}} |D_x^{\alpha+1} D_t^N \mathbf{v}|^2\, dQ \leqslant C_8\left(\frac{1}{\varkappa_N}, C_6 + C_7; A_j\right), \quad \forall t^* < t_1. \tag{4.12}$$

Adding (4.9), (4.11) and (4.12), we obtain, for a generalized solution of the Cauchy problems (0.23), (0.24) of order $j \geq 1$, the estimate (4.1) in which the constant C_1 is equal to $C_6 + C_7 + C_8$.

The uniqueness of a generalized solution of the Cauchy problem (0.23), (0.24) of order $j \geq 2$ may be proved in exactly the same way as was done in Theorem 1.1, since a generalized solution of the Cauchy problem (0.23), (0.24) of order $j \geq 2$ is a generalized solution of this problem in the sense of Theorem 1.1.

We return now to the two-dimensional Cauchy problem (0.28), (0.29). For this problem we define a generalized solution of order $j \geq 1$ to be a function $\psi(x, t)$, which, together with the derivatives $D_x^\alpha D_t^l \Delta\psi$, $\alpha = 0, 1, \ldots, j$; $l = 0, 1, \ldots, N-1$, is bounded in $E_2 \times (0, T)$, has generalized derivatives $D_x^{j+1} D_t^l \Delta\psi$ and $D_x^j D_t^l \Delta^2\psi$ (and all derivatives subordinate to them) belonging to $L_2(E_2)$, $0 \leq t \leq T$, $l = 0, 1, \ldots, N-1$, and generalized derivatives $D_x^{\alpha+1} D_t^N \Delta\psi \in L_2(E_2 \times (0, T))$, $\alpha = 0, 1, \ldots, j$, satisfies the initial conditions (0.29) and also, almost everywhere in $E_2 \times (0, T)$, the equation (0.28). The following theorem is then valid.

Theorem 4.2. *Let $g(x, t) \in W_2^{j,0}(E_2)$, $0 \leq t \leq T$, and $\int_0^T \|g\|_{W_2^{j,0}(E_2)}\, dt \equiv B_j < \infty$, and let $\psi_{0m}(x) \in W_2^{j+5}(E_2)$, $m = 0, 1, \ldots, N-1$. Then for all $t^* < t_1$ (the* *quantity* t_1 *is determined only by the data of the problem) the following estimate holds for a generalized solution of the Cauchy problem* (0.28), (0.29) *of order* $j \geq 1$:

$$\sum_{\alpha=0}^{j}\left\{\iint_{E_2\times(0,\,t^*)} |D_x^{\alpha+1}D_t^N\Delta\psi|^2\,dQ+\sum_{l=0}^{N-1}\left[\max_{0\leqslant t\leqslant t^*}\int_{E_2}(|D_x^{\alpha+1}D_t^l\Delta\psi|^2\right.\right.$$
$$\left.\left.+|D_x^{\alpha}D_t^l\Delta^2\psi|^2)\,dx+\max_{E_2\times(0,\,t^*)}|D_x^{\alpha}D_t^l\Delta\psi|\right]\right\}\leqslant C_9, \quad (4.13)$$

where the constant C_9 *depends only on* B_j, *the norms* $\|\psi_{0m}\|_{W_2^{j+5}(E_2)}$, $m = 0, 1, \dots, N-1$; κ_l, $l = 1, \dots, N$; *and* T *and* $t^* < t_1$; *moreover,* $C_9 \to \infty$ *if* $t^* \to t_1$ *or* $\kappa_N \to 0$.

A generalized solution of the Cauchy problem (0.28), (0.29) *of order* $j \geq 2$ *is unique for all* $T > 0$.

The proof of the Theorem 4.2 is similar to the proof of Theorem 4.1 (in connection with the proof of uniqueness for $j \geq 2$, see the proof of Theorem 1.4). The proof of (4.13) for a generalized solution of the Cauchy problem (0.28), (0.29) of order $j \geq 1$ depends on the following analytical fact: In the integral

$$\left(\frac{\partial\psi}{\partial x_1}\frac{\partial}{\partial x_2}\sum_{l=1}^{N} l\varkappa_l\frac{\partial^{l-1}\Delta^2\psi}{\partial t^{l-1}}-\frac{\partial\psi}{\partial x_2}\frac{\partial}{\partial x_1}\sum_{l=1}^{N} l\varkappa_l\frac{\partial^{l-1}\Delta^2\psi}{\partial t^{l-1}},\ \sum_{l=1}^{N} l\varkappa_l\frac{\partial^{l-1}\Delta^2\psi}{\partial t^{l-1}}\right)_{W_2^j(E_2)} \quad (4.14)$$

the terms containing derivatives with respect to x of the highest order are equal to zero:

$$\int_{E_2}\left[\frac{\partial\psi}{\partial x_1}\frac{\partial}{\partial x_2}\sum_{l=1}^{N} l\varkappa_l D_x^{\alpha}\frac{\partial^{l-1}\Delta^2\psi}{\partial t^{l-1}}-\frac{\partial\psi}{\partial x_2}\frac{\partial}{\partial x_1}\sum_{l=1}^{N} l\varkappa_l D_x^{\alpha}\frac{\partial^{l-1}\Delta^2\psi}{\partial t^{l-1}}\right]$$
$$\times\sum_{l=1}^{N} l\varkappa_l D_x^{\alpha}\frac{\partial^{l-1}\Delta^2\psi}{\partial t^{l-1}}\,dx=0, \quad (4.15)$$
$$\alpha=0,\ 1,\dots,j;\quad t\geqslant 0,$$

while the remaining nonlinear terms for $j \geq 1$, by virtue of the Sobolev imbedding theorems, may be estimated (in the small) in terms of positive integrals obtainable from the fundamental linear terms

$$\sum_{l=1}^{N}\varkappa_l\frac{\partial^l\Delta^2\psi}{\partial t^l}.$$

This, in fact, makes it possible to obtain the estimate (4.13) (in the small).

§5. Small-amplitude gravitational waves on the surface of an unbounded non-Newtonian fluid

The three-dimensional problem concerning small-amplitude gravitational waves on the surface of an unbounded non-Newtonian fluid with the constitutive relation (0.7) may be reduced (cf. [17], Chapter 2, or [18], Chapter 8) to the solution in the lower half-plane $-\infty < x,\ y < \infty,\ z \leq 0$, for $t \leq 0$, of the linearized system of equations

$$\frac{\partial v_x}{\partial t}-\sum_{l=0}^{N}\varkappa_l\frac{\partial^l\Delta v_x}{\partial t^l}=-\frac{\partial p}{\partial x},\quad \frac{\partial v_y}{\partial t}-\sum_{l=0}^{N}\varkappa_l\frac{\partial^l\Delta v_y}{\partial t^l}=-\frac{\partial p}{\partial y},$$

$$\frac{\partial v_z}{\partial t}-\sum_{l=0}^{N}\varkappa_l\frac{\partial^l\Delta v_z}{\partial t^l}=-\frac{\partial p}{\partial z}-g,\quad \frac{\partial v_x}{\partial x}+\frac{\partial v_y}{\partial y}+\frac{\partial v_z}{\partial z}=0,\quad \varkappa_0\equiv\nu>0, \tag{5.1}$$

subject to the following boundary conditions on the free surface $z = 0$ $(t \geq 0)$:

$$\frac{\partial}{\partial t}\left[p-2\sum_{l=0}^{N}\varkappa_l\frac{\partial^l}{\partial t^l}\frac{\partial v_z}{\partial z}\right]=gv_z, \tag{5.2}$$

$$\sum_{l=0}^{N}\varkappa_l\frac{\partial^l}{\partial t^l}\left(\frac{\partial v_x}{\partial z}+\frac{\partial v_z}{\partial x}\right)=0,\quad \sum_{l=0}^{N}\varkappa_l\frac{\partial^l}{\partial t^l}\left(\frac{\partial v_y}{\partial z}+\frac{\partial v_z}{\partial y}\right)=0, \tag{5.3}$$

where g is the gravitational acceleration. Seeking a solution of the system (5.1) which depends on t, x and y by means of the factors

$$e^{-i\omega t+ikx}\sin ly \quad\text{or}\quad e^{-i\omega t+ikx}\cos ly$$

(where k and $l > 0$ are specified wave numbers) and decays as $z \to -\infty$, we obtain

$$v_x=e^{-i\omega t+ikx}\cos ly\left(ikAe^{\sqrt{k^2+l^2}z}+mBe^{mz}-lCe^{mz}\right),$$
$$v_y=e^{-i\omega t+ikx}\sin ly\left(-lAe^{\sqrt{k^2+l^2}z}+mCe^{mz}+ikCe^{mz}\right),$$
$$v_z=e^{-i\omega t+ikx}\cos ly\left(\sqrt{k^2+l^2}Ae^{\sqrt{k^2+l^2}z}-ikBe^{mz}-lCe^{mz}\right), \tag{5.4}$$
$$p=-gz+i\omega Ae^{-i\omega t+ikx}\cos ly\,e^{\sqrt{k^2+l^2}z},$$

where

$$m=\sqrt{k^2+l^2-\frac{i\omega}{\nu_N}},\quad \nu_N(\omega)=\sum_{l=0}^{N}\varkappa_l(-i\omega)^l, \tag{5.5}$$

and A, B and C are arbitrary constants. Substituting the solution (5.4), (5.5) into the boundary conditions (5.2), (5.3), we obtain a system of linear algebraic equations for determining A, B and C, and from the compatibility condition for this system we obtain a dispersion equation for determining the complex frequency ω of the oscillations:

$$\Delta(\omega)\equiv\begin{vmatrix}\omega^2+2i\nu_N(k^2+l^2)\omega-g\sqrt{k^2+l^2} & ikg+2\nu_Nkm\omega & gl-2i\nu_Nlm\omega\\ 2ik\sqrt{k^2+l^2} & m^2+k^2 & -l(ik+m)\\ -2l\sqrt{k^2+l^2} & ikl & m^2+ikm+l^2\end{vmatrix}=0. \tag{5.6}$$

If the kinematic coefficient of viscosity $\nu \equiv \kappa_0$, and if the shear coefficients κ_j, $j = 1, \ldots, N$, are small in the sense that

$$\nu n_{kl}^2 \ll \sqrt{g n_{kl}}, \quad \varkappa_j \sim \nu (\sqrt{g n_{kl}})^{-j}, \quad j=1, \ldots, N, \quad n_{kl} = \sqrt{k^2+l^2} \tag{5.7}$$

(the first of the conditions (5.7) represents a well-known condition for smallness of the viscosity in the theory of gravitational waves in a viscous fluid; see [17] or [18]), then (5.6) reduces to the equation

$$\omega^2 - g n_{kl} = 2 i g \nu_N(\omega)\, n_{kl}^2 \left(\frac{g n_{kl}}{\omega} - \omega\right), \tag{5.8}$$

whose roots (within the limits of accuracy adopted) are of the same form as in the case of a Newtonian fluid of small viscosity [17], [18]:

$$\omega = \pm \sqrt{g n_{kl}} - 2 i \nu n_{kl}.^{(2)} \tag{5.9}$$

This same result can be obtained for the problem (5.1)–(5.3), subject to the conditions (5.7), through use of the asymptotic method presented in Chapter 8 of [18].

In the two-dimensional case ($v_y \equiv 0$, $v_x = v_x(x, z, t)$, $v_z = v_z(x, z, t)$) the problem (5.1)–(5.3) has the solution (cf. [17], Chapter 2)

$$v_x = e^{-i\omega t + ikx}\left(Ae^{kz} + Be^{mz}\right), \quad v_z = -ie^{-i\omega t + ikx}\left(Ae^{kz} + \frac{k}{m} Be^{mz}\right),$$

$$p = -qz + \frac{\omega}{k} Ae^{-i\omega t + ikx + kz}, \tag{5.10}$$

where, as before, m and $\nu_N(\omega)$ are given by (5.9). The dispersion equation for the complex frequency ω of the oscillations has the form

$$\left[2 - \frac{i\omega}{k^2 \nu_N(\omega)}\right]^2 + \frac{g}{k^3 \nu_N^2(\omega)} = 4\sqrt{1 - \frac{i\omega}{k^2 \nu_N(\omega)}}. \tag{5.11}$$

Under the conditions (5.7) on the viscosity and shear coefficients the roots of (5.11) may be expressed, as before, by the well-known formula (5.9).

In some simple situations the initial-boundary value problems for the equations (0.23) may, as in the case of the Navier-Stokes equations, be substantially simplified and solved in closed form. One such problem is the problem concerning the nonstationary motion of a non-Newtonian fluid in the half-space $-\infty < y, z < \infty$, $x > 0$, due to harmonic oscillations of the plane bounding the fluid. This

(2) More exactly,

$$\omega = [\pm \sqrt{g n_{kl}} - 2 i \nu n_{kl}]\left(1 - 2 n_{kl}^2 \sum_{j=1}^{N} \frac{\varkappa_j}{(\sqrt{g n_{kl}})^{j-1}}\right),$$

but, by virtue of the conditions (5.7),

$$2 n_{kl}^2 \sum_{j=i}^{N} \varkappa_j (\sqrt{g n_{kl}})^{1-j} \approx \frac{2 \nu n_{kl}^2}{\sqrt{g n_{kl}}}\left(1 + o\left(\frac{1}{g n_{kl}}\right)\right) = o(1).$$

problem was solved for the Navier-Stokes equations, for example, in [17], and for the equations (0.23) for $N = 1$ in [9].

Let us assume that the non-Newtonian fluid whose motion is described by (0.23) is in contact with the plane $-\infty < y,\ z < \infty$, which is oscillating harmonically with frequency ω in the direction of the y axis. The problem then reduces to solving the equations (0.23) with $f(x, t) \equiv 0$ in the half-space $-\infty < y,\ z < \infty$, $x > 0$, subject to the boundary conditions

$$v_x = 0,\quad v_z = 0,\quad v_y = u_0 e^{-i\omega t} \tag{5.12}$$

for $x = 0$, $-\infty < t < \infty$.

From the conditions of symmetry for the problem (0.23), (5.12) we have the same simplifications as for the analogous problem in the case of the Navier-Stokes equations [17], and we are confronted with solving the following linear problem for the only velocity component different from zero, namely, $v_y \equiv v(x, t)$:

$$\frac{\partial v}{\partial t} - \sum_{l=1}^{N} \varkappa_l \frac{\partial^{l+2} v}{\partial t^l \partial x^2} = \nu \frac{\partial^2 v}{\partial x^2},\quad x > 0;\quad v|_{x=0} = u_0 e^{-i\omega t},\quad -\infty < t < \infty. \tag{5.13}$$

We seek a solution of the problem (5.13) which is periodic in x and t and of the form $v(x, t) = u_0 e^{i(kx - \omega t)}$. Substituting this solution into (5.13), we obtain

$$k^2 = \frac{i\omega}{\nu + \sum_{l=1}^{N} \varkappa_l (-i\omega)^l}, \tag{5.14}$$

whence $k = \pm(a_N + ib_N)$; here

$$a_N = \sqrt{\frac{\omega}{2\nu}} \cdot \frac{\sqrt{-\gamma_N + \sqrt{1 + 2\gamma_N^2}}}{\sqrt{1 + \gamma_N^2}},\quad b_N = \sqrt{\frac{\omega}{2\nu}} \cdot \frac{1}{\sqrt{1 + \gamma_N^2}\sqrt{-\gamma_N + \sqrt{1 + 2\gamma_N^2}}}, \tag{5.15}$$

where

$$\gamma_N = \frac{\alpha_N}{\beta_N},\quad \alpha_N = \sum_{l=1,3,\ldots,2\left[\frac{N}{2}\right]+1} \varkappa_l (-1)^{\frac{l-1}{2}} \omega^l,$$
$$\beta_N = \sum_{l=2,4,\ldots,2\left[\frac{N}{2}\right]} \varkappa_l (-1)^{l/2} \omega^l + \nu. \tag{5.16}$$

Then

$$v(x, t) = u_0 e^{-b_N x} e^{i(a_N x - \omega t)}. \tag{5.17}$$

Just as in the case of the Navier-Stokes equations the solution (5.17) represents a transverse wave, directed perpendicular to the direction of motion of the plane and decaying exponentially with the depth $x > 0$.

The frictional force acting per unit area of the plane and making harmonic oscillations along the y axis is directed along this axis and is calculated from the constitutive relation (0.7). Moreover, the only nonzero component of the stress tensor turns out to be

$$T_{xy} = \left(\nu \frac{\partial v}{\partial x} + \sum_{l=1}^{N} \varkappa_l \frac{\partial^{l+1} v}{\partial t^l \partial x} \right) \Bigg|_{x=0}. \tag{5.18}$$

Assuming u_0 to be real, we substitute the solution (5.17) into the stress tensor component (5.18). Separating out the real part, we obtain

$$T_{xy} = \nu u_0 A_N \left\{ \gamma_N \cos(\omega t - \varphi_N(x)) - \cos\left(\omega t - \varphi_N(x) + \frac{\pi}{2}\right) \right\}, \tag{5.19}$$

where

$$A_N = \sqrt{\frac{\omega}{2\nu}} \cdot \frac{\sqrt{1 + [-\gamma_N + \sqrt{1 + 2\gamma_N^2}]^2}}{\sqrt{1+\gamma_N^2}\sqrt{-\gamma_N + \sqrt{1+2\gamma_N^2}}},$$

$$\cot \varphi_N(x) = -\gamma_N + \sqrt{1 + 2\gamma_N^2}. \tag{5.20}$$

When $\kappa_1 = \kappa_2 = \cdots = \kappa_N = 0$, the solution obtained coincides with the solution of the analogous problem for the Navier-Stokes equations [17].

BIBLIOGRAPHY

1. James B. Serrin, Jr., *Mathematical principles of classical fluid mechanics*, Handbuch der Physik, Band 8/1, Strömungsmechanik I, Springer-Verlag, Berlin, 1959, pp. 125–263. MR 21 #6836b.

2. O. A. Ladyženskaja, *Mathematical problems in the dynamics of a viscous incompressible fluid*, 2nd rev. aug. ed., "Nauka", Moscow, 1970; English transl. of 1st ed., *The mathematical theory of viscous incompressible flow*, Gordon and Breach, New York, 1963; rev. 1969. MR 42 #6442; 27 #5034b; 40 #7610.

3. ———, *Local solvability of nonstationary problems for incompressible ideal and viscous fluids and for vanishing viscosity*, Zap Naučn. Sem. Leningrad. Otdel. Mat. Inst. Steklov. (LOMI) 21 (1971), 65–78 = J. Soviet Math. 1 (1973), 441–451. MR 45 #2979.

4. ———, *The dynamical system that is generated by the Navier-Stokes equations*, Zap. Naučn. Sem. Leningrad. Otdel. Mat. Inst. Steklov. (LOMI) 27 (1972), 91–115 = J. Soviet Math. 3 (1975), 458–479. MR 48 #6720.

5. L. I. Sedov, *Mechanics of continuous media*. Vol. 2, "Nauka", Moscow, 1973. (Russian)

6. W. L. Wilkinson, *Non-Newtonian fluids. Fluid mechanics, mixing and heat transfer*, Pergamon Press, New York, 1960. MR 22 #1272.

7. Frederick R. Eirich, editor, *Rheology. Theory and applications*. Vol. I, Academic Press, New York, 1956.

8. Alfred M. Freudenthal and Hilda Geiringer, *The mathematical theories of the inelastic continuum*, Handbuch der Physik. Band 6: Elastizität und Plastizität, Springer-Verlag, Berlin, 1958, pp. 229–433. MR 20 #483.

9. A. P. Oskolkov, *The uniqueness and solvability in the large of boundary value problems for the equations of motion of aqueous solutions of polymers*, Zap. Naučn. Sem. Leningrad. Otdel. Mat. Inst. Steklov. (LOMI) 38 (1973), 98–136 = J. Soviet Math. (to appear).

10. ———, *A certain nonstationary quasilinear system with a small parameter that regularizes the system of Navier-Stokes equations*, Problemy Mat. Anal., vyp. 4, Izdat. Leningrad. Univ., Leningrad, 1973, pp. 78–87 = J. Soviet Math. 6 (1976), 51–57.

11. Ja. I. Voĭtkunskiĭ, V. B. Amfilohiev and V. A. Pavlovskiĭ, *The equations of motion of a fluid with regard for its relaxation properties*, Trudy Leningrad. Korablestr. Inst. vyp. 69 (1970), 19–26. (Russian) R. Ž. Meh. **1971** #9Б916.

12. V. A. Pavlovskiĭ, *Theoretical description of weak aqueous polymer solutions*, Dokl. Akad. Nauk SSSR **200** (1971), 809–812 = Soviet Phys. Dokl. **16** (1971/72), 853–855.

13. A. P. Oskolkov, *Solvability in the large of the first boundary value problem for a certain quasilinear third-order system that is encountered in the study of the motion of a viscous fluid*, Zap. Naučn. Sem. Leningrad. Otdel. Mat. Inst. Steklov. (LOMI) **27** (1972), 145–160 = J. Soviet Math. **3** (1975), 508–521. MR **48** #6722.

14. ——, *On the uniqueness and global solvability of the first boundary value problem for systems of equations describing the motion of an aqueous solution of polymers*, Trudy Leningrad. Korablestr. Inst. vyp. **80** (1972), 51–54. (Russian) RŽ Meh. **1973** #8Б798.

15. ——, *On certain two-dimensional degenerate quasilinear equations that arise in the theory of non-Newtonian fluids*, Problemy Mat. Anal., vyp. 5, Izdat. Leningrad. Univ., Leningrad, 1975, pp. 122–132 = J. Soviet Math. (to appear).

16. O. A. Ladyženskaja and N. N. Ural′ceva, *Linear and quasi-linear equations of elliptic type*, 2nd ed., "Nauka", Moscow, 1973; English transl. of 1st ed., Academic Press, New York, 1968. MR **39** #5941.

17. L. M. Landau and E. M. Lifšic, *Mechanics of continuous media*, 2nd ed., GITTL, Moscow, 1953; English transl., *Course of theoretical physics*, vols. 6, 7, Pergamon Press, Oxford; Addison-Wesley, Reading, Mass., 1959. MR **16**, 412; **21** #6839.

18. N. N. Moiseev and V. V. Rumjancev, *Dynamic stability of bodies containing fluid*, "Nauka", Moscow, 1965; English transl., Springer-Verlag, New York, 1968.

Translated by J. F. HEYDA

ESTIMATION OF THE SOLUTION OF AN OVERDETERMINED PARABOLIC INITIAL-BOUNDARY VALUE PROBLEM

UDC 517.946

Š. SAHAEV

Abstract. An investigation is made of the differential properties of the solution of the initial-boundary value problem of finding a solenoidal vector in a domain $\Omega \subset E^3$ that (i) satisfies a system of three equations of heat conduction, (ii) has prescribed tangential components on the boundary of Ω and (iii) initially vanishes. Estimates of this vector are established in the norms of $W_p^{2,1}$ and in the Hölder norms.

Bibliography: 9 titles

In the present paper we investigate the solution of the problem

$$c\frac{\partial \mathbf{E}}{\partial t} - \Delta \mathbf{E} = 0, \quad \operatorname{div} \mathbf{E} = 0,$$
$$\mathbf{E}(x, t)|_{t=0} = 0, \quad \mathbf{E}_\tau|_{x\in S} = \psi_\tau|_{x\in S}, \tag{1}$$

where $\mathbf{E}(x, t) = (E_1(x, t), E_2(x, t), E_3(x, t))$ is an unknown vector to be determined in a bounded cylinder $Q_t = \Omega \times [0, T]$ $(x \in \Omega \subset R^3,\ t \in [0, T])$: $\mathbf{E}_\tau = \mathbf{E} - \mathbf{n}(\mathbf{E} \cdot \mathbf{n})$ is its tangential component on the boundary S of the domain Ω, $\mathbf{n}(x)$ is the outward unit normal to S at a point $x \in S$, ψ is a vector prescribed on $S_T = S \times [0, T]$ with zero normal component $\psi_n = \psi \cdot \mathbf{n}$, and c is a positive constant.

The investigation of problem (1) was begun in the paper [1], in which it was shown that a solution can be written in the form of a heat potential whose density $\lambda(x, t)$ is defined in terms of ψ via an integral equation. In the present paper this solution is estimated in the norms of $W_p^{2,1}(Q_T)$ and $C^{1+\alpha,1/2+\alpha/2}(\overline{Q}_T)$ defined below.

Definition 1. $W_p^{2,1}(Q_T)$ is the space consisting of those elements of $L_p(Q_T)$ whose generalized derivatives of first and second orders with respect to x and of first order with respect to t belong to $L_p(Q_T)$. Its norm is defined by

$$\|u\|_{W_p^{2,1}(Q_T)} = \sum_{|\mu|\leqslant 2} \|D^\mu u\|_{p, Q_T} + \|D_t u\|_{p, Q_T},$$

where $\|v(x, t)\|_{p,Q_T} = (\int_0^T dt \int_\Omega |v(x, t)|^p dx)^{1/p}$ is the norm in $L_p(Q_T)$.

For an exact description of the traces on S_T of the functions from $W_p^{2,1}(Q_T)$

AMS (MOS) subject classifications (1970). Primary 35K50, 35B45, 35N05.

and their first derivatives, we require the spaces $W_p^{2r,r}(S_T)$ with nonintegral $r < 1$, which were introduced by L. N. Slobodeckiĭ [2]. Suppose given a covering $\{S^{(k)}\}$ $(k = 1, \ldots, N)$ of S such that an explicit equation can be written for each $S^{(k)}$ in local coordinates $\{z_1^{(k)}, z_2^{(k)}, z_3^{(k)}\}$ whose origin is a point $x^{(k)} \in S^{(k)}$, whose $z_3^{(k)}$ axis is directed along the outward normal to S at $x^{(k)}$, and whose $z_1^{(k)}$ and $z_2^{(k)}$ axes lie in the tangent plane to S at $x^{(k)}$.

Definition 2. $W_p^{2r,r}(S_T)$ is the Banach space of functions defined on S_T with finite norm

$$\|u\|^p_{W_p^{2r,r}(S_T)} = \|u\|^p_{L_p(S_T)} + \int_0^T dt \int_S \int_S \frac{|u(x, t) - u(y, t)|^p}{|x - y|^{2+2pr}} dS_x dS_y + \int_0^T \int_0^T \int_S \frac{|u(x, t) - u(x, t')|^p}{|t - t'|^{1+rp}} dS_x dt dt', \tag{2}$$

if $2r < 1$, or

$$\|u\|^p_{W_p^{2r,r}(S_T)} = \|u\|^p_{L_p(S_T)} + \sum_{k=1}^N \sum_{m=1}^2 \|u^{(m)}\|^p_{L_p(S_T^{(k)})} + \sum_{k=1}^N \sum_{m=1}^2 \int_0^T dt \int_{S^{(k)}} \int_{S^{(k)}} |u^{(m)}(x, t) - u^{(m)}(y, t)|^p \frac{dS_x dS_y}{|x - y|^{2+p(2r-1)}} + \int_0^T \int_0^T \int_S |u(x, t) - u(x, t')|^p \frac{dS_x dt dt'}{|t - t'|^{1+rp}}, \tag{3}$$

if $2 > 2r > 1$. Here $u^{(m)} = \partial u / \partial z_m^{(k)}$ and $S_T^{(k)} = S^{(k)} \times [0, T]$.

Definition 3. $\overset{\circ}{W}{}_p^{2,1}(Q_T)$ is the subspace of $W_p^{2,1}(Q_T)$ consisting of those functions that vanish when $t = 0$.

The space $\overset{\circ}{W}{}_p^{2,1}(Q_T)$ can also be defined as the subspace of those functions belonging to $W_p^{2,1}(Q_T)$ whose extensions by zero into the cylinder $\tilde{Q}_T = \Omega \times [-T, T]$ belong to $W_p^{2,1}(\tilde{Q}_T)$.

Definition 4. By $\overset{\circ}{W}{}_p^{2r,r}(S_T)$ with nonintegral $r < 1$ is meant the subspace of $W_p^{2r,r}(S_T)$ in which the norm $\|u^{(0)}\|^p_{W_p^{2r,r}(\tilde{S}_T)}$ is finite. Here $\tilde{S}_T = S \times [-T, T]$, $u^{(0)} = u(x, t)$ when $t > 0$, $u^{(0)} = 0$ when $t < 0$, and the norm $\|u^{(0)}\|^p_{W_p^{2r,r}(\tilde{S}_T)}$ is defined by (2) and (3), in which the integration with respect to t is carried out from $-T$ to T.

It can be shown [3], [4] that the space $\overset{\circ}{W}{}_p^{2r,r}(S_T)$ has the following properties: when $r < 1/p$ it coincides with $W_p^{2r,r}(S_T)$ and the norms $\|u^{(0)}\|^p_{W_p^{2r,r}(\tilde{S}_T)}$ and $\|u\|^p_{W_p^{2r,r}(S_T)}$ are equivalent; when $r > 1/p$ it is the subspace of $W_p^{2r,r}(S_T)$ consisting of those functions that vanish when $t = 0$, the norms $\|u^{(0)}\|^p_{W_p^{2r,r}(\tilde{S}_T)}$ and $\|u\|^p_{W_p^{2r,r}(S_T)}$ being also equivalent on this subspace; and when $r = 1/p$ it is the subspace of $W_p^{2/p,1/p}(S_T)$ in which the norm

$$|u|^p_{W_p^{2/p,\,1/p}(S_T)} = \|u\|^p_{W_p^{2/p,\,1/p}(S_T)} + \int_0^T \|u(x,\,t)\|^p_{L_p(S)} \frac{dt}{t}\,, \tag{4}$$

which is equivalent to $\|u^{(0)}\|^p_{W_p^{2/p,1/p}(\tilde{S}_T)}$, is finite.

In addition, it can be shown that if $u \in \overset{\circ}{W}{}_p^{2,1}(Q_T)$, then

$$u|_{S_T} \in \overset{\circ}{W}{}_p^{2-1/p,\,1-1/2p}(S_T),\quad \left.\frac{\partial u}{\partial x_i}\right|_{S_T} \in \overset{\circ}{W}{}_p^{1-1/p,\,1/2-1/2p}(S_T)$$

and

$$\|u^{(0)}\|_{W_p^{2-1/p,\,1-1/2p}(\tilde{S}_T)} + \sum_{i=1}^3 \|(u_{x_i})^{(0)}\|_{W_p^{1-1/p,\,1/2-1/2p}(\tilde{S}_T)} \leqslant C\|u\|_{W_p^{2,\,1}(Q_T)}. \tag{5}$$

Definition 5. By $C^{\alpha,\,\alpha/2}(\overline{Q}_T)$ for $\alpha < 1$ is meant the space of Hölder continuous functions with finite norm

$$|u|_{\alpha,\,\bar{Q}_T} = \sup_{\bar{Q}_T}|u| + \sup_{\bar{Q}_T}\frac{|u(x,\,t) - u(y,\,t)|}{|x-y|^\alpha} + \sup_{\bar{Q}_T}\frac{|u(x,\,t)-u(x,\,t')|}{|t-t'|^{\alpha/2}};$$

the space $C^{\alpha,\alpha/2}(S_T)$ is defined analogously.

Definition 6. By $C^{1+\alpha,1/2+\alpha/2}(\overline{Q}_T)$ is meant the space of functions with finite norm

$$|u|_{1+\alpha,\,\bar{Q}_T} = \sup_{\bar{Q}_T}|u| + \sum_{i=1}^3 |u_{x_i}|_{\alpha,\,\bar{Q}_T} + \sup_{\bar{Q}_T}\frac{|u(x,\,t)-u(x,\,t')|}{|t-t'|^{\alpha/2+1/2}}.$$

Definition 7. By $C^{1+\alpha,1/2+\alpha/2}(S_T)$ is meant the space of functions defined on S_T with finite norm

$$|u|_{1+\alpha,\,S_T} = \sup_{S_T}|u| + \max_k \sum_{m=1}^2 \sup_{S_T^{(k)}}|u^{(m)}|$$
$$+ \max_k \sum_{m=1}^2 \sup_{S_T^{(k)}} \frac{|u^{(m)}(x,\,t) - u^{(m)}(y,\,t)|}{|x-y|^\alpha} + \sup_{S_T}\frac{|u(x,\,t)-u(x,\,t')|}{|t-t'|^{\frac{1+\alpha}{2}}}.$$

The spaces introduced above can also be defined for vectors $\mathbf{u} = (u_1, u_2, u_3)$ with the obvious changes in the definitions of the norms; for example, $\|D^\mu \mathbf{u}\|_{p,\Omega} = \sum_i \|D^\mu u_i\|_{p,\Omega}$). The notation for the spaces remains the same.

§1. Estimation of the solution of problem (1) in $W_p^{2,1}(Q_T)$

Theorem 1. *If $S \in C^{(3)}$ and in problem* (1) $\psi \in W_p^{2-1/p,1-1/2p}(S_T)$, *then* $\mathbf{E} \in W_p^{2,1}(Q_T)$ *and*

$$\sum_{j=1}^3 \|E_j\|_{W_p^{2,\,1}(Q_T)} \leqslant C \sum_{j=1}^3 \|\psi_j\|_{W_p^{2-1/p,\,1-1/2p}(S_T)},$$

where C denotes here and below a constant quantity.

The following two facts are used in the proof of Theorem 1:

1) the estimate

$$\sum_{j=1}^{3} \|E_j\|_{W_p^{2,1}(Q_T)} \leqslant C \sum_{j=1}^{3} \|\lambda_j\|_{W_p^{2-1/p,\,1-1/2p}(S_T)}, \tag{6}$$

the validity of which follows from the representation of the solution of problem (1) in the form of a heat potential,

$$\mathbf{E}(x, t) = \int_0^t d\tau \int_S \operatorname{curl}_x [\Gamma(x, \eta, t-\tau) \cdot \boldsymbol{\lambda}(\eta, \tau)]\, dS_\eta, \tag{7}$$

where $\Gamma(x, t)$ is a fundamental solution of the equation $(c\partial/\partial t - \Delta)v(x, t) = 0$ and has the form

$$\Gamma(x, t) = \begin{cases} -\dfrac{2\sqrt{c}}{(4\pi t)^{3/2}} \exp\left\{-\dfrac{cx^2}{4t}\right\}, & t > 0, \\ 0, & t < 0, \end{cases} \tag{8}$$

while the density λ is determined from the system of integral equations [1]

$$\begin{aligned} \boldsymbol{\lambda}(\xi, t) + \int_0^t d\tau \int_S \Big\{ & (\mathbf{n}(\xi) - \mathbf{n}(\eta), \boldsymbol{\lambda}(\eta, \tau)) \operatorname{grad}_\xi \Gamma(\xi, \eta, t-\tau) \\ & - \boldsymbol{\lambda}(\eta, \tau) \frac{\partial}{\partial n_\xi} \Gamma(\xi, \eta, t-\tau) \Big\}\, dS_\eta = [\mathbf{n}(\xi) \times \boldsymbol{\psi}] \equiv \boldsymbol{\omega}(\xi, t), \end{aligned} \tag{9}$$

in which $\mathbf{n}(\xi)$ and $\mathbf{n}(\eta)$ are the inward normals at the points ξ and η respectively;

2) the following theorem, which provides an estimate of the solution of system (9).

Theorem 2. *If* $S \in C^{(2)}$, $\omega \cdot \mathbf{n} = 0$ *and* $\omega \in \overset{\circ}{W}{}_p^{2r,r}(S_T)$ *with* $r < 1/2$, *then* $\lambda \in \overset{\circ}{W}{}_p^{2r,r}(S_T)$ *and*

$$\sum_{j=1}^{3} \|\lambda_j^{(0)}\|_{W_p^{2r,r}(\tilde{S}_T)} \leqslant C \sum_{j=1}^{3} \|\omega_j^{(0)}\|_{W_p^{2r,r}(\tilde{S}_T)}. \tag{10}$$

If $S \in C^{(3)}$, *then the above assertions are valid for* $r < 1$.

The proof of the theorem will be carried out in stages.

Extending λ and ω by zero onto the half-line $t < 0$, we write system (9) in terms of projections:

$$\lambda_j^{(0)}(\xi, t) + \int_{-\infty}^{t} d\tau \int_S \sum_{i=1}^{3} K_{ij}(\xi, \eta, t-\tau)\, \lambda_i^{(0)}(\eta, \tau)\, dS_\eta = \omega_j^{(0)}(\xi, t); \tag{11}$$

the kernels $K_{ij}(\xi, \eta, t-\tau)$ have the form

$$\begin{aligned} K_{ij}(\xi, \eta, t-\tau) = {} & \left(\frac{c}{4\pi}\right)^{3/2} \frac{\exp\left\{-\dfrac{c|\xi-\eta|^2}{4(t-\tau)}\right\}}{(t-\tau)^{5/2}} \\ & \times [(n_i(\xi) - n_i(\eta))(\xi_j - \eta_j) - \delta_{ij}|\xi - \eta| \cos\varphi], \end{aligned} \tag{12}$$

where $\phi = \angle(\xi - \eta, \mathbf{n}(\xi))$, the $n_i(\xi)$ and the $n_i(\eta)$ are the direction cosines of the vectors $\mathbf{n}(\xi)$ and $\mathbf{n}(\eta)$ respectively, and δ_{ij} is the Kronecker delta. The following estimates of the kernels $K_{ij}(\xi, \eta, t-\tau)$ will be used in the proof of Theorem 2.

Lemma 1. *If $S \in C^{(2)}$, then*

$$|K_{ij}(\xi, \eta, t-\tau)| \leqslant \frac{C}{(t-\tau)^{3/2}} \exp\left\{-c_1 \frac{|\xi-\eta|^2}{t-\tau}\right\}, \tag{13}$$

$$\left|\frac{\partial}{\partial \xi_m} K_{ij}(\xi, \eta, t-\tau)\right| \leqslant \frac{C}{(t-\tau)^2} \exp\left\{-c_1 \frac{|\xi-\eta|^2}{t-\tau}\right\}, \tag{14}$$

$$\left|\frac{\partial}{\partial t} K_{ij}(\xi, \eta, t-\tau)\right| \leqslant \frac{C}{(t-\tau)^{5/2}} \exp\left\{-c_1 \frac{|\xi-\eta|^2}{t-\tau}\right\}, \tag{15}$$

while for $S \in C^{(2+\alpha)}$

$$\left|\frac{\partial}{\partial \xi_m} K_{ij}(\xi, \eta, t-\tau) - \frac{\partial}{\partial \zeta_m} K_{ij}(\zeta, \eta, t-\tau)\right| \leqslant C \frac{|\xi-\zeta|^\alpha}{(t-\tau)^{2+\alpha/2}} \exp\left\{-c_1 \frac{|\eta''-\eta|^2}{t-\tau}\right\}, \tag{16}$$

where η'' is the nearest to η of the points ξ and ζ. Finally, if $S \in C^{(3)}$, estimate (16) *is valid for any $\alpha \in [0, 1]$.*

All of these estimates are easily established on the basis of (12).

Lemma 2. *Let $K(\xi, \eta, t-\tau)$ be a function that satisfies inequalities* (13)–(15) *and vanishes for $t-\tau < 0$. If $\delta < 1/3$, then*

$$\int_{-\infty}^{t} d\tau \int_S |K(\xi, \eta, t-\tau) - K(\zeta, \eta, t-\tau)|^{1+\delta} dS_\eta \leqslant C|\xi-\zeta|^{1-3\delta}, \tag{17}$$

$$\int_{-\infty}^{2t'-t} d\tau \int_S |K(\xi, \eta, t-\tau) - K(\xi, \eta, t'-\tau)|^{1+\delta} dS_\eta \leqslant C|t-t'|^{\frac{1-3\delta}{2}} \tag{18}$$

for $t > t' > \tau$. In addition, if $1/2 < a < 1$ and $b < 2 + a$, then

$$\int_{-\infty}^{t} d\tau \int_S \int_S \frac{|K(\xi, \eta, t-\tau) - K(\zeta, \eta, t-\tau)|^a}{|\xi-\zeta|^b} dS_\xi dS_\zeta \leqslant C, \tag{19}$$

$$\int_S dS_\xi \int_{-T}^{T} \int_{-T}^{t} \frac{|K(\xi, \eta, t-\tau) - K(\xi, \eta, t'-\tau)|^a}{|t-t'|^{b/2}} dt dt' \leqslant C. \tag{20}$$

Finally, if $2s - d > 1$ and $c > -4 + (4s - 2d)$, then

$$\int_{-\infty}^{t} d\tau \int_{\sigma_\xi(\rho)} \left|\frac{\partial}{\partial \xi_m} K(\xi, \eta, t-\tau)\right|^s |\xi-\eta|^c (t-\tau)^d dS_\eta \leqslant C\rho^{c+4+2d-4s}, \tag{21}$$

where $\sigma_\xi(\rho)$ is the part of the surface S interior to the sphere with center ξ and radius ρ.

Remark 1. By taking a sufficiently large number for ρ in (21), we see that

$$\int_{-\infty}^{t} d\tau \int_S \left|\frac{\partial}{\partial \xi_m} K(\xi, \eta, t-\tau)\right|^s |\xi-\eta|^c (t-\tau)^d dS_\eta \leqslant C.$$

Proof. Using inequalities (13) and (14), we have

$$\int_{-\infty}^{t} d\tau \int_S |K(\xi, \eta, t-\tau) - K(\zeta, \eta, t-\tau)|^{1+\delta} dS_\eta$$

$$= \int_{-\infty}^{t} d\tau \int_{\sigma_\xi(2|\xi-\zeta|)} |K(\xi, \eta, t-\tau) - K(\zeta, \eta, t-\tau)|^{1+\delta} dS_\eta$$

$$+ \int_{-\infty}^{t} d\tau \int_{S\setminus\sigma_\xi(2|\xi-\zeta|)} |K(\xi, \eta, t-\tau) - K(\zeta, \eta, t-\tau)|^{1+\delta} dS_\eta$$

$$\leqslant C\Bigg[\int_{\sigma_\xi(2|\xi-\zeta|)} \frac{dS_\eta}{|\xi-\eta|^{1+\delta}} + \int_{\sigma_\xi(2|\xi-\zeta|)} \frac{dS_\eta}{|\zeta-\eta|^{1+\delta}}$$

$$+ |\xi-\zeta|^{1+\delta} \int_{S\setminus\sigma_\xi(2|\xi-\zeta|)} \frac{dS_\eta}{|\zeta-\eta|^{2+4\delta}}\Bigg] \leqslant C|\xi-\zeta|^{1-3\delta},$$

i.e. estimate (17) is obtained.

Further, using the fact that $(t-\tau) \leq 3(t'-\tau)$ for $\tau \in (-\infty, 2t'-t)$ and $|\xi-\eta| \leq \sqrt{t-\tau}$ in $\sigma_\xi(\rho)$, we have

$$\int_{-\infty}^{2t'-t} d\tau \int_S |K(\xi, \eta, t-\tau) - K(\xi, \eta, t'-\tau)|^{1+\delta} dS_\eta$$

$$\leqslant C\Bigg[\int_{-\infty}^{2t'-t} d\tau \int_{\sigma_\xi(\rho)} \exp\left\{-c_0 \frac{|\xi-\eta|^2}{t-\tau}\right\} \frac{dS_\eta}{(t-\tau)^{3(1+\delta)/2}}$$

$$+ \int_{-\infty}^{2t'-t} d\tau \int_{\sigma_\xi(\rho)} \exp\left\{-c_0 \frac{|\xi-\eta|^2}{t-\tau}\right\} \cdot \frac{dS_\eta}{(t'-\tau)^{3(1+\delta)/2}}$$

$$+ (t-t')^{(1+\delta)/2} \int_{-\infty}^{2t'-t} d\tau \int_{S\setminus\sigma_\xi(\rho)} \exp\left\{-c_0 \frac{|\xi-\eta|^2}{t-\tau}\right\}$$

$$\times \frac{dS_\eta}{(t'-\tau)^{2(1+\delta)}}\Bigg] \leqslant C(t-t')^{(1-3\delta)/2}$$

Using (14) and (15), we get

$$\int_{-\infty}^{t} d\tau \int_S \int_S \frac{|K(\xi, \eta, t-\tau) - K(\zeta, \eta, t-\tau)|^a}{|\xi-\zeta|^b} dS_\xi dS_\zeta$$

$$\leqslant C \int_{-\infty}^{t} d\tau \int_S \int_S \left[\exp\left\{-c_0 \frac{|\xi-\eta|^2}{t-\tau}\right\} + \exp\left\{-c_0 \frac{|\zeta-\eta|^2}{t-\tau}\right\}\right] \frac{dS_\xi dS_\zeta}{(t-\tau)^{2a}|\xi-\zeta|^{b-a}}$$

$$\leqslant C \int_S \int_S |\xi-\zeta|^{a-b} (|\xi-\eta|^{2-4a} + |\zeta-\eta|^{2-4a}) dS_\xi dS_\zeta \leqslant C,$$

if $1/2 < a < 1$ and $b < 2 + a$. Estimate (20) is established analogously. In the same way, using (14), we have

$$\int_{-\infty}^{t} d\tau \int_{\sigma_\xi(\rho)} \left|\frac{\partial}{\partial \xi_m} K(\xi, \eta, t-\tau)\right|^s |\xi-\eta|^c (t-\tau)^d \, dS_\eta$$

$$\leqslant C \int_{\sigma_\xi(\rho)} |\xi-\eta|^c dS_\eta \int_{-\infty}^{t} \exp\left\{-c_0 \frac{|\xi-\eta|^2}{t-\tau}\right\} \frac{d\tau}{(t-\tau)^{2s-d}}$$

$$\leqslant C \int_{\sigma'_\xi(\rho)} |\xi-\eta|^{c+2+2d-4s} dS_\eta \leqslant C\rho^{c+4+2d-4s},$$

i.e. inequality (21) is proved. Lemma 2 is completely proved. The following lemma is proved analogously.

Lemma 3. *If $K(\xi, \eta, t-\tau)$ satisfies* (13) *and* (15), *then*

$$\int_{t-\rho}^{t} d\tau \int_{S} |K(\xi, \eta, t-\tau)|^s (t-\tau)^d |\eta-\xi|^c dS_\eta \leqslant C\rho^{\frac{1}{2}(c-3s+2d+4)}$$

for $c > -2$, and $c > -4 + 3s - 2d$, while

$$\int_{-\infty}^{2t'-t} d\tau \int_{S} |K(\xi, \eta, t-\tau) - K(\xi, \eta, t'-\tau)|^s (t-\tau)^d |\eta-\xi|^c dS_\eta$$

$$\leqslant C(t-t')^{\frac{1}{2}(c-3s+2d+4)}$$

for $c > -2$ and $-4 + 3s - 2d < c < -4 + 5s - 2d$.

Lemma 4. *Suppose $S \in C^{(3)}$ and $K(\xi, \eta, t-\tau)$ is a function satisfying* (16) *with $\alpha = 1$. If $5s/2 - d > 1$, and $-4 + (4s - 2d) < c < -4 + (5s - 2d)$, then*

$$\int_{-\infty}^{t} d\tau \int_{S \setminus \sigma_\xi(2|\xi-\zeta|)} \left|\frac{\partial}{\partial \xi_m} K(\xi, \eta, t-\tau) - \frac{\partial}{\partial \zeta_m} K(\zeta, \eta, t-\tau)\right|^s |\xi-\eta|^c (t-\tau)^d dS_\eta$$

$$\leqslant C|\xi-\eta|^{c-4s+2d+4}. \tag{22}$$

Proof. Since $|\eta - \xi| \geq 2|\xi - \zeta|$ and $|\eta - \zeta| \geq 2|\xi - \zeta|$, in the integrand, we have

$$\frac{2}{3}|\xi-\eta| \leqslant \frac{2}{3}|\zeta-\eta| + \frac{2}{3}|\zeta-\xi| \leqslant |\zeta-\eta| \leqslant |\xi-\eta|$$

$$+|\xi-\zeta| \leqslant \frac{3}{2}|\xi-\eta|. \tag{23}$$

It follows that for $\eta \in S\setminus\sigma_\xi(2|\xi-\zeta|)$

$$\exp\left\{-c_1 \frac{|\eta''-\eta|^2}{t-\tau}\right\} \leqslant \exp\left\{-c_2 \frac{|\xi-\eta|^2}{t-\tau}\right\}, \quad c_2 = \frac{4}{9}c_1 \tag{24}$$

and the integral in the left side of (22) is bounded by

$$C|\xi-\zeta|^s \int_{S\setminus\sigma_\xi} |\xi-\eta|^c dS_\eta \int_{-\infty}^{t} (t-\tau)^{d-\frac{5s}{2}} \exp\left\{-c_2 \frac{|\xi-\eta|^2}{t-\tau}\right\} d\tau$$

$$\leqslant C|\xi-\zeta|^s \int_{S\setminus\sigma_\xi} |\xi-\eta|^{c+2+2d-5s} dS_\eta \leqslant C|\xi-\zeta|^{c-4s+2d+4}.$$

In addition to (13)–(16), we make use of the fact that the potential with kernel $K(\xi, \eta, t-\tau)$ and density 1 is a smooth function on S_T.

Lemma 5. *The functions*

$$k_i(\xi)=\int_{-\infty}^{t} d\tau \int_S K_{ij}(\xi,\ \eta,\ t-\tau)\,dS_\eta$$

belong to the space $C^{1+\alpha}(S)$ *for any* $\alpha \in [0, 1]$.

In fact, since

$$\int_{-\infty}^{t} \Gamma(\xi,\ \eta,\ t-\tau)\,d\tau=\frac{1}{2\pi\cdot|\xi-\eta|},$$

we have

$$k_i(\xi)=\frac{1}{2\pi}\int_S\left[(n_i(\xi)-n_i(\eta))\frac{\partial}{\partial\xi_j}\frac{1}{|\xi-\eta|}-\delta_{ij}\frac{\partial}{\partial n_\xi}\frac{1}{|\xi-\eta|}\right]dS_\eta$$

and the assertion of the lemma follows from the properties of the single- and double-layer potentials with a differentiable density [5].

We proceed to the proof of Theorem 2. We will use the scheme of argument of V. A. Solonnikov in [6] that was developed for estimating the solution of a steady-flow problem of hydrodynamics in the space $W_p^2(\Omega)$.

The solution of system (9) can be represented by the method of successive approximations in terms of a resolvent kernel [1]:

$$\boldsymbol{\lambda}(\xi,\ t)=\int_0^t d\tau\int_S R(\xi,\ \eta,\ t-\tau)\,\boldsymbol{\omega}(\eta,\ \tau)\,dS_\eta+\boldsymbol{\omega}(\xi,\ t),$$

where $R(\xi,\ \eta,\ t-\tau)$ is a matrix whose elements R_{ij} have a weak singularity (they satisfy (13)). Therefore

$$\|\boldsymbol{\lambda}\|_{p,\ S_T}\leqslant C\,\|\boldsymbol{\omega}\|_{p,\ S_T}. \tag{25}$$

We will show that the integral operator in the left side of (11) acts from $L_p(S_T)$ into $\overset{\circ}{W}{}_p^{r,r/2}(S_T)$ for any $r<1$, and from $\overset{\circ}{W}{}_p^{r,r/2}(S_T)$ into $\overset{\circ}{W}{}_p^{2r,r}(S_T)$.

Suppose $\lambda\in L_p(S_T)$ and $K(\xi,\ \eta,\ t-\tau)$ is any function satisfying (13)–(15). We will prove that the function

$$\psi^{(0)}(\xi,\ t)=\int_{-\infty}^{t} d\tau\int_S K(\xi,\ \eta,\ t-\tau)\,\lambda^{(0)}(\eta,\ \tau)\,dS_\eta \tag{26}$$

satisfies the estimate

$$\|\psi^{(0)}\|_{W_p^{r,\,r/2}(\tilde S_T)}\leqslant C\,\|\lambda^{(0)}\|_{p,\ \tilde S_T} \tag{27}$$

for any $r<1$. Suppose $p'=p/(p-1)$ and $0<\epsilon<1/3p'$. With the use of Hölder's inequality, we get

$$|\psi^{(0)}(\xi,\ t)-\psi^{(0)}(\zeta,\ t)|$$
$$\leqslant\left(\int_{-\infty}^{t} d\tau\int_S|K(\xi,\ \eta,\ t-\tau)-K(\zeta,\ \eta,\ t-\tau)|^{1-\epsilon p}\,|\lambda^{(0)}(\eta,\ \tau)|^p\,dS_\eta\right)^{1/p}$$
$$\times\left(\int_{-\infty}^{t} d\tau\int_S|K(\xi,\ \eta,\ t-\tau)-K(\zeta,\ \eta,\ t-\tau)|^{1+\epsilon p'}dS_\eta\right)^{1/p'};$$

but the latter integral is bounded by $C|\xi-\zeta|^{1-3\epsilon p'}$ by virtue of (17), so that

$$\int_{-T}^{T} dt \int_S \int_S \frac{|\psi^{(0)}(\xi,t)-\psi^{(0)}(\zeta,t)|^p}{|\xi-\zeta|^{2+rp}}\, dS_\xi dS_\zeta \leqslant C \int_{-T}^{T} d\tau \int_S |\lambda^{(0)}(\eta,\tau)|^p\, dS_\eta$$
$$\times \int_{-\infty}^{t} dt \int_S \int_S \frac{|K(\xi,\eta,t-\tau)-K(\zeta,\eta,t-\tau)|^{1-\varepsilon p}}{|\xi-\zeta|^{3-p+3\varepsilon p+rp}}\, dS_\xi dS_\zeta .$$

As a result of (19), the inner integral is bounded by a constant, and therefore

$$\int_{-T}^{T} dt \int_S \int_S \frac{|\psi^{(0)}(\xi,t)-\psi^{(0)}(\zeta,t)|^p}{|\xi-\zeta|^{2+rp}}\, dS_\xi dS_\zeta \leqslant C\|\lambda^{(0)}\|^p_{p,\tilde{S}_T}. \tag{28}$$

We now estimate $\psi^{(0)}(\xi,t)-\psi^{(0)}(\xi,t')$ for $t>t'>0$ (the case $t'>t>0$ is considered analogously). We have

$$|\psi^{(0)}(\xi,t)-\psi^{(0)}(\xi,t')|^p \leqslant \int_{-\infty}^{2t'-t} d\tau \int_S |K(\xi,\eta,t-\tau)-K(\xi,\eta,t'-\tau)|^{1-\varepsilon p}$$
$$\times |\lambda^{(0)}(\eta,\tau)|^p\, dS_\eta \left(\int_{-\infty}^{2t'-t} d\tau \int_S |K(\xi,\eta,t-\tau)-K(\xi,\eta,t'-\tau)|^{1+\varepsilon p'} dS_\eta \right)^{p-1}$$
$$+ \int_{2t'-t}^{t} d\tau \int_S |K(\xi,\eta,t-\tau)|\,|\lambda^{(0)}(\eta,\tau)|^p\, dS_\eta \left(\int_{2t'-t}^{t} d\tau \int_S |K(\xi,\eta,t-\tau)|\, dS_\eta \right)^{p-1}$$
$$+ \int_{2t'-t}^{t'} d\tau \int_S |K(\xi,\eta,t'-\tau)|\,|\lambda^{(0)}(\eta,\tau)|^p\, dS_\eta \left(\int_{2t'-t}^{t'} d\tau \int_S |K(\xi,\eta,t'-\tau)|\, dS_\eta \right)^{p-1}$$

and hence, using (18), Lemma 3 and (20), we get

$$\int_{-T}^{T} dt \int_{-T}^{t} dt' \int_S \frac{|\psi^{(0)}(\xi,t)-\psi^{(0)}(\xi,t')|^p}{|t-t'|^{1+\frac{r}{2}p}}\, dS_\xi \leqslant C\|\lambda^{(0)}\|^p_{L_p(\tilde{S}_T)}.$$

This equality together with (25) and (28) implies (27).

Applying our result to the integral in the left side of (11), we obtain for the solution of equation (9) the estimate

$$\sum_{j=1}^{3} \|\lambda_j^{(0)}\|_{W_p^{r,r/2}(\tilde{S}_T)} \leqslant C \sum_{j=1}^{3} \|\omega_j^{(0)}\|_{W_p^{r,r/2}(\tilde{S}_T)} \quad (r<1). \tag{29}$$

The first part of Theorem 2 is thereby proved.

In order to prove the second part, we again consider the function (26), assuming that $\lambda \in \overset{\circ}{W}{}_p^{r,r/2}(S_T)$ and $K(\xi,\eta,t-\tau)$ satisfies (14) and (21). In addition, we assume that

$$k(\xi)=\int_{-\infty}^{t} d\tau \int_S K(\xi,\eta,t-\tau)\, dS_\eta \in C^{1+\alpha}(S) \tag{29a}$$

(this was established in Lemma 5 for $K(\dots)=K_{ij}(\xi,\eta,t-\tau)$). We write $\psi^{(0)}(\xi,t)$ in the form

$$\psi^{(0)}(\xi, t) = \int_{-\infty}^{t} d\tau \int_S K(\xi, \eta, t-\tau)[\lambda^{(0)}(\eta, \tau) - \lambda^{(0)}(\xi, t)]\, dS_\eta + \lambda^{(0)}(\xi, t)\, k(\xi)$$

and differentiate both sides on the surface S. The corresponding operator of differentiation is denoted by D (it can be assumed that $\xi \in S^{(k)}$ and $D = \partial/\partial z_m^{(k)}$, $m = 1, 2$). We have

$$D\psi^{(0)}(\xi, t) = \int_{-\infty}^{t} d\tau \int_S DK(\xi, \eta, t-\tau)[\lambda^{(0)}(\eta, \tau) - \lambda^{(0)}(\xi, t)]\, dS_\eta + \lambda^{(0)}(\xi, t)\, Dk(\xi) \tag{30}$$

and we will show that $D\psi^{(0)} \in L_p(\widetilde{S}_T^{(k)})$, where $\widetilde{S}_T^{(k)} = S^{(k)} \times [-T, T]$. The last term in the right side of (30) obviously belongs to $L_p(\widetilde{S}_T^{(k)})$ by virtue of (29a); so that it remains to estimate the integral

$$\varphi(\xi, t) = \int_{-\infty}^{t} d\tau \int_S DK(\xi, \eta, t-\tau)[\lambda^{(0)}(\eta, \tau) - \lambda^{(0)}(\xi, t)]\, dS_\eta = \varphi_1(\xi, t) + \varphi_2(\xi, t),$$

where

$$\varphi_1(\xi, t) = \int_{-\infty}^{t} d\tau \int_S DK(\xi, \eta, t-\tau)[\lambda^{(0)}(\eta, \tau) - \lambda^{(0)}(\xi, \tau)]\, dS_\eta,$$

$$\varphi_2(\xi, t) = \int_{-\infty}^{t} d\tau \int_S DK(\xi, \eta, t-\tau)[\lambda^{(0)}(\xi, \tau) - \lambda^{(0)}(\xi, t)]\, dS_\eta.$$

Let us show that $\phi_i(\xi, t) \in L_p(\widetilde{S}_T^{(k)})$. We have

$$|\varphi_1(\xi, t)| \leqslant \left(\int_{-\infty}^{t} \frac{d\tau}{(t-\tau)^{1-\frac{r}{4}p}} \int_S |\lambda^{(0)}(\eta, \tau) - \lambda^{(0)}(\xi, \tau)|^p \frac{dS_\eta}{|\xi-\eta|^{2+rp}} \right)^{1/p}$$
$$\times \left(\int_{-\infty}^{t} d\tau \int_S |DK(\xi, \eta, t-\tau)|^{p'} |\xi-\eta|^{2p'/p+rp'} (t-\tau)^{p'/p-\frac{r}{4}p'}\, dS_\eta \right)^{1/p'},$$

$$|\varphi_2(\xi, t)| \leqslant \left(\int_{-\infty}^{t} \frac{d\tau}{(t-\tau)^{1+\frac{r}{2}p}} \int_S |\lambda^{(0)}(\xi, \tau) - \lambda^{(0)}(\xi, t)|^p \frac{dS_\eta}{|\xi-\eta|^{2-\frac{r}{2}p}} \right)^{1/p}$$
$$\times \left(\int_{-\infty}^{t} d\tau \int_S |DK(\xi, \eta, t-\tau)|^{p'} |\xi-\eta|^{2p'/p-\frac{r}{2}p'} (t-\tau)^{p'/p+\frac{r}{2}p'}\, dS_\eta \right)^{1/p'}.$$

Since the numbers

$$p' = s_1,\ \frac{2p'}{p} + rp' = c_1,\ \frac{p'}{p} - \frac{r}{4}p' = d_1$$

and

$$p' = s_2,\ \frac{2p'}{p} - \frac{r}{2}p' = c_2,\ \frac{p'}{p} + \frac{r}{2}p' = d_2$$

satisfy the conditions of Lemma 2, we have

$$\|\varphi(\xi, \zeta, t)\|^p_{L_p(S_T^{(k)})} \leqslant C\left[\int_{-T}^{T} dt \int_{S}\int_{S^{(k)}} |\lambda^{(0)}(\eta, \tau) - \lambda^{(0)}(\xi, \tau)|^p \frac{dS_\xi dS_\eta}{|\xi - \eta|^{2+rp}}\right.$$

$$\left. + \int_{S^{(k)}} dS_\xi \int_{-T}^{T} dt \int_{-\infty}^{t} |\lambda^{(0)}(\xi, \tau) - \lambda^{(0)}(\xi, t)|^p \frac{d\tau}{(t-\tau)^{1+\frac{r}{2}p}}\right] \leqslant C\|\lambda^{(0)}\|^p_{W_p^{r, r/2}(\tilde{S}_T)}.$$

Hence

$$\sum_{m=1}^{2}\sum_{k=1}^{N}\int_{-T}^{T} dt \int_{S^{(k)}} \left|\frac{\partial\psi^{(0)}}{\partial z_m^{(k)}}\right|^p dS \leqslant C\|\lambda^{(0)}\|^p_{W_p^{r, r/2}(\tilde{S}_T)}. \tag{31}$$

We will next estimate the expression

$$\sum_{k=1}^{N}\int_{-T}^{T} dt \int_{S^{(k)}} dS_\xi \int_{S^{(k)}} |D\psi^{(0)}(\xi, t) - D\psi^{(0)}(\zeta, t)|^p \frac{dS_\zeta}{|\xi - \zeta|^{2+p(2r-1)}}, \tag{32}$$

assuming that the integration with respect to dS_ξ and dS_ζ has been carried out over only one of the $S^{(k)}$. This does not constitute a loss of generality, since it is always possible to take overlapping $S^{(k)}$ and to conduct the estimate following below only for sufficiently close ξ and ζ. Estimate (31) can be taken in place of it when ξ and ζ are farther apart.

We thus take $\xi, \zeta \in S^{(k)}$ and distinguish a domain σ_ξ from S by means of the sphere of radius $2|\xi - \zeta|$ with center at ξ, assuming $|\xi - \zeta| \leq d/4$.

On the basis of (30) we compose the expression

$$\begin{aligned} D\psi^{(0)}(\xi, t) - D\psi^{(0)}(\zeta, t) &= \int_{-\infty}^{t} d\tau \int_{\sigma_\xi} DK(\xi, \eta, t-\tau)[\lambda^{(0)}(\eta, \tau) - \lambda^{(0)}(\eta, t]dS_\eta \\ &- \int_{-\infty}^{t} d\tau \int_{\sigma_\xi} DK(\zeta, \eta, t-\tau)[\lambda^{(0)}(\eta, \tau) - \lambda^{(0)}(\eta, t)]dS_\eta \\ &+ \int_{-\infty}^{t} d\tau \int_{S\setminus\sigma_\xi} [DK(\xi, \eta, t-\tau) - DK(\zeta, \eta, t-\tau)][\lambda^{(0)}(\eta, \tau) \\ &- \lambda^{(0)}(\eta, t)]dS_\eta + \int_{-\infty}^{t} d\tau \int_{\sigma_\xi} DK(\xi, \eta, t-\tau)[\lambda^{(0)}(\eta, t) - \lambda^{(0)}(\xi, t)]dS_\eta \\ &- \int_{-\infty}^{t} d\tau \int_{\sigma_\xi} DK(\zeta, \eta, t-\tau)[\lambda^{(0)}(\eta, t) - \lambda^{(0)}(\zeta, t)]dS_\eta \\ &+ \int_{-\infty}^{t} d\tau \int_{S\setminus\sigma_\xi} [DK(\xi, \eta, t-\tau) - DK(\zeta, \eta, t-\tau)][\lambda^{(0)}(\eta, t) - \lambda^{(0)}(\xi, t)]dS_\eta \\ &+ \lambda^{(0)}(\zeta, t)[Dk(\xi) - Dk(\zeta)] + [\lambda^{(0)}(\xi, t) - \lambda^{(0)}(\zeta, t)] \\ &\times\left[Dk(\xi) - \int_{-\infty}^{t} d\tau \int_{S\setminus\sigma_\xi} DK(\zeta, \eta, t-\tau)dS_\eta\right] = \sum_{i=1}^{8}\varphi_i(\xi, \zeta, t). \end{aligned} \tag{33}$$

By virtue of (29a) (we can take $\alpha = r$)

$$\left.\begin{aligned} &|Dk(\xi) - Dk(\zeta)| \leqslant C|\xi - \zeta|^r \quad (r < 1), \\ &\left| Dk(\xi) - \int_{-\infty}^{t} d\tau \int_{S\setminus\sigma_\xi} DK(\zeta, \eta, t-\tau)\, dS_\eta \right| \leqslant C(1 + |\ln|\xi - \zeta||), \end{aligned}\right\} \quad (33a)$$

and we therefore have for the last two summands in the right side of (33) the estimate

$$\int_{-T}^{T} dt \int_{S^{(k)}} dS_\xi \int_{S^{(k)}} \sum_{i=7}^{8} |\varphi_i(\xi, \zeta, t)|^p \frac{dS_\zeta}{|\xi - \zeta|^{2+p(2r-1)}} \leqslant C\Bigg[\int_{S^{(k)}} \frac{dS_\xi}{|\xi - \zeta|^{2+p(r-1)}} \int_{-T}^{T} dt$$

$$\times \int_{S^{(k)}} |\lambda^{(0)}(\zeta, t)|^p \, dS_\zeta + \int_{-T}^{T} dt \int_{S^{(k)}} dS_\xi \int_{S^{(k)}} |\lambda^{(0)}(\xi, t) - \lambda^{(0)}(\zeta, t)|^p \quad (34)$$

$$\times \frac{|\xi - \zeta|^{p(1-r)}(1 + |\ln|\xi - \zeta||)}{|\xi - \zeta|^{2+rp}} dS_\zeta\Bigg],$$

so that it remains to estimate the first six summands in the right side of (33). We estimate them with the use of Hölder's inequality. We have

$$|\varphi_1(\xi, \zeta, t)| \leqslant \left(\int_{\sigma_\xi} \frac{dS_\eta}{|\xi - \eta|^{2 - \frac{\varepsilon p}{2}}} \int_{-\infty}^{t} \frac{|\lambda^{(0)}(\eta, \tau) - \lambda^{(0)}(\eta, t)|^p}{(t-\tau)^{1 + \frac{r}{2}p}} d\tau \right)^{1/p}$$

$$\times \left(\int_{-\infty}^{t} d\tau \int_{\sigma_\xi} |DK(\xi, \eta, t-\tau)|^{p'} |\xi - \eta|^{2p'/p - \frac{\varepsilon p'}{2}} (t-\tau)^{p'/p + \frac{r}{2}p'} dS_\eta \right)^{1/p'},$$

$$|\varphi_2(\xi, \zeta, t)| \leqslant \left(\int_{\sigma_\xi} \frac{dS_\eta}{|\zeta - \eta|^{2 - \frac{\varepsilon p}{2}}} \int_{-\infty}^{t} \frac{|\lambda^{(0)}(\eta, \tau) - \lambda^{(0)}(\eta, t)|^p}{(t-\tau)^{1 + \frac{r}{2}p}} d\tau \right)^{1/p}$$

$$\times \left(\int_{-\infty}^{t} d\tau \int_{\sigma_\xi} |DK(\zeta, \eta, t-\tau)|^{p'} |\zeta - \eta|^{2p'/p - \frac{\varepsilon p'}{2}} (t-\tau)^{p'/p + \frac{r}{2}p'} dS_\eta \right)^{1/p'},$$

$$|\varphi_3(\xi, \zeta, t)| \leqslant \left(\int_{S\setminus\sigma_\xi} \frac{dS_\eta}{|\zeta - \eta|^2} \int_{-\infty}^{t} \frac{|\lambda^{(0)}(\eta, \tau) - \lambda^{(0)}(\eta, t)|^p}{(t-\tau)^{1 + \frac{r}{2}p}} d\tau \right)^{1/p}$$

$$\times \Bigg(\int_{-\infty}^{t} d\tau \int_{S\setminus\sigma_\xi} |DK(\xi, \eta, t-\tau) - DK(\zeta, \eta, t-\tau)|^{p'} |\zeta - \eta|^{2p'/p}$$

$$\times (t-\tau)^{p'/p + \frac{r}{2}p'} dS_\eta \Bigg)^{1/p'},$$

$$|\varphi_4(\xi, \zeta, t)| \leqslant \left| \int_{\sigma_\xi} [\lambda^{(0)}(\eta, t) - \lambda^{(0)}(\xi, t)]\, dS_\eta \int_{-\infty}^{t} DK(\xi, \eta, t-\tau)\, d\tau \right|$$

$$\leqslant C \int_{\sigma_\xi} \frac{|\lambda^{(0)}(\eta, t) - \lambda^{(0)}(\xi, t)|}{|\xi - \eta|^2} dS_\eta \leqslant C|\xi - \zeta|^r \left(\int_{\sigma_\xi} \frac{|\lambda^{(0)}(\eta, t) - \lambda^{(0)}(\xi, t)|^p}{|\xi - \eta|^{2+rp}} dS_\eta \right)^{1/p},$$

$$|\varphi_5(\xi, \zeta, t)| \leqslant C|\xi-\zeta|^r \left(\int\limits_{\sigma_\xi} \frac{|\lambda^{(0)}(\eta, t) - \lambda^{(0)}(\zeta, t)|^p}{|\zeta-\eta|^{2+rp}} dS_\eta \right)^{1/p};$$

and, finally, using the fact that

$$\int\limits_{-\infty}^{t} |DK(\xi, \eta, t-\tau) - DK(\zeta, \eta, t-\tau)|\, d\tau \leqslant \frac{C}{|\xi-\eta|^2}$$

when $\eta \in S \backslash \sigma_\xi$, we get

$$|\varphi_6(\xi, \zeta, t)| \leqslant C|\xi-\zeta|^r \left(\int\limits_{S\backslash\sigma_\xi} \frac{|\lambda^{(0)}(\eta, t) - \lambda^{(0)}(\xi, t)|^p}{|\xi-\eta|^{2+rp}} dS_\eta \right)^{1/p}.$$

Since the numbers

$$p' = s_1 = s_2, \quad \frac{2p'}{p} - \varepsilon\frac{p}{2} = c_1 = c_2, \quad \frac{p'}{p} + \frac{r}{2}p' = d_1 = d_2$$

satisfy the conditions of Lemma 2 (see (21)) and the numbers

$$p' = s_3, \frac{2p'}{p} = c_3, \quad \frac{p'}{p} + \frac{r}{2}p' = d_3$$

satisfy the conditions of Lemma 4 (all of this for $\phi_1(\xi, \zeta, t)$, $\phi_2(\xi, \zeta, t)$, $\phi_3(\xi, \zeta, t)$), we get

$$\int\limits_{-T}^{T} dt \int\limits_{S^{(k)}} dS_\xi \int\limits_{S^{(k)}} \sum_{i=1}^{6} |\varphi_i(\xi, \zeta, t)|^p \frac{dS_\zeta}{|\xi-\zeta|^{2+p(2r-1)}}$$

$$\leqslant C\Bigg[\int\limits_{S^{(k)}} \frac{dS_\zeta}{|\xi-\zeta|^{2+p(r-1)+\varepsilon p/2}} \int\limits_{S^{(k)}} \frac{dS_\xi}{|\xi-\eta|^{2-\varepsilon p/2}} \int\limits_{-T}^{T} dt \int\limits_{-\infty}^{t} d\tau$$

$$\times \int\limits_{\sigma_\xi} \frac{|\lambda^{(0)}(\eta, \tau) - \lambda^{(0)}(\eta, t)|^p}{(t-\tau)^{1+\frac{r}{2}p}} dS_\eta + \int\limits_{S^{(k)}} \frac{dS_\zeta}{|\zeta-\eta|^{2-\varepsilon p/2}} \int\limits_{-T}^{T} dt \int\limits_{-\infty}^{t} d\tau$$

$$\times \int\limits_{\sigma_\xi} \frac{|\lambda^{(0)}\eta, \tau) - \lambda^{(0)}(\eta, t)|^p}{(t-\tau)^{1+\frac{r}{2}p}} dS_\eta + \int\limits_{S^{(k)}} \frac{|\ln|\xi-\zeta||}{|\xi-\zeta|^{2+p(r-1)}} dS_\xi \int\limits_{-T}^{T} dt \int\limits_{-\infty}^{t} d\tau \tag{35}$$

$$\times \int\limits_{S\backslash\sigma_\xi} \frac{|\lambda^{(0)}(\eta, \tau) - \lambda^{(0)}(\eta, t)|^p}{(t-\tau)^{1+\frac{r}{2}p}} dS_\eta + \int\limits_{-T}^{T} dt \int\limits_{S^{(k)}} dS_\xi \int\limits_{S^{(k)}} \frac{dS_\zeta}{|\xi-\zeta|^{2+p(r-1)}}$$

$$\times \int\limits_{S} \frac{|\lambda^{(0)}(\eta, t) - \lambda^{(0)}(\xi, t)|^p}{|\xi-\eta|^{2+rp}} dS_\eta + \int\limits_{-T}^{T} dt \int\limits_{S^{(k)}} dS_\zeta \int\limits_{S^{(k)}} \frac{dS_\xi}{|\xi-\zeta|^{2+p(r-1)}}$$

$$\times \int\limits_{\sigma_\xi} \frac{|\lambda^{(0)}(\eta, t) - \lambda^{(0)}(\zeta, t)|^p}{|\zeta-\eta|^{2+rp}} dS_\eta \Bigg].$$

Combining (34) and (35), we have for the quantity (32) an estimate in the form

$$\sum_{k=1}^{N}\int_{-T} dt \int_{S^{(k)}} dS_\xi \int_{S^{(k)}} |D\psi^{(0)}(\xi, t) - D\psi^{(0)}(\zeta, t)|^p \frac{dS_\zeta}{|\xi-\zeta|^{2+p(2r-1)}} \leqslant C\|\lambda^{(0)}\|^p_{W_p^{r,r/2}(\bar{S}_T)}. \tag{36}$$

We will finally estimate the quantity

$$\int_{-T}^{T} dt \int_{-T}^{T} dt' \int_S |\psi^{(0)}(\xi, t) - \psi^{(0)}(\xi, t')|^p \frac{dS_\xi}{(t-t')^{1+rp}}. \tag{37}$$

To this end we compose on the basis of (26) the relation

$$\begin{aligned}\psi^{(0)}(\xi, t) - \psi^{(0)}(\xi, t') = &\int_{-\infty}^{2t'-t} d\tau \int_S [K(\xi, \eta, t-\tau) - K(\xi, \eta, t'-\tau)][\lambda^{(0)}(\eta, \tau)\\ &- \lambda^{(0)}(\eta, t')]\,dS_\eta + \int_{2t'-t}^{t} d\tau \int_S K(\xi, \eta, t-\tau)[\lambda^{(0)}(\eta, \tau) - \lambda^{(0)}(\eta, t)]\,dS_\eta\\ &- \int_{2t'-t}^{t'} d\tau \int_S K(\xi, \eta, t'-\tau)[\lambda^{(0)}(\eta, \tau) - \lambda^{(0)}(\eta, t')]\,dS_\eta\\ &+ \int_S [\lambda^{(0)}(\eta, t) - \lambda^{(0)}(\eta, t')]\,dS_\eta \int_0^{2(t-t')} K(\xi, \eta, \tau)\,dS_\eta = \sum_{i=1}^{4} f_i(\xi, t, t').\end{aligned} \tag{38}$$

We estimate (38) with the use of Hölder's inequality. We have

$$\begin{aligned}|f_1(\xi, t, t')| \leqslant &\left(\int_{-\infty}^{2t'-t} d\tau \int_S \frac{|\lambda^{(0)}(\eta, \tau) - \lambda^{(0)}(\eta, t')|^p}{|\xi-\eta|^{2-\frac{r}{2}p}(t'-\tau)^{1+\frac{r}{2}p}}\,dS_\eta\right)^{1/p}\\ &\times\left(\int_{-\infty}^{2t'-t} d\tau \int_S |K(\xi, \eta, t-\tau) - K(\xi, \eta, t'-\tau)|^{p'} (t'-\tau)^{p'/p+\frac{r}{2}p'}\right.\\ &\left.\times|\xi-\eta|^{2p'/p-\frac{r}{2}p'}\,dS_\eta\right)^{1/p'},\end{aligned}$$

$$\begin{aligned}|f_2(\xi, t, t')| \leqslant &\left(\int_{2t'-t}^{t} d\tau \int_S \frac{|\lambda^{(0)}(\eta, t) - \lambda^{(0)}(\eta, \tau)|^p}{|\xi-\eta|^{2-\frac{r}{2}p}(t-\tau)^{1+\frac{r}{2}p}}\,dS_\eta\right)^{1/p}\\ &\times\left(\int_{2t'-t}^{t} d\tau \int_S |K(\xi, \eta, t-\tau)|^{p'} |\xi-\eta|^{2p'/p-\frac{r}{2}p'} (t-\tau)^{p'/p+\frac{r}{2}p'}\,dS_\eta\right)^{1/p'},\end{aligned}$$

$$\begin{aligned}|f_3(\xi, t, t')| \leqslant &\left(\int_{2t'-t}^{t'} d\tau \int_S \frac{|\lambda^{(0)}(\eta, \tau) - \lambda^{(0)}(\eta, t')|^p}{|\xi-\eta|^{2-\frac{r}{2}p}(t'-\tau)^{1+\frac{r}{2}p}}\,dS_\eta\right)^{1/p}\\ &\times\left(\int_{2t'-t}^{t'} d\tau \int_S |K(\xi, \eta, t'-\tau)|^{p'} |\xi-\eta|^{2p'/p-\frac{r}{2}p'} (t'-\tau)^{p'/p+\frac{r}{2}p'}\,dS_\eta\right)^{1/p'},\end{aligned}$$

$$|f_4(\xi, t, t')| \leqslant \left(\int_S \frac{|\lambda^{(0)}(\eta, t) - \lambda^{(0)}(\eta, t')|^p}{|\xi - \eta|^{2-\frac{r}{2}p}} dS_\eta \int_0^{2(t-t')} \frac{d\tau}{\tau^{1-\frac{r}{4}p}} \right)^{1/p}$$

$$\times \left(\int_0^{2(t-t')} d\tau \int_S |K(\xi, \eta, \tau)|^{p'} \tau^{p'/p - \frac{r}{4}p'} |\xi - \eta|^{2p'/p - \frac{r}{2}p'} dS_\eta \right)^{1/p'}.$$

Since the numbers

$$p' = s_1, \quad \frac{2p'}{p} - \frac{r}{2} p' = c_1, \quad \frac{p'}{p} + \frac{r}{2} p' = d_1$$

satisfy the conditions of Lemma 3 (see the second inequality) and the numbers

$$p' = s_2 = s_3, \quad \frac{2p'}{p} - \frac{r}{2} p' = c_2 = c_3, \quad \frac{p'}{p} + \frac{r}{2} p' = d_2 = d_3$$

and

$$p' = s_4, \quad \frac{2p'}{p} - \frac{r}{2} p' = c_4, \quad \frac{p'}{p} - \frac{r}{4} p' = d_4$$

also satisfy the conditions of Lemma 3 (see the first inequality), we get

$$\int_{-T}^{T} dt \int_{-T}^{T} dt' \int_S |\psi^{(0)}(\xi, t) - \psi^{(0)}(\xi, t')|^p \frac{dS_\xi}{(t-t')^{1+rp}} \leqslant \int_{-T}^{T} dt \int_{-T}^{T} dt'$$

$$\times \int_S \sum_{i=1}^{4} |f_i(\xi, t, t')|^p \frac{dS_\xi}{(t-t')^{1+rp}} \leqslant C \Bigg[\int_{-T}^{T} \frac{dt}{(t-t')^{1+rp-\frac{p}{2}\left(1+\frac{r}{2}\right)}} \int_{-T}^{T} dt' \int_{-\infty}^{2t'-t} d\tau$$

$$\times \int_S dS_\xi \int_S \frac{|\lambda^{(0)}(\eta, \tau) - \lambda^{(0)}(\eta, t')|^p}{|\xi - \eta|^{2-\frac{r}{2}p} (t' - \tau)^{1+\frac{r}{2}p}} dS_\eta + \int_{-T}^{T} \frac{dt'}{(t-t')^{1+rp-\frac{p}{2}\left(1+\frac{r}{2}\right)}}$$

$$\times \int_{-T}^{T} dt \int_S dS_\xi \int_{2t'-t}^{t} d\tau \int_S \frac{|\lambda^{(0)}(\eta, t) - \lambda^{(0)}(\eta, \tau)|^p}{|\xi - \eta|^{2-\frac{r}{2}p} (t-\tau)^{1+\frac{r}{2}p}} dS_\eta$$

$$+ \int_{-T}^{T} \frac{dt}{(t-t')^{1+rp-\frac{p}{2}\left(1+\frac{r}{2}\right)}} \int_{-T}^{T} dt' \int_S dS_\xi \int_{2t'-t}^{t'} d\tau \int_S \frac{|\lambda^{(0)}(\eta, t') - \lambda^{(0)}(\eta, \tau)|^p}{|\xi - \eta|^{2-\frac{p}{2}r} (t'-\tau)^{1+\frac{r}{2}p}} dS_\eta$$

$$+ \int_{-T}^{T} dt \int_{-T}^{T} dt' \int_S dS_\xi \int_S \frac{|\lambda^{(0)}(\eta, t) - \lambda^{(0)}(\eta, t')|^p}{|\xi - \eta|^{2-\frac{r}{2}p} (t-t')^{1+\frac{r}{2}p}} dS_\eta$$

$$\times \frac{(t-t')^{\frac{p}{2}(1-r)}}{(t-t')^{\frac{r}{2}p}} \int_0^{2(t-t')} \frac{d\tau}{\tau^{1-\frac{r}{4}p}} \Bigg],$$

i.e. an estimate for the quantity (37) in the form

$$\int_{-T}^{T} dt \int_{-T}^{T} dt' \int_S |\psi^{(0)}(\xi, t) - \psi^{(0)}(\xi, t')|^p \frac{dS_\xi}{|t-t'|^{1+rp}} \leqslant C \|\lambda^{(0)}\|^p_{W_p^{r, r/2}(S_T)}. \quad (39)$$

As a result of the established inequalities (31), (36) and (39), we have obtained along with (27) the estimate

$$\|\psi^{(0)}\|_{W_p^{2r,r}(\tilde{S}_T)} \leqslant C\|\lambda^{(0)}\|_{W_p^{r,r/2}(\tilde{S}_T)}. \tag{40}$$

Applying this result to the integral in the left side of (11), we obtain for the solution of (9) the estimate

$$\sum_{j=1}^{3}\|\lambda^{(0)}\|_{W_p^{2r,r}(\tilde{S}_T)} \leqslant C\sum_{j=1}^{3}\|\omega_j^{(0)}\|_{W_p^{2r,r}(\tilde{S}_T)} \quad (1<2r<2); \tag{41}$$

the second part of Theorem 2 is thereby proved. Estimates (29) and (41) together with (25) provide a complete proof of Theorem 2.

§2. Estimation of the solution of problem (1) in $C^{1+\alpha,1/2+\alpha/2}(\overline{Q}_T)$

Theorem 3. *If* $S \in C^{(2+\alpha)}$, $\psi \cdot \mathbf{n} = 0$ *and in problem* (1) $\psi \in C^{1+\alpha,1/2+\alpha/2}(S_T)$, *then* $\mathbf{E} \in C^{1+\alpha,1/2+\alpha/2}(\overline{Q}_T)$ *and*

$$\sum_{j=1}^{3}|E_j|_{1+\alpha,\,\bar{Q}_T} \leqslant C\sum_{j=1}^{3}|\psi_j|_{1+\alpha,\,S_T}.$$

The validity of this theorem rests on the proof of the following two theorems.

Theorem 4. *If* $S \in C^{(2+\alpha)}$ *and the density* λ *is such that* $\lambda \cdot \mathbf{n} = 0$ *and* $\lambda \in C^{1+\alpha,1/2+\alpha/2}(S_T)$, *then* $\mathbf{E} \in C^{1+\alpha,1/2+\alpha/2}(\overline{Q}_T)$ *and*

$$\sum_{j=1}^{3}|E_j|_{1+\alpha,\,\bar{Q}_T} \leqslant C\sum_{j=1}^{3}|\lambda_j|_{1+\alpha,\,S_T}.$$

The validity of Theorem 4 follows from the representation (7) of the solution of problem (1) in the form of a heat potential, and from the results of [3], [7] and [8].

Theorem 5. *If* $S \in C^{(2+\alpha)}$ *and the free term in the system of integral equations* (9) *is such that* $\omega \cdot \mathbf{n} = 0$ *and* $\omega \in C^{1+\alpha,1/2+\alpha/2}(S_T)$, *then*

$$\lambda \in C^{1+\alpha,\,1+\alpha/2}(S_T)$$

and

$$\sum_{j=1}^{3}|\lambda_j|_{1+\alpha,\,S_T} \leqslant C\sum_{j=1}^{3}|\omega_j|_{1+\alpha,\,S_T}.$$

We will prove Theorem 5 in stages. The estimate

$$\sup_{S_T}|\lambda| \leqslant C\sup_{S_T}|\omega| \tag{42}$$

follows from the representation of the solution of (9) in terms of a resolvent kernel. We will show that the integral operator in the left side of (11) acts from $C(S_T)$ (a space of continuous functions) into $C^{\alpha,\alpha/2}(S_T)$ for $\alpha < 1$, and from $C^{\alpha,\alpha/2}(S_T)$ into $C^{1+\alpha,\,1/2+\alpha/2}(S_T)$.

Suppose $\lambda \in C(S_T)$ and $K(\xi, \eta, t-\tau)$ is any function satisfying (13)–(15). We will prove that (26) satisfies

$$|\psi|_{\alpha, S_T} \leqslant C \sup_{S_T} |\lambda|. \tag{43}$$

To this end we will first establish that

$$\sup_{S_T} \frac{|\psi^{(0)}(\xi, t) - \psi^{(0)}(\zeta, t)|}{|\xi - \zeta|^\alpha} \leqslant C \sup_{S_T} |\lambda| \tag{44}$$

for any $\alpha < 1$.

Estimate (44) follows from the inequality (obtained from (26))

$$|\psi^{(0)}(\xi, t) - \psi^{(0)}(\zeta, t)| \leqslant \int_{-\infty}^{t} d\tau \int_S |K(\xi, \eta, t-\tau) - K(\zeta, \eta, t-\tau)|$$

$$\times |\lambda^{(0)}(\eta, \tau)|\, dS_\eta \leqslant C|\xi - \zeta|^\alpha \int_{-\infty}^{t} \frac{d\tau}{(t-\tau)^{\frac{1}{2}(3+\alpha)}} \int_S |\lambda^{(0)}(\eta, \tau)|$$

$$\times \left(\exp\left\{-c_0 \frac{|\xi-\eta|^2}{t-\tau}\right\} + \exp\left\{-c_0 \frac{|\zeta-\eta|^2}{t-\tau}\right\}\right) dS_\eta.$$

We will now estimate $\psi^{(0)}(\xi, t) - \psi^{(0)}(\xi, t')$ for $t > t'$ (the case $t < t'$ is estimated analogously). To this end we compose on the basis of (26) the expression

$$\psi^{(0)}(\xi, t) - \psi^{(0)}(\xi, t') = \int_{-\infty}^{2t'-t} d\tau \int_S [K(\xi, \eta, t-\tau) - K(\xi, \eta, t'-\tau)]$$

$$\times \lambda^{(0)}(\eta, \tau)\, dS_\eta + \int_{2t'-t}^{t} d\tau \int_S K(\xi, \eta, t-\tau)\, \lambda^{(0)}(\eta, \tau)\, dS_\eta$$

$$- \int_{2t'-t}^{t'} d\tau \int_S K(\xi, \eta, t'-\tau)\, \lambda^{(0)}(\eta, \tau)\, dS_\eta.$$

Hence, using (13), (18) and Lemma 3, we have

$$\sup_{S_T} \frac{|\psi^{(0)}(\xi, t) - \psi^{(0)}(\xi, t')|}{(t-t')^{\alpha/2}} \leqslant C \sup_{S_T} |\lambda|. \tag{45}$$

Inequalities (44) and (45) together with (42) confirm the validity of (43). Applying (43) to the integral in the left side of (11), we obtain for the solution of (9) the estimate

$$\sum_{j=1}^{3} |\lambda_j|_{\alpha, S_T} \leqslant C \sum_{j=1}^{3} |\omega_j|_{\alpha, S_T}; \tag{46}$$

the first part of Theorem 5 is thereby proved.

In order to prove the second part of Theorem 5, we consider the function (26) under the assumption that $\lambda \in C^{\alpha, \alpha/2}(S_T)$ and $K(\xi, \eta, t-\tau)$ satisfies (14) and (21). In addition, we assume that $k(\xi) \in C^{1+\alpha}(S)$ (see Lemma 5). Then from (30) we have

$$\max_k \sup_{S_T^{(k)}} |D\psi^{(0)}| \leqslant C|\lambda|_{\alpha, S_T}. \tag{47}$$

We must now estimate the expression

$$\max_k \sup_{S_T^{(k)}} \left[\frac{1}{|\xi-\zeta|^\alpha} |D\psi^{(0)}(\xi, t) - D\psi^{(0)}(\zeta, t)| \right]. \tag{48}$$

To this end we make use of (33). We have

$$|D\psi^{(0)}(\xi, t) - D\psi^{(0)}(\zeta, t)| \leqslant \left\{ \int_{-\infty}^{t} d\tau \int_S |DK(\xi, \eta, t-\tau) \right.$$

$$- DK(\zeta, \eta, t-\tau)| (t-\tau)^{\alpha/2} \cdot \frac{|\lambda^{(0)}(\eta, \tau) - \lambda^{(0)}(\eta, t)|}{(t-\tau)^{\alpha/2}} dS_\eta$$

$$+ \int_{-\infty}^{t} d\tau \int_{\sigma_\xi} |DK(\xi, \eta, t-\tau)| |\xi-\eta|^\alpha \cdot \frac{|\lambda^{(0)}(\eta, t) - \lambda^{(0)}(\xi, t)|}{|\xi-\eta|^\alpha} dS_\eta$$

$$+ \int_{-\infty}^{t} d\tau \int_{\sigma_\xi} |DK(\zeta, \eta, t-\tau)| \cdot |\zeta-\eta|^\alpha \frac{|\lambda^{(0)}(\eta, t) - \lambda^{(0)}(\zeta, t)|}{|\zeta-\eta|^\alpha} dS_\eta$$

$$+ \int_{-\infty}^{t} d\tau \int_{S\setminus\sigma_\xi} |DK(\xi, \eta, t-\tau) - DK(\zeta, \eta, t-\tau)| |\xi-\eta|^\alpha$$

$$\times \frac{|\lambda^{(0)}(\eta, t) - \lambda^{(0)}(\xi, t)|}{|\xi-\eta|^\alpha} dS_\eta + |\lambda^{(0)}(\xi, t) - \lambda^{(0)}(\zeta, t)|$$

$$\times \left| \int_{-\infty}^{t} d\tau \int_{S\setminus\sigma_\xi} |DK(\zeta, \eta, t-\tau)| dS_\eta + |Dk(\xi)| \right| + |\lambda^{(0)}(\zeta, t)| |Dk(\xi) - Dk(\zeta)|.$$

Hence, taking into account the fact that the last two summands in the right side admit (with the use of (33a)) the estimate

$$C[|\lambda^{(0)}(\xi, t) - \lambda^{(0)}(\zeta, t)| (|\ln|\xi-\zeta|| + 1) + |\xi-\zeta|^\alpha |\lambda^{(0)}(\zeta, t)|] \leqslant$$

$$\leqslant C|\xi-\zeta|^\alpha \left[\frac{|\lambda^{(0)}(\xi, t) - \lambda^{(0)}(\zeta, t)|}{|\xi-\zeta|^{\alpha_1}} \cdot |\xi-\zeta|^{\alpha'} (|\ln|\xi-\zeta|| + 1) + |\lambda^{(0)}(\zeta, t)| \right],$$

where $\alpha_1 = \alpha + \alpha' < 1$, and using inequality (21) from Lemma 2 (for $s = 1$, $c = \alpha$, $d = 0$), Lemma 4 (for $e_1 = 0$, $s_1 = 1$, $d_1 = \alpha/2$ and $c_2 = \alpha$, $s_2 = 1$, $d_s = 0$) and Lemma 5, we have an estimate for (48) in the form

$$\max_k \sup_{S_T^k} \frac{|D\psi^{(0)}(\xi, t) - D\psi^{(0)}(\zeta, t)|}{|\xi-\zeta|^\alpha} \leqslant C|\lambda|_{\alpha_1, S_T}, \quad \alpha_1 > \alpha. \tag{49}$$

Finally, let us esimtate the quantity

$$\sup_{S_T} \left[\frac{1}{(t-t')^{\frac{1}{2}(1+\alpha)}} |\psi^{(0)}(\xi, t) - \psi^{(0)}(\xi, t')| \right]. \tag{50}$$

To this end we make use of (38). Then

$$|\psi^{(0)}(\xi, t) - \psi^{(0)}(\xi, t')| \leqslant \left\{ \int_{-\infty}^{2t'-t} d\tau \int_S |K(\xi, \eta, t-\tau) - K(\xi, \eta, t'-\tau)| \right.$$

$$\times (t'-\tau)^{\alpha/2} \cdot \frac{|\lambda^{(0)}(\eta, \tau) - (\lambda^{(0)}(\eta, t')|}{(t'-\tau)^{\alpha/2}} dS_\eta + \int_{2t'-t}^{t} d\tau \int_S |K(\xi, \eta, t-\tau)| (t-\tau)^{\alpha/2}$$

$$\times \frac{|\lambda^{(0)}(\eta, \tau) - \lambda^{(0)}(\eta, t)|}{(t-\tau)^{\alpha/2}} dS_\eta + \int_{2t'-t}^{t'} d\tau \int_S |K(\xi, \eta, t'-\tau)| (t'-\tau)^{\alpha/2}$$

$$\times \frac{|\lambda^{(0)}(\eta, \tau) - \lambda^{(0)}(\eta, t')|}{(t'-\tau)^{\alpha/2}} dS_\eta +$$

$$+ (t-t')^{\alpha/2} \int\limits_S \frac{|\lambda^{(0)}(\eta, t) - \lambda^{(0)}(\eta, t')|}{(t-t')^{\alpha/2}} dS_\eta \int\limits_0^{2(t-t')} |K(\xi, \eta, \tau)| d\tau \Big\}.$$

Hence, using Lemma 3 for $s_1 = s_2 = s_3 = 1$, $d_1 = d_2 = d_3 = \alpha/2$ and $c_1 = c_2 = c_3 = 0$, we have an estimate for (50) in the form

$$\sup_{S_T} \left[\frac{1}{(t-t')^{\frac{1}{2}(1+\alpha)}} |\psi^{(0)}(\xi, t) - \psi^{(0)}(\xi, t')| \right] \leqslant C |\lambda|_{\alpha, S_T}. \tag{51}$$

As a result of (47), (49) and (51), we obtain along with (43) the estimate

$$|\psi|_{1+\alpha;\, S_T} \leqslant C |\lambda|_{\alpha,\, S_T}. \tag{52}$$

Applying (52) to the integral in the left side of (11), we obtain for the solution of (9) the estimate

$$\sum_{j=1}^{3} |\lambda_j|_{1+\alpha,\, S_T} \leqslant C \sum_{j=1}^{3} |\omega_j|_{1+\alpha,\, S_T}; \tag{53}$$

the second part of Theorem 5 is thereby proved. Collecting estimates (42), (46) and (53), we confirm that Theorem 5 is completely proved.

Theorem 3 can now be established with the use of Theorems 4 and 5.

Remark 2. Some of the results of §1 were announced in the author's note [9].

In conclusion, I wish to express my gratitude to V. A. Solonnikov for his advice and comments on the present work.

BIBLIOGRAPHY

1. Š. Sahaev, *Solution of the first initial-boundary-value problem for the nonstationary system of Maxwell's equations*, Akad. Nauk Kazah. SSR Trudy Inst. Mat. Meh. 2 (1972), 69–77. (Russian)

2. L. N. Slobodeckiĭ, *Estimates of solutions of elliptic and parabolic systems*, Dokl. Akad. Nauk SSSR 120 (1958), 468–471. (Russian) MR 21 #5060.

3. V. A. Solonnikov, *A priori estimates for second-order parabolic equations*, Trudy Mat. Inst. Steklov. **70** (1964), 133–212; English transl., Amer. Math. Soc. Transl. (2) 65 (1967), 51–137. MR **28** #5267.

4. O. A. Ladyženskaja, V. A. Solonnikov and N. N. Ural′ceva, *Linear and quasilinear equations of parabolic type*, "Nauka", Moscow, 1967; English transl., Amer. Math. Soc., Providence, R. I., 1968. MR **39** #3159a, b.

5. N. M. Gunther, *La théorie du potentiel et ses applications aux problèmes fondamentaux de la physique mathématique*, Gauthier-Villars, Paris, 1934.

6. O. A. Ladyženskaja, *Mathematical problems in the dynamics of a viscous incompressible fluid*, 2nd rev. aug. ed., "Nauka", Moscow, 1970; English transl. of 1st ed., Gordon and Breach, New York, 1963; rev., 1969. MR **42** #6442; **27** #5034b; **40** #7610.

7. Avner Friedman, *Boundary estimates for second order parabolic equations and their applications*, J. Math. Mech. **7** (1958), 771–791. MR **21** #7363.

8. L. I. Kamynin and V. N. Maslennikova, *Schauder type boundary estimates of the solution of a problem with directional derivative for a parabolic equation in a noncylindrical region*, Sibirsk. Mat. Ž. 7 (1966), 83–128 = Siberian Math. J. 7 (1966), 65–103. MR 33 #4480.

9. Š. Sahaev, *An estimate of the solution of the first initial-boundary-value problem for Maxwell's equation in the space* $W^{2,1}_{pxt}(Q_T)$, $p > 1$, Izv. Akad. Nauk Kazah. SSR Ser. Fiz.-Mat. **1973**, no. 1, 83–85. (Russian) MR 51 #13484.

Translated by S. SMITH

Trudy Mat. Inst. Steklov. 127 (1975)

Proc. Steklov Inst. Math. 127 (1975)

ESTIMATES FOR SOLUTIONS OF A BOUNDARY VALUE PROBLEM IN MAGNETOHYDRODYNAMICS

UDC 517.946

Š. SAHAEV AND V. A. SOLONNIKOV

Abstract. A nonstationary, nonlinear problem is considered for the system of magnetohydrodynamic equations which describes the motion of a conducting fluid in a vessel with ideally conducting walls. The problem is solved "in the small" with respect to time in classes of functions whose derivatives are Hölder continuous or integrable with power $p > 1$.

Bibliography: 15 titles.

§1. Formulation of the problem and preliminary remarks

Suppose that a liquid is contained in a finite vessel Ω whose walls S are made of an ideally conducting material, and that the liquid is subject to hydrodynamic and electromagnetic forces. It is required to determine the motion of the fluid and the electromagnetic field in Ω if initially the fluid velocity and the magnetic field intensity are known. This problem reduces to finding vectors $\mathbf{v}$, $\mathbf{H}$, $\mathbf{E}$ and a function $p(x, t)$ which in the cylinder $Q_T = \Omega \times [0, T]$ satisfy the system of equations

$$\begin{gathered}\frac{\partial \mathbf{v}}{\partial t} - \nu\Delta\mathbf{v} + \sum_{k=1}^{3} v_k \frac{\partial \mathbf{v}}{\partial x_k} - \frac{\mu}{\rho}\sum_{k=1}^{3} H_k \frac{\partial \mathbf{H}}{\partial x_k} + \frac{1}{\rho}\operatorname{grad}\left(p + \frac{\mu \mathbf{H}^2}{2}\right) = \mathbf{f}(x, t),\\ \operatorname{div}\mathbf{v} = 0, \quad \operatorname{div}\mathbf{H} = 0, \quad \operatorname{curl}\mathbf{E} + \mu\frac{\partial \mathbf{H}}{\partial t} = 0,\\ \operatorname{curl}\mathbf{H} - \sigma(\mathbf{E} + \mu[\mathbf{v}, \mathbf{H}]) = \mathbf{j}(x, t),\end{gathered} \tag{1.1}$$

the initial conditions

$$\mathbf{v}(x, 0) = \mathbf{v}_0(x), \quad \mathbf{H}(x, 0) = \mathbf{H}_0(x) \tag{1.2}$$

and the boundary conditions

$$\mathbf{v}|_S = \mathbf{a}, \quad H_n|_S = 0, \quad \mathbf{E}_\tau|_S = 0. \tag{1.3}$$

Here $\mathbf{v} = (v_1, v_2, v_3)$ is the fluid velocity; $\mathbf{E}$ and $\mathbf{H}$ are the vectors of electric and magnetic intensity, $\mathbf{E}_\tau = \mathbf{E} - \mathbf{n}E_n$; $\mathbf{n}$ is the outer normal to the surface S; $H_n = \mathbf{H} \cdot \mathbf{n}$; p is the pressure; $\mathbf{f}$ represents external hydrodynamic forces; $\mathbf{j}(x, t)$ gives the prescribed currents; μ is the magnetic permeability; σ is the conductivity; ρ is the density; ν is the kinematic viscosity of the fluid. We have neglected displacement currents.

AMS (MOS) subject classifications (1970). Primary 76W05; Secondary 35Q10.

This problem can be reformulated by eliminating the vector $\mathbf{E}(x, t)$. The following equations are then obtained for $\mathbf{v}$, $\mathbf{H}$ and p:

$$\frac{\partial \mathbf{v}}{\partial t} - \nu\Delta \mathbf{v} + \sum_{k=1}^{3} v_k \frac{\partial \mathbf{v}}{\partial x_k} - \frac{\mu}{\rho}\sum_{k=1}^{3} H_k \frac{\partial \mathbf{H}}{\partial x_k} + \frac{1}{\rho}\operatorname{grad}\left(p + \frac{\mu \mathbf{H}^2}{2}\right) = \mathbf{f}, \tag{1.4}$$

$$\frac{\partial \mathbf{H}}{\partial t} - \frac{1}{\sigma\mu}\Delta \mathbf{H} - \operatorname{curl}[\mathbf{v} \times \mathbf{H}] = \frac{1}{\sigma\mu}\operatorname{curl}\mathbf{j}, \tag{1.5}$$

$$\operatorname{div}\mathbf{v} = 0, \quad \operatorname{div}\mathbf{H} = 0, \tag{1.6}$$

$$\left.\begin{aligned} &\mathbf{v}|_{t=0} = \mathbf{v}_0(x), \quad \mathbf{H}|_{t=0} = \mathbf{H}_0(x), \\ &\mathbf{v}|_S = \mathbf{a}, \quad H_n|_S = 0, \ \operatorname{curl}_\tau \mathbf{H}|_S = \mathbf{j}_\tau|_S. \end{aligned}\right\} \tag{1.7}$$

We shall assume that $a_n = 0$.

The problem (1.1)–(1.3) in a generalized formulation was studied in [1], where results were obtained of the same type as for the nonstationary, nonlinear Navier-Stokes system (cf., for example, [2]): solvability "in the large" with respect to t in a class of vectors with finite energy integral (but with no uniqueness theorem), and solvability "in the small" in a more restricted class of vectors in which there is a uniqueness theorem and smoothness of the solution in the interior of the domain. In [3] the same problem was studied in the formulation (1.4)–(1.7) in the class of vectors $W_2^{2,1}(Q_T)$ with derivatives entering in the system (1.4)–(1.6) which are square integrable. Among the results of that work were precise estimates in the norms of $W_2^{2,1}(Q_T)$ both of the solution of problem (1.4)–(1.7) itself as well as of solutions of a similar initial-value problem for systems differing from (1.4)–(1.6) by additional lower order terms. In the present work this result is extended to the norms of $W_p^{2,1}(Q_T)$ for any $p > 1$ and to the norms of $C^{2+\alpha, 1+\alpha/2}(Q_T)$ with $\alpha < 1$. Classical solvability of the problem (1.4)–(1.7) and smoothness of the solution in the entire domain up to the boundary are also proved.

We list the norms used in the present paper (the space is denoted by a subscript in the notation for the norm):

$$\|u(x)\|_{L_p(\Omega)} = \left(\int_\Omega |u(x)|^p\,dx\right)^{1/p}, \quad \|u(x, t)\|_{L_p(Q_T)} = \left(\int_{Q_T} |u(x, t)|^p\,dxdt\right)^{1/p},$$

$$\|u(x)\|_{W_p^l(\Omega)} = \sum_{|\alpha| \leqslant l} \|D^\alpha u\|_{L_p(\Omega)},$$

$$\|u(x, t)\|_{W_p^{2l, l}(Q_T)} = \sum_{|\alpha| \leqslant 2l} \|D_x^\alpha u\|_{L_p(Q_T)} + \sum_{j=1}^{l} \left\|\frac{\partial^j u}{\partial t^j}\right\|_{L_p(Q_T)}.$$

Here $l > 0$ is an integer; the derivatives $\partial^{|\alpha|} u/\partial x_1^{\alpha_1} \dots \partial x_n^{\alpha_n}$ are interpreted as generalized derivatives in the sense of Sobolev [4]. The spaces

$W_p^l(\Omega)$ with integer l were introduced by Sobolev and those with fractional l by Slobodeckiĭ [5]; the norm in the latter is given by

$$\|u(x)\|_{W_p^l(\Omega)} = \sum_{|\alpha|<l} \|D^\alpha u\|_{L_p(\Omega)} + \sum_{|\alpha|=[l]} \left(\int_\Omega \int_\Omega |D^\alpha u(x) - D^\alpha u(y)|^p \frac{dx\,dy}{|x-y|^{n+p\lambda}}\right)^{1/p}, \quad (1.8)$$

where $l = \lambda + [l]$, λ being the fractional part and $[l]$ the integral part of l.

We now define the spaces $W_p^{l,l/2}(S_T)$ with nonintegral $l > 0$ on the surface $S_T = S \times [0, T]$. Let $S^{(k)}$ be subsets of the surface S which cover it and map into domains $D^{(k)} \subset R^{n-1}$ under smooth, invertible mappings Z_k, and let $S_T^{(k)} = S^{(k)} \times [0, T]$ and $D_T^{(k)} = D^{(k)} \times [0, T]$. We introduce the norms

$$\|u^{(k)}\|^p_{W_p^{l,l/2}(D_T^{(k)})} = \int_0^T \|u^{(k)}(z, t)\|^p_{W_p^l(D_T^{(k)})} dt + \int_{D^{(k)}} \|u^{(k)}(z, t)\|^p_{W_p^{l/2}(0, T)} dz$$

(the norms on the right side are defined by means of (1.8); $z = Z_k \circ x$ are local coordinates on $S^{(k)}$, and $u^{(k)} = u \circ Z_k^{-1}$) and

$$\|u\|^p_{W_p^{l,l/2}(S_T)} = \sum_k \|u^{(k)}(z, t)\|^p_{W_p^{l,l/2}(D_T^{(k)})}.$$

In addition, we shall make use of the Hölder norms

$$|u(x)|_{C^\alpha(\Omega)} \equiv |u|_\Omega^{(\alpha)} = \sup_{x\in\Omega} |u(x)| + \sup_{x,x'\in\Omega} \frac{|u(x) - u(x')|}{|x - x'|^\alpha}, \quad 0 < \alpha < 1,$$

$$|u(x)|_{C^l(\Omega)} \equiv |u|_\Omega^{(l)} = \sum_{|\beta|<l} \sup_{x\in\Omega} |D^\beta u(x)| + \sum_{|\beta|=[l]} |D^\beta u(x)|_\Omega^{(\lambda)},$$

$$|u(x, t)|_{C^{l,l/2}(Q_T)} \equiv |u|_{Q_T}^{(l)} = \sup_{t\in[0,T]} |u(x, t)|_\Omega^{(l)} + \sup_{x\in\Omega} |u(x, t)|_{(0,T)}^{(l/2)},$$

$$|u(x, t)|_{C^{l,l/2}(S_T)} = \sup_{t\in[0,T]} |u(x, t)|_S^{(l)} + \sup_{x\in S} |u(x, t)|_{(0,T)}^{(l/2)},$$

where $|v(x)|_S^{(l)} = \max_k |v^{(k)}(z)|_{D^{(k)}}^{(l)}$.

Let us consider the space $W_p^{2,1}(Q_T)$, $p > 1$. It is known that if $u(x, t) \in W_p^{2,1}(Q_T)$, then $v = u|_{t=0} \in W_p^{2-2/p}(\Omega)$, $w = u|_{x\in S} \in W_p^{2-1/p,1-1/2p}(S_T)$ and $u_i = u_{x_i}|_S \in W_p^{1-1/p,1/2-1/2p}(S_T)$. The functions v, w and u_i satisfy certain consistency conditions on the face $S^{(0)} = \{x \in S,\ t = 0\} \subset \partial Q_T$ if they are meaningful there (according to the imbedding theorems, $v, w \in W_p^{2-3/p}(S^{(0)})$ for $p > 3/2$ and $u_i \in W_p^{1-3/p}(S^{(0)})$ for $p > 3$). We have the equalities $v|_S = w|_{t=0}$ if $p > 3/2$ and $v_{x_i}|_S = u_i|_{t=0}$ if $p > 3$. If $p < 3/2$ or $p < 3$ there are no consistency conditions; in the extreme cases $p = 3/2$ or $p = 3$ the conditions acquire a different formulation.

Theorem 1. *For any function* $u \in W_{3/2}^{2,1}(Q_T)$

$$\int_0^T\int_S\int_\Omega \frac{|u(x,t)-u(y,0)|^{3/2}}{(|x-y|^2+t)^{5/2}}\,dS_x dy dt \leqslant C\|u\|^{3/2}_{W^{2,1}_{3/2}(Q_T)}. \tag{1.9}$$

Moreover, if $u \in W_3^{2,1}(Q_T)$, *then*

$$\int_0^T\int_S\int_\Omega \frac{|u_{x_i}(x,t)-u_{y_i}(y,0)|^3}{(|x-y|^2+t)^{5/2}}\,dS_x dy dt \leqslant C\|u\|^3_{W_3^{2,1}(Q_T)}. \tag{1.10}$$

Proof. It suffices to prove both inequalities for the half space $\Omega = R^3_+ : x_3 > 0$, $S = R^2$, since the problem can always be reduced to this case by a local straightening of the boundary. Moreover, it can be assumed that the function $u(x, t)$ is smooth and that $T = \infty$. We have

$$I = \left(\int_0^\infty\int_{R^2}\int_{R^3_+} \frac{|u(x',0,t)-u(y,0)|^{3/2}}{(|x'-y|^2+t)^{5/2}}\,dx' dy dt\right)^{2/3}$$

$$\leqslant C_1\left(\int_0^\infty\int_0^\infty\int_{R^3_+} \frac{|u(x',0,t)-u(x',x_3,0)|^{3/2}}{(x_3^2+t)^{3/2}}\,dx' dx_3 dt\right)^{2/3}$$

$$+ C_2\left(\int_{R^2}\int_0^\infty\int_{R^2_+} \frac{|u(x',y_3,0)-u(y',y_3,0)|^{3/2}}{|x'-y'|^3}\,dx' dy' dy_3\right)^{2/3}.$$

Since the last integral can be estimated by the principal part of the $W^{2/3}_{3/2}(R^3_+)$ norm of the function $u(x, 0)$ and hence, by the imbedding theorem, since

$$\sum_{|\alpha|=2}\|D^\alpha u\|_{L_{3/2}(Q_T)} + \|u_t\|_{L_{3/2}(Q_T)},$$

it remains to estimate the first integral I on the right side. We use the identity

$$\begin{aligned} u(x',0,t)-u(x',x_3,0) &= \tfrac{1}{2}(u(x',0,4t)-u(x',2x_3,0)) \\ &+ [u(x',0,t)-2u(x',x_3,t)+u(x',2x_3,t)] - \tfrac{1}{2}[u(x',0,4t) \\ &- 2u(x',x_3,4t)+u(x',2x_3,4t)] - [u(x',x_3,0)-u(x,t)] \\ &+ [u(x,t)-u(x,4t)] + \tfrac{1}{2}[u(x',2x_3,0)-u(x',2x_3,t)] \\ &- \tfrac{1}{2}[u(x',2x_3,t)-u(x',2x_3,4t)]. \end{aligned}$$

Expressing the differences in square brackets by $\partial^2 u/\partial x_3^2$ or u_t and using an inequality of Hardy, we obtain

$$I \leqslant \frac{1}{2} I + C \left\{ \left[\int_0^\infty \int_0^\infty \int_{R^2} \frac{dx' dx_3 dt}{x_3^3} \left(\int_0^{2x_3} \xi \,| u_{\xi\xi}(x', \xi, t) | \, d\xi \right)^{3/2} \right]^{2/3} + \left[\int_0^\infty \int_0^\infty \int_{R^2} \frac{dx' dx_3 dt}{t^{3/2}} \left(\int_0^{4t} | u_\tau (x', x_3, \tau) | \, d\tau \right)^{3/2} \right]^{2/3} \right\}$$

$$\leqslant \frac{1}{2} I + C (\| u_{x_3 x_3} \|_{L_{3/2}(Q_\infty)} + \| u_t \|_{L_{3/2}(Q_\infty)})$$

and after combining all estimates we have

$$\left(\int_0^\infty \int_{R^2} \int_{R_+^3} \frac{| u(x', 0, t) - u(y, 0) |^{3/2}}{(|x' - y|^2 + t)^{5/2}} dx' dy dt \right)^{2/3} \leqslant C \Big(\sum_{|\alpha|=2} \| D^\alpha u \|_{L_{3/2}(Q_\infty)} + \| u_t \|_{L_{3/2}(Q_\infty)} \Big).$$

This completes the proof of (1.9).

To prove (1.10) we use the identity

$$u_i(x, 0, t) - u_i(y', y_3, 0) = u_i(x', y_3, t) - u_i(x', y_3, 0) - \int_0^{y_3} u_{i3}(x', z, t) \, dz + [u_i(x', y_3, 0) - u_i(y', y_3, 0)],$$

in which $u_i = \partial u / \partial x_i$ and $u_{ik} = \partial^2 u / \partial x_i \partial x_k$. From this it follows that

$$\left(\int_0^\infty \int_{R^2} \int_{R_+^3} \frac{| u_{x_i}(x, t) - u_{y_i}(y, 0) |^3}{(|x' - y|^2 + t)^{5/2}} dx' dy dt \right)^{1/3} \leqslant C_1 \left(\int_0^\infty \int_{R_+^3} \frac{| u_{y_i}(y, t) - u_{y_i}(y, 0) |^3}{(y_3^2 + t)^{3/2}} dy dt \right)^{1/3} + C_2 \sum_{|\alpha|=2} \| D^\alpha u \|_{L_3(Q_\infty)}.$$

Furthermore let $\zeta \in C_0^\infty(q)$, where q is the cube $0 \le z_i \le 1$, and $\int_q \zeta(z) dz = 1$. We set $q_t = \{ z | 0 \le z_i \le \sqrt{t} \}$ and $\zeta_i = \partial \zeta / \partial z_i$. Then

$$u_{y_i}(y, t) - u_{y_i}(y, 0) = -\frac{1}{t^2} \int_{q_t} [u(y+z, t) - u(y+z, 0)] \zeta_i \left(\frac{z}{\sqrt{t}} \right) dz + \frac{1}{t^{3/2}} \int_{q_t} [u_{y_i}(y, t) - u_{y_i}(y+z, t)] \zeta \left(\frac{z}{\sqrt{t}} \right) dz - \frac{1}{t^{3/2}} \int_{q_t} [u_{y_i}(y, 0) - u_{y_i}(y+z, 0)] \zeta \left(\frac{z}{\sqrt{t}} \right) dz,$$

whence by means of Hölder's inequality and the imbedding theorem for $W_p^{2,1}(Q_T)$ we obtain

$$\left(\int_0^\infty \int_{R_+^3} \frac{|u_{y_i}(y,t)-u_{y_i}(y,0)|^3}{(y_3^2+t)^{3/2}}\,dydt\right)^{1/3} \leqslant C\left(\sum_{|\alpha|=2}\|D_x^\alpha u\|_{L_3(Q_\infty)}+\left\|\frac{\partial u}{\partial t}\right\|_{L_3(Q_\infty)}\right)$$

$$+C\left(\int_{R_+^3}\int_{R_+^3} \frac{|u_{y_i}(y,0)-u_{y_i}(y+z,0)|^3}{|z|^4}\,dydz\right)^{1/3}$$

$$\leqslant C\left(\sum_{|\alpha|=2}\|D^\alpha u\|_{L_3(Q_\infty)}+\left\|\frac{\partial u}{\partial t}\right\|_{L_3(Q_\infty)}\right).$$

Combining the estimates, we obtain (1.10). The proof of the theorem is complete.

The case $u(y, 0) = 0$ is considered in [6]; on the left side of (1.9) and (1.10) in this case there are the integrals

$$\int_0^T\int_S |u(x,t)|^{3/2} t^{-1}dSdt \quad \text{and} \quad \int_0^T\int_S |u_{x_i}|^3\, t^{-1}dSdt.$$

If $u|_S = 0$ they are replaced by

$$\int_\Omega |u(y,0)|^{3/2}\rho^{-1}(y)\,dy \quad \text{and} \quad \int_\Omega |u_{y_i}(y,0)|^3\rho^{-1}(y)\,dy,$$

where $\rho(y)$ is the distance from y to S.

In addition to functions, we shall consider three-component vector-valued functions $\mathbf{u} = (u_1, u_2, u_3)$. We say that a vector $\mathbf{u}$ belongs to one of the spaces listed above if each of its components belongs to this space, and we set $\|\mathbf{u}\| = \Sigma_1^3\|u_i\|$.

We now introduce subspaces of solenoidal vectors. As is known [2], $\mathbf{L}_2(\Omega) = \mathbf{G}(\Omega) \oplus \mathring{\mathbf{J}}(\Omega) = \mathring{\mathbf{G}}(\Omega) \oplus \mathbf{J}(\Omega)$, where $\mathbf{G}(\Omega)$ and $\mathring{\mathbf{G}}(\Omega)$ are spaces of vectors of the form $\mathbf{u} = \operatorname{grad}\phi$ and $\mathbf{u} = \operatorname{grad}\psi$, where $\phi, \psi \in W_2^1(\Omega)$, $\psi|_S = 0$, $\mathbf{J}(\Omega)$ is the closure in $\mathbf{L}_2$ of the set of smooth solenoidal vectors (i.e. such that $\operatorname{div}\mathbf{u} = \Sigma_1^3 \partial u_k/\partial x_k = 0$), and $\mathring{\mathbf{J}}(\Omega)$ is the closure of the set of smooth solenoidal vectors which satisfy the condition $u_n|_S = 0$.

We denote by $\mathring{\mathbf{J}}(Q_T)$ and $\mathbf{J}(Q_T)$ the subspaces of vectors of $\mathbf{L}_2(Q_T)$ which belong to $\mathring{\mathbf{J}}(\Omega)$ and $\mathbf{J}(\Omega)$ for almost all t, and we define the following spaces:

$$\mathring{\mathbf{J}}_p(\Omega) = \mathring{\mathbf{J}}(\Omega)\cap\mathbf{L}_p(\Omega), \quad \mathring{\mathbf{J}}_p(Q_T) = \mathring{\mathbf{J}}(Q_T)\cap\mathbf{L}_p(Q_T),$$
$$\mathbf{J}_p(\Omega) = \mathbf{J}(\Omega)\cap\mathbf{L}_p(\Omega), \quad \mathbf{J}_p(Q_T) = \mathbf{J}(Q_T)\cap\mathbf{L}_p(Q_T),$$
$$\mathring{\mathbf{J}}^\alpha(\Omega) = \mathring{\mathbf{J}}(\Omega)\cap\mathbf{C}^\alpha(\Omega), \quad \mathring{\mathbf{J}}^\alpha(Q_T) = \mathring{\mathbf{J}}(Q_T)\cap\mathbf{C}^{\alpha,\alpha/2}(Q_T),$$
$$\mathbf{J}^\alpha(\Omega) = \mathbf{J}(\Omega)\cap\mathbf{C}^\alpha(\Omega), \quad \mathbf{J}^\alpha(Q_T) = \mathbf{J}(Q_T)\cap\mathbf{C}^{\alpha,\alpha/2}(Q_T).$$

§2. The linear problem

If the nonlinear terms are dropped in (1.4) and (1.5), then problem (1.4)–(1.7) decomposes into two problems: an initial boundary value problem for the Navier-Stokes equation

$$\frac{\partial \mathbf{v}}{\partial t} - \nu\Delta\mathbf{v} + \frac{1}{\rho}\operatorname{grad} p = \mathbf{f}, \quad \operatorname{div}\mathbf{v} = 0,$$
$$\mathbf{v}\big|_S = \mathbf{a}, \quad \mathbf{v}\big|_{t=0} = \mathbf{v}_0(x) \tag{2.1}$$

and an overdetermined initial boundary value problem

$$\sigma\mu\frac{\partial \mathbf{H}}{\partial t} - \Delta\mathbf{H} = \operatorname{curl}\mathbf{j}, \quad \operatorname{div}\mathbf{H} = 0,$$
$$H_n\big|_S = 0, \ \operatorname{curl}_\tau\mathbf{H}\big|_S = \mathbf{j}_\tau\big|_S, \quad \mathbf{H}\big|_{t=0} = \mathbf{H}_0(x). \tag{2.2}$$

The following assertions are proved regarding problem (2.1) in [7]–[9].

Theorem 2. *Let* $S \in C^2$. *For any* $\mathbf{f} \in \overset{\circ}{\mathbf{J}}_p(Q_T)$, $p > 1$, $\mathbf{v}_0 \in \mathbf{W}_p^{2-2/p}(\Omega)$ *and* $\mathbf{a} \in \mathbf{W}_p^{2-1/p,\,1-1/2p}(S_T)$, $a_n = 0$, *satisfying the consistency conditions*

$$\mathbf{v}_0\big|_S = \mathbf{a}\big|_{t=0}, \quad \text{if} \quad p > \frac{3}{2},$$
$$\int_0^T\int_S\int_\Omega \frac{|\mathbf{v}_0(y) - \mathbf{a}(x, t)|^{3/2}}{(|x-y|^2 + t)^{5/2}}\, dS_x\,dy\,dt < +\infty \quad \text{for} \quad p = 3/2, \tag{2.3}$$

the problem (2.1) *has a unique solution* $\mathbf{v} \in \mathbf{W}_p^{2,1}(Q_T)$, $\operatorname{grad} p \in \mathbf{L}_p(Q_T)$, *which if* $p \neq 3/2$ *satisfies the inequality*

$$\|\mathbf{v}\|_{\mathbf{W}_p^{2,1}(Q_T)} + \|\operatorname{grad} p\|_{\mathbf{L}_p(Q_T)} \leqslant C(T)\Big(\|\mathbf{f}\|_{\mathbf{L}_p(Q_T)} + \|\mathbf{v}_0\|_{\mathbf{W}_p^{2-2/p}(\Omega)} + \|\mathbf{a}\|_{\mathbf{W}_p^{2-1/p,\,1-1/2p}(S_T)}\Big). \tag{2.4}$$

If $p = 3/2$, *the integral* (2.3) *to the power* $2/3$ *with a constant factor is added to the right side of* (2.4).

In [9] the consistency conditions were formulated as follows: "there exists a vector $\mathbf{v}(x, t) \in \mathbf{W}_p^{2,1}(Q_T)$ such that $\mathbf{v}|_S = \mathbf{a}$ and $\mathbf{v}|_{t=0} = \mathbf{v}_0$." Here we have made use of Theorem 1.

Theorem 3. *Let* $S \in C^{2+\alpha}$, $0 < \alpha < 1$. *For any* $\mathbf{f} \in \overset{\circ}{\mathbf{J}}{}^{\alpha}(Q_T)$, $\mathbf{v}_0 \in \mathbf{C}^{2+\alpha}(\Omega)$ *and* $\mathbf{a} \in \mathbf{C}^{2+\alpha,\,1+\alpha/2}(S_T)$ *satisfying the consistency conditions*

$$\left.\begin{aligned} \mathbf{v}_0|_S &= \mathbf{a}|_{t=0}, \\ \frac{\partial \mathbf{a}}{\partial t}\Big|_{t=0} &= \mathbf{P}\nu\Delta\mathbf{v}_0 + \mathbf{f}(x, 0)|_S, \end{aligned}\right\} \tag{2.5}$$

where $\mathbf{P}$ *is the orthogonal projection in* $\mathbf{L}_2(\Omega)$ *onto* $\overset{\circ}{\mathbf{J}}(\Omega)$, *there exists a unique solution of problem* (2.1)

$$\mathbf{v} \in \mathbf{C}^{2+\alpha,\, 1+\alpha/2}(Q_T), \quad \operatorname{grad} p \in \mathbf{C}^{\alpha,\, \alpha/2}(Q_T),$$

satisfying the inequality

$$|\mathbf{v}|_{Q_T}^{(2+\alpha)} + |\operatorname{grad} p|_{Q_T}^{(\alpha)} \leqslant C(T)\left(|\mathbf{f}|_{Q_T}^{(\alpha)} + |\mathbf{v}_0|_{\Omega}^{(2+\alpha)} + |\mathbf{a}|_{S_T}^{(2+\alpha)}\right). \tag{2.6}$$

In this section analogous theorems are established for problem (2.2), which is reduced to the auxiliary problem

$$\begin{gathered} \sigma\mu\frac{\partial \mathbf{E}}{\partial t} - \Delta\mathbf{E} = \mathbf{0}, \quad \operatorname{div}\mathbf{E} = 0, \\ \mathbf{E}_\tau|_S = \psi, \quad \mathbf{E}|_{t=0} = \mathbf{0}, \end{gathered} \tag{2.7}$$

studied in [10] and [11]. It is shown in [10] that for any vector $\psi(x, t)$ with $\psi_n = 0$ which satisfies a Hölder condition in both arguments the problem (2.7) has a unique classical solution representable in the form of a potential

$$\mathbf{E}(x, t) = \operatorname{curl}\mathbf{A}(x, t), \tag{2.8}$$

$$\mathbf{A}(x, t) = \int_0^t d\tau \int_S \Gamma(x - \eta, t - \tau)\,\boldsymbol{\lambda}(\eta, \tau)\,dS_\eta, \tag{2.9}$$

where

$$\Gamma(x, t) = \begin{cases} -\dfrac{\sqrt{\sigma\mu}}{(4\pi t)^{3/2}}\, e^{-\frac{\sigma\mu x^2}{4t}} & \text{if } t > 0, \\ 0 & \text{if } t < 0 \end{cases}$$

is the fundamental solution of the heat equation and $\boldsymbol{\lambda}$ is a vector determined by a system of integral equations of Volterra-Fredholm type (similar to those found in solving initial-boundary value problems for a single heat equation) with right side $\omega = [\psi \times \mathbf{n}]$.

Moreover, in [11] the following theorem is proved.

Theorem 4. *If* $S \in C^2$, *then for any* $\psi \in \mathbf{W}_p^{1-1/p,\, 1/2-1/2p}(S_T)$ *with* $\psi_n = 0$ *satisfying the consistency conditions* $\psi|_{t=0} = 0\omega$ *if* $p > 3$ *and*

$$\int_{S_T} |\psi(x, t)|^3 t^{-1}\, dS\, dt < +\infty \quad \textit{for} \quad p = 3,$$

if $p = 3$, *the vector* $\mathbf{A}$ *of* (2.9) *belongs to* $\mathbf{W}_p^{2,1}(Q_T)$ *and*

$$\|\mathbf{A}\|_{\mathbf{W}_p^{2,1}(Q_T)} \leqslant C\|\boldsymbol{\psi}\|_{\mathbf{W}_p^{1-1/p,\,1/2-1/2p}(S_T)},\quad p\neq 3,$$
$$\|\mathbf{A}\|_{\mathbf{W}_3^{2,1}(Q_T)} \leqslant C\left[\|\boldsymbol{\psi}\|_{\mathbf{W}_3^{2/3,\,1/3}(S_T)} + \left(\int\limits_{S_T}|\boldsymbol{\psi}|^3 t^{-1} dS dt\right)^{1/3}\right]. \tag{2.10}$$

If $S \in C^{2+\alpha}$, $\psi \in \mathbf{C}^{1+\alpha,(1+\alpha)/2}(S_T)$ *and* $\psi|_{t=0} = 0$, *then* $\mathbf{A} \in \mathbf{C}^{2+\alpha,1+\alpha/2}(Q_T)$ *and*

$$|\mathbf{A}|_{Q_T}^{(2+\alpha)} \leqslant C|\boldsymbol{\lambda}|_{S_T}^{(1+\alpha)} \leqslant C|\boldsymbol{\psi}|_{S_T}^{(1+\alpha)}.$$

In order to make use of all these results, we must first of all reduce the system and initial conditions (2.2) to a homogeneous system.

Lemma 1. *Any vector* $\mathbf{u}(x) \in \mathbf{W}_p^r(\Omega) \cap \mathbf{J}_p(\Omega)$ $(r \geq 0;\ \mathbf{W}_p^r = \mathbf{L}_p$ *if* $r = 0)$ *can be continued to all of* R^3 *in such a manner that* $\mathbf{u} \in \mathbf{W}_p^r(R^3) \cap \mathbf{J}_p(R^3)$ *and*

$$\|\mathbf{u}\|_{\mathbf{W}_p^r(R^3)} \leqslant C\|\mathbf{u}\|_{\mathbf{W}_p^r(\Omega)} \tag{2.11}$$

with constant not depending on $\mathbf{u}$. *It is here assumed that* $S \in C^{1+r'}$ *with* $r' > r$.

Proof. Let $\mathbf{v}(x)$ be a solution of the problem

$$\operatorname{curl}\mathbf{v} = \mathbf{u},\quad \operatorname{div}\mathbf{v} = 0,\quad v_n|_S = 0. \tag{2.12}$$

It is solvable for any $\mathbf{u} \in \mathbf{J}_p(\Omega)$ and, as is shown in [12], the solution satisfies the inequality

$$\|\mathbf{v}\|_{\mathbf{W}_p^{r+1}(\Omega)} \leqslant C\|\mathbf{u}\|_{\mathbf{W}_p^r(\Omega)}. \tag{2.13}$$

We now extend $\mathbf{v}$ to the entire space R^3 with preservation of class, i.e. so that

$$\|\mathbf{v}\|_{\mathbf{W}_p^{r+1}(R^3)} \leqslant C\|\mathbf{v}\|_{\mathbf{W}_p^{r+1}(\Omega)} \tag{2.14}$$

(if $p = 2$ this is done in [13], and for $p \neq 2$ it is done similarly). The vector $\operatorname{curl}\mathbf{v}$ is the desired extension of the vector $\mathbf{u}$.

Theorem 5. *If* $S \in C^3$, *then problem* (2.2) *has a unique solution in the class* $\mathbf{W}_p^{2,1}(Q_T)$ *for any* $\mathbf{H}_0 \in \mathbf{W}_p^{2-2/p}(\Omega) \cap \overset{\circ}{\mathbf{J}}_p(\Omega)$, $\operatorname{curl}\mathbf{j} \in \mathbf{J}_p(Q_T)$, $\mathbf{j}_\tau \in \mathbf{W}_p^{1-1/p,1/2-1/2p}(S_T)$, *satisfying the consistency conditions*: $\operatorname{curl}_\tau\mathbf{H}_0|_S = \mathbf{j}_\tau|_{t=0}$ *if* $p > 3$ *and*

$$\int\limits_0^T\int\limits_S\int\limits_\Omega \frac{|\operatorname{curl}_\tau\mathbf{H}_0(y) - \mathbf{j}_\tau(x,t)|^3}{(|x-y|^2+t)^{5/2}}\, dS_x dy dt < +\infty,\quad p = 3 \tag{2.15}$$

(here $\mathrm{curl}_\tau \mathbf{H}_0(y) = \mathrm{curl}\, \mathbf{H}_0(y) - \mathbf{n}(y)\mathrm{curl}_n \mathbf{H}_0(y)$; *the vector* $\mathbf{n}$ *is extended from* S *to the interior of* Ω *in such a manner that* $\mathbf{n} \in C^2(\Omega)$).

The solution of problem (2.1) *satisfies the inequality*

$$\|\mathbf{H}\|_{W_p^{2,1}(Q_T)} \leqslant C\Big[\|\mathrm{rot}\,\mathbf{j}\|_{L_p(Q_T)} + \|\mathbf{H}_0\|_{W_p^{2-2/p}(\Omega)} + \|\mathbf{j}_\tau\|_{W_p^{1-1/p,\,1/2-1/2p}(S_T)}\Big], \quad p \neq 3. \tag{2.16}$$

If $p = 3$ *the integral* (2.15) *to the power* $1/3$ *with a constant factor is added to the right side.*

Proof. We extend the vectors $\mathbf{H}_0(x)$ and $\mathbf{g} = \mathrm{curl}\,\mathbf{j}$ with preservation of class and the solenoidal property to R^3 in accordance with Lemma 1 with $r = 2 - 2/p$ and $r = 0$. Let $\mathbf{h}_0(x, t)$ be a solution of the Cauchy problem

$$\sigma\mu \frac{\partial \mathbf{h}_0}{\partial t} - \Delta \mathbf{h}_0 = \mathbf{g}, \quad \mathbf{h}_0\big|_{t=0} = \mathbf{H}_0(x). \tag{2.17}$$

Since $\mathrm{div}\,\mathbf{h}_0$ satisfies the homogeneous heat equation and is zero at $t = 0$, it follows that

$$\mathrm{div}\,\mathbf{h}_0(x, t) = 0. \tag{2.18}$$

Let $\mathbf{E}$ be a solution of the auxiliary problem

$$\begin{gathered}\sigma\mu \frac{\partial \mathbf{E}}{\partial t} - \Delta \mathbf{E} = 0, \quad \mathrm{div}\,\mathbf{E} = 0,\\ \mathbf{E}\big|_{t=0} = 0, \quad \mathbf{E}_\tau\big|_S = (-\mathrm{curl}_\tau \mathbf{h}_0 + \mathbf{j}_\tau)\big|_S,\end{gathered} \tag{2.19}$$

and let $\mathbf{h}$ be defined in terms of $\mathbf{E}$ as a solution of the problem

$$\mathrm{curl}\,\mathbf{h} = \mathbf{E}, \quad \mathrm{div}\,\mathbf{h} = 0, \quad h_n\big|_S = -h_{0n}\big|_S. \tag{2.20}$$

It is solvable, since $\mathbf{E} \in \mathbf{J}_p(Q_T)$; moreover, (2.8) holds for $\mathbf{E}$, so that

$$\mathbf{h}(x, t) = \mathbf{A}(x, t) + \mathrm{grad}\,\varphi(x, t), \tag{2.21}$$

where ϕ is a solution of the Neumann problem

$$\Delta\varphi = -\mathrm{div}\,\mathbf{A}, \quad \frac{\partial\varphi}{\partial n}\Big|_S = -(A_n + h_{0n})\big|_S. \tag{2.22}$$

We shall show that $\mathbf{h}(x, t)$ satisfies the relations

$$\sigma\mu \frac{\partial \mathbf{h}}{\partial t} - \Delta \mathbf{h} = 0, \quad \mathrm{div}\,\mathbf{h} = 0, \tag{2.23}$$

$$\mathbf{h}\big|_{t=0} = 0, \mathrm{curl}_\tau \mathbf{h}\big|_S = -(\mathrm{curl}_\tau \mathbf{h}_0 - \mathbf{j}_\tau)\big|_S, \quad h_n\big|_S = -h_{0n}\big|_S. \tag{2.24}$$

All of these are obvious with the exception of the heat equation. We shall show that this equation is also satisfied. The vector

$$\boldsymbol{\omega} = \sigma\mu\mathbf{h} - \int_0^t \Delta\mathbf{h}(x, \tau)\, d\tau \tag{2.25}$$

satisfies the equations

$$\begin{gathered} \operatorname{curl}\boldsymbol{\omega} = \sigma\mu\mathbf{E} - \int_0^t \Delta\mathbf{E}\, d\tau = \mathbf{0}, \quad \operatorname{div}\boldsymbol{\omega} = 0, \\ \omega_n|_S = \left[-\sigma\mu h_{0n} + \int_0^t \operatorname{curl}_n \operatorname{curl}\mathbf{h}\, d\tau\right]\Big|_S. \end{gathered} \tag{2.26}$$

But since $\operatorname{curl}_n \phi|_S$ depends only on $\omega_T|_S$, using (2.24) we obtain

$$\begin{aligned} \omega_n|_S &= -\sigma\mu h_{0n} - \int_0^t \operatorname{curl}_n \operatorname{curl}\mathbf{h}_0\, d\tau + \int_0^t \operatorname{curl}_n \mathbf{j}\, d\tau\,|_S \\ &= -\sigma\mu H_{0n} - \int_0^t \left(\sigma\mu\frac{\partial h_{0n}}{\partial\tau} + \operatorname{curl}_n \operatorname{curl}\mathbf{h}_0 - \operatorname{curl}_n \mathbf{j}\right) d\tau = 0. \end{aligned} \tag{2.27}$$

It follows from (2.26) and (2.27) that $\omega = 0$; differentiating (2.25) with respect to t, we obtain the desired equation (2.23). It follows from (2.17), (2.18), (2.23) and (2.24) that $\mathbf{H} = \mathbf{h}_0 + \mathbf{h}$ is a solution of problem (2.2).

We estimate the vectors $\mathbf{h}_0$, $\mathbf{h}$ and $\mathbf{H}$. For the solution of the Cauchy problem (2.17) there is the known estimate

$$\begin{aligned} \|\mathbf{h}_0\|_{\mathbf{W}_p^{2,1}(\Pi_T)} &\leqslant C\left(\|\mathbf{H}_0\|_{\mathbf{W}_p^{2-2/p}(R^3)} + \|\mathbf{g}\|_{\mathbf{L}_p(\Pi_T)}\right) \\ &\leqslant C\left(\|\mathbf{H}_0\|_{\mathbf{W}_p^{2-2/p}(\Omega)} + \|\operatorname{curl}\mathbf{j}\|_{\mathbf{L}_p(Q_T)}\right). \end{aligned} \tag{2.28}$$

Further, as a consequence of Theorem 4,

$$\begin{aligned} \|\mathbf{A}\|_{\mathbf{W}_p^{2,1}(Q_T)} &\leqslant C\left(\|\operatorname{curl}\mathbf{h}_0\|_{\mathbf{W}_p^{1-1/p,\,1/2-1/2p}(S_T)} + \|\mathbf{j}_\tau\|_{\mathbf{W}_p^{1-1/p,\,1/2-1/2p}(S_T)}\right) \\ &\leqslant C\left(\|\mathbf{h}_0\|_{\mathbf{W}_p^{2,1}(Q_T)} + \|\mathbf{j}_\tau\|_{\mathbf{W}_p^{1-1/p,\,1/2-1/2p}(S_T)}\right) \end{aligned} \tag{2.29}$$

(if $p = 3$ the integral (2.15) to the power 1/3 with a constant factor must be added to the right side). We estimate $\operatorname{grad}\phi$. Since ϕ is a solution of problem (2.22), it follows, as shown in [14], that

$$\begin{aligned} \|\operatorname{grad}\varphi\|_{\mathbf{W}_p^2(\Omega)} \leqslant \|\varphi\|_{W_p^3(\Omega)} &\leqslant C\left(\|A_n\|_{W_p^{2-1/p}(S)} + \|h_{0n}\|_{W_p^{2-1/p}(S)}\right) \\ &\leqslant C\left(\|\mathbf{A}\|_{\mathbf{W}_p^2(\Omega)} + \|\mathbf{h}_0\|_{\mathbf{W}_p^2(\Omega)}\right). \end{aligned} \tag{2.30}$$

In order to estimate grad ϕ_t, we represent ϕ_t as a sum $\phi_t = \phi_t' + \phi_t''$, where ϕ_t' and ϕ_t'' are solutions of the problems

$$\Delta\varphi_t' = -\operatorname{div} A_t, \quad \frac{\partial \varphi_t'}{\partial n}\Big|_S = -A_{tn}|_S,$$

$$\Delta\varphi_t'' = 0, \quad \frac{\partial \varphi_t''}{\partial n}\Big|_S = -h_{0nt}|_S = \operatorname{curl}_n \operatorname{curl} \mathbf{h}_0|_S,$$

and we then make use of estimates obtained in [9] and [14]:

$$\|\operatorname{grad} \varphi_t'\|_{\mathbf{L}_p(\Omega)} \leqslant C\|\mathbf{A}_t\|_{\mathbf{L}_p(\Omega)},$$

$$\|\operatorname{grad} \varphi_t''\|_{\mathbf{L}_p(\Omega)} \leqslant C\|\mathbf{h}_0\|_{\mathbf{W}_p^2(\Omega)}.$$

From these estimates and (2.30) it follows that

$$\|\operatorname{grad} \varphi\|_{\mathbf{W}_p^{2,1}(Q_T)} \leqslant C(\|\mathbf{A}\|_{\mathbf{W}_p^{2,1}(Q_T)} + \|\mathbf{h}_0\|_{\mathbf{W}_p^{2,1}(Q_T)}),$$

which together with (2.28) and (2.29) gives (2.16). The solution found is unique, since the solution w of problem (2.2) with j = 0 and $\mathbf{H}_0 = 0$ is at the same time a solution of the parabolic initial value problem

$$\sigma\mu\mathbf{w}_t - \Delta\mathbf{w} = \mathbf{0},$$

$$\mathbf{w}|_{t=0} = \mathbf{0}, \quad w_n|_S = 0, \quad \operatorname{curl}_\tau \mathbf{w}|_S = \mathbf{0}$$

(the complementarity condition is easily checked).

It is shown in [15] that w = 0. This completes the proof of the theorem. We proceed to study problem (2.2) in the class $\mathbf{C}^{2+\alpha, 1+\alpha/2}(Q_T)$.

Lemma 2. *Any solenoidal vector* $\mathbf{u} \in \mathbf{C}^{k+\alpha}(\Omega)$ *can be extended to* R^3 *with preservation of class and the solenoidal property in such a manner that* $|\mathbf{u}|_{R^3}^{(k+\alpha)} \leq C|\mathbf{u}|_\Omega^{(k+\alpha)}$. *It is here assumed that* $S \in C^{k+1+\alpha}$.

The proof is similar to that of Lemma 1; in place of (2.13) the estimate $|\mathbf{v}|_\Omega^{(k+1+\alpha)} \leq C|\mathbf{u}|_\Omega^{(k+\alpha)}$ established in [12] is used.

Lemma 3. *For any* $f \in C^{\alpha, \alpha/2}(Q_T)$ *and* $\psi \in C^{1+\alpha,(1+\alpha)/2}(S_T)$ *satisfying the conditions*

$$f(x, 0) = 0, \quad \psi(x, 0) = 0, \tag{2.31}$$

the potentials

$$v(x, t) = \int_0^t d\tau \int_\Omega \Gamma(x-y, t-\tau) f(y, \tau)\, dy,$$

$$w(x, t) = \int_0^t d\tau \int_S \Gamma(x-y, t-\tau) \psi(y, \tau)\, dS$$

belong to $C^{2+\alpha,1+\alpha/2}(Q_T)$ *and satisfy the inequalities*

$$|v|_{Q_T}^{(2+\alpha)} \leqslant C|f|_{Q_T}^{(\alpha)}, \quad |w|_{Q_T}^{(2+\alpha)} \leqslant C|\psi|_{S_T}^{(1+\alpha)}. \tag{2.32}$$

Proof. In virtue of (2.31) the functions f and ψ can be extended by zero to the region $t<0$ with preservation of class. We consider the heat potential of the simple layer w. We have

$$\frac{\partial w}{\partial t} = \int_{-\infty}^{t} d\tau \int_S \frac{\partial \Gamma(x-y,\ t-\tau)}{\partial t} [\psi_0(y,\tau)-\psi_0(y,t)]\,dS,$$

$$\frac{\partial^2 w}{\partial x_i \partial x_j} = \int_{-\infty}^{t} d\tau \int_S \frac{\partial^2 \Gamma(x-y,\ t-\tau)}{\partial x_i \partial x_j} [\psi_0(y,\tau)-\psi_0(y,t)]\,dS + \frac{1}{4\pi}\int_S \frac{\partial^2}{\partial x_i \partial x_j}\frac{1}{(x-y)}\psi(y,t)\,dS \equiv w'_{ij}+w''_{ij},$$

where $\psi_0=\psi$ for $\tau>0$ and $\psi_0=0$ for $\tau<0$. In the case $S=R^{n-1}$ potentials of this form have been considered in [15]. In particular, repeating the computations of pp. 65 and 68 (= 73 and 76–77) of [15], it can be established that

$$\left|\frac{\partial w(x,t)}{\partial t} - \frac{\partial w(x,t')}{\partial t'}\right| \leqslant C|\psi|_{S_T}^{(1+\alpha)}|t-t'|^{\alpha/2},$$

$$|w'_{ij}(x,t)-w'_{ij}(z,t)| \leqslant C|\psi|_{S_T}^{(1+\alpha)}|x-z|^{\alpha}.$$

It is known that the latter inequality holds also for functions w''_{ij}, the derivatives of the electrostatic simple-layer potential.

All of these inequalities prove the estimate (2.32) for w. The volume heat potential v is estimated similarly.

Theorem 6. *If* $S \in C^{3+\alpha}$, $0<\alpha<1$, *then for any* $\mathbf{g}=\operatorname{curl}\mathbf{j} \in \mathbf{J}^{\alpha}(Q_T)$, $\mathbf{H}_0 \in \mathbf{C}^{2+\alpha}(\Omega)\cap \overset{\circ}{\mathbf{J}}{}^{\alpha}(\Omega)$ *and* $\mathbf{j}_\tau \in \mathbf{C}^{1+\alpha,(1+\alpha)/2}(S_T)$ *satisfying the consistency condition* $\operatorname{curl}_\tau \mathbf{H}_0|_S = \mathbf{j}_\tau|_{t=0,x\in S}$ *problem* (2.2) *has a unique solution*

$$\mathbf{H} \in \mathbf{C}^{2+\alpha,\,1+\alpha/2}(Q_T)$$

and

$$|\mathbf{H}|_{Q_T}^{(2+\alpha)} \leqslant C\left[|\mathbf{H}_0|_{\Omega}^{(2+\alpha)} + |\operatorname{curl}\mathbf{j}|_{Q_T}^{(\alpha)} + |\mathbf{j}_\tau|_{S_T}^{(1+\alpha)}\right]. \tag{2.33}$$

Proof. We extend the vectors $\mathbf{H}_0(x) \in \mathbf{C}^{2+\alpha}(\Omega)\cap \mathbf{J}^{\alpha}(\Omega)$ and $\mathbf{g}(x,0) = \operatorname{curl}\mathbf{j}(x,0) \in \mathbf{J}^{\alpha}(\Omega)$ with preservation of class and the solenoidal property to the whole space R^3. Let

$$\mathbf{h}_1(x,\ t) \in \mathbf{C}^{2+\alpha,\ 1+\alpha/2}(Q_T)\cap \mathbf{J}^{\alpha}(Q_T)$$

be a solution of the Cauchy problem

$$\sigma\mu \frac{\partial \mathbf{h}_1}{\partial t} - \Delta \mathbf{h}_1 = \mathbf{g}(x, 0), \quad \mathbf{h}_1|_{t=0} = \mathbf{H}_0(x), \quad x \in R^3. \tag{2.34}$$

It is easy to verify that the vector $\mathbf{h}_0(x, t) = \mathbf{h}_1(x, t) + \mathbf{h}_2(x, t) + \mathbf{h}_3(x, t)$, where

$$\mathbf{h}_2(x, t) = \int_0^t d\tau \int_\Omega \Gamma(x-y, t-\tau)[\mathbf{g}(y, \tau) - \mathbf{g}(y, 0)]\, dy,$$

$$\mathbf{h}_3(x, t) = \int_0^t d\tau \int_S \Gamma(x-y, t-\tau)[(\mathbf{j}(y, \tau) - \mathbf{j}(y, 0)) \times \mathbf{n}(y)]\, dS,$$

satisfies (2.17) and (2.18) in Q_T. Invoking known estimates for the solution of the Cauchy problem (2.34) (cf., for example, [15]) and using Lemma 3 to estimate the heat potentials $\mathbf{h}_2$ and $\mathbf{h}_3$, we obtain

$$|\mathbf{h}_0|_{Q_T}^{(2+\alpha)} \leqslant [C_1(|\mathbf{g}(x, 0)|_{R^3}^{(\alpha)} + |\mathbf{H}_0|_{R^3}^{(2+\alpha)}) + C_2|\mathbf{g}(y, t)|_{Q_T}^{(\alpha)} + C_3|\mathbf{j}(y, t)|_{S_T}^{(1+\alpha)}$$
$$\leqslant C(|\mathbf{H}_0|_\Omega^{(2+\alpha)} + |\operatorname{rot} \mathbf{j}|_{Q_T}^{(\alpha)} + |\mathbf{j}|_{S_T}^{(1+\alpha)}).$$

The vector $\mathbf{h} = \mathbf{H} - \mathbf{h}_0$ is constructed just as in Theorem 5, and it remains to estimate it in the norm of $C^{2+\alpha, 1+\alpha/2}(Q_T)$. By Theorem 4

$$|\mathbf{A}|_{Q_T}^{(2+\alpha)} \leqslant C(|\operatorname{curl}_\tau \mathbf{h}_0|_{S_T}^{(1+\alpha)} + |\mathbf{j}_\tau|_{S_T}^{(1+\alpha)}) \leqslant C(|\mathbf{h}_0|_{Q_T}^{(2+\alpha)} + |\mathbf{j}_\tau|_{S_T}^{(1+\alpha)}).$$

Further, for vectors grad ϕ, grad ϕ_t' and grad ϕ_t'' the following estimates are valid:

$$\sup_t |\operatorname{grad} \varphi|_\Omega^{(2+\alpha)} \leqslant C \sup_t (|\mathbf{A}|_S^{(2+\alpha)} + |\mathbf{h}_0|_S^{(2+\alpha)}),$$
$$|\operatorname{grad} \varphi_t'|_{Q_T}^{(\alpha)} \leqslant C|\mathbf{A}|_{Q_T}^{(2+\alpha)},$$
$$|\operatorname{grad} \varphi_t''|_{Q_T}^{(\alpha)} \leqslant C|\operatorname{curl} \mathbf{h}_0|_{S_T}^{(1+\alpha)} \leqslant C|\mathbf{h}_0|_{Q_T}^{(2+\alpha)}$$

(the first estimate for a solution of problem (2.22) is well known; the second and third are proved in §6 of [9]). Thus

$$|\operatorname{grad} \varphi|_{Q_T}^{(2+\alpha)} \leqslant C(|\mathbf{A}|_{Q_T}^{(2+\alpha)} + |\mathbf{h}_0|_{Q_T}^{(2+\alpha)}).$$

Combining the above inequalities, we arrive at (2.33). This completes the proof of the theorem.

§3. The nonlinear problem

In this section we consider the nonlinear problem (1.4)–(1.7), where for brevity we write the system (1.4)–(1.6) as a single differential equation. Let $\mathbf{f} \in \mathring{\mathbf{J}}(\Omega)$, and let $\mathbf{P}$ be the orthogonal (in $L_2(\Omega)$) projection onto $\overset{\circ}{\mathbf{J}}(\Omega)$. We project both sides of (1.4) onto the space $\overset{\circ}{\mathbf{J}}(\Omega)$ in order to eliminate the term grad $(p + \mu \mathbf{H}^2/2)$. The equation so obtained together with (1.5) can be written as

$$\frac{d\mathbf{u}}{dt} + A_0\mathbf{u} + K(\mathbf{u}, \mathbf{u}) = \mathbf{F}, \tag{3.1}$$

where

$$\mathbf{u} = \begin{pmatrix} \mathbf{v} \\ \mathbf{H} \end{pmatrix}, \quad \mathbf{F} = \begin{pmatrix} \mathbf{f} \\ (\sigma\mu)^{-1} \operatorname{curl} \mathbf{j} \end{pmatrix}, \quad A_0\mathbf{u} = -\begin{pmatrix} \nu \mathbf{P} \Delta \mathbf{v} \\ (\sigma\mu)^{-1} \Delta \mathbf{H} \end{pmatrix},$$

$$K(\mathbf{u}, \mathbf{u}') = \frac{1}{2} \begin{pmatrix} \mathbf{P} \sum_{k=1}^{3} \left(v_k \mathbf{v}'_{x_k} + v'_k \mathbf{v}_{x_k} - \frac{\mu}{\rho} H_k \mathbf{H}_{x_k} - \frac{\mu}{\rho} H'_k \mathbf{H}_{x_k} \right) \\ -\operatorname{curl}([\mathbf{v} \times \mathbf{H}'] + [\mathbf{v}' \times \mathbf{H}]) \end{pmatrix}.$$

We do not write down (1.6), since we shall always work in the space of solenoidal $\mathbf{u}$ (i.e. such that $\mathbf{v}, \mathbf{H} \in \overset{\circ}{\mathbf{J}}(\Omega)$).

The boundary and initial conditions (1.7) can be reduced to homogeneous conditions; new linear terms thereby appear on the left side of (3.1). Let $\mathbf{u}' = \binom{\mathbf{v}'}{\mathbf{H}'}$, where $\mathbf{v}'$ and $\mathbf{H}'$ are solenoidal vectors satisfying conditions (1.7) (they can be constructed, for example, by solving problem (2.1), (2.2)). Then for $\mathbf{u} - \mathbf{u}' = \mathbf{U} = \binom{\mathbf{w}}{\mathbf{h}}$ we obtain

$$\frac{d\mathbf{U}}{dt} + A_0\mathbf{U} + 2K(\mathbf{u}', \mathbf{U}) + K(\mathbf{U}, \mathbf{U}) = \mathbf{G}, \tag{3.2}$$

where

$$\mathbf{G} = \mathbf{F} - \frac{\partial \mathbf{u}'}{\partial t} - A_0\mathbf{u}' - K(\mathbf{u}', \mathbf{u}') \equiv \begin{pmatrix} \mathbf{g} \\ \boldsymbol{\omega} \end{pmatrix}.$$

Since $[\mathbf{v}' \times \mathbf{H}']_\tau|_S = [a' \times \mathbf{H}']_\tau|_S = 0$ and $\operatorname{curl}_n \operatorname{curl} \mathbf{H}'|_S = \operatorname{curl}_n \operatorname{curl}_\tau \mathbf{H}'|_S = \operatorname{curl}_n \mathbf{j}|_S$, it follows that the vector

$$\boldsymbol{\omega} = \frac{1}{\sigma\mu} \operatorname{curl} \mathbf{j} - \mathbf{H}'_t - \frac{1}{\sigma\mu} \operatorname{curl} \operatorname{curl} \mathbf{H}' + \operatorname{curl}[\mathbf{v}' \times \mathbf{H}']$$

satisfies the condition

$$\omega_n|_S = 0. \tag{3.3}$$

It is also obviously solenoidal; thus $\omega \in \overset{\circ}{\mathbf{J}}(\Omega)$. Moreover, the vector $\mathbf{U}$ satisfies a zero initial condition

$$\mathbf{U}|_{t=0} = 0, \tag{3.4}$$

and its components $\mathbf{w}$ and $\mathbf{h}$ satisfy the boundary conditions

$$\mathbf{w}|_S = \mathbf{0}, \ \mathrm{curl}_\tau \mathbf{h}|_S = \mathbf{0}, \ h_n|_S = 0. \tag{3.5}$$

We consider the problem (3.2), (3.4) in the class $\mathbf{W}_p^{2,1}(Q_T)$. We assume that the vectors $\mathbf{f}$, $\mathbf{v}_0$ and $\mathbf{a}$ satisfy the conditions of Theorem 2 and the vectors $\mathbf{j}$ and $\mathbf{H}_0$ those of Theorem 4. Solving problems (2.1), (2.2), we obtain vectors $\mathbf{v}' \in \mathbf{W}_p^{2,1}(Q_T)$ and $\mathbf{H}' \in \mathbf{W}_p^{2,1}(Q_T)$ satisfying (1.7), and (2.4) and (2.16). Therefore the vector $\mathbf{G} = \binom{\mathbf{g}}{\omega}$ in (3.2) belongs to $\mathbf{L}_p(Q_T) \times \mathbf{L}_p(Q_T)$; moreover, from the definition of $\mathbf{g}$ and $\mathbf{w}$ and also from (3.3) it is evident that $\mathbf{g}, \mathbf{w} \in \mathring{\mathbf{J}}_p(Q_T)$, and thus $\mathbf{G} \in \mathring{\mathbf{J}}_p(Q_T) \times \mathring{\mathbf{J}}_p(Q_T)$.

We define the operators A_0 and K on the set $\mathbf{D}_p(\Omega)$ of vectors of $\mathbf{W}_p^2(\Omega) \times \mathbf{W}_p^2(\Omega)$ which are solenoidal and satisfy the homogeneous boundary conditions (3.5).

We denote by $\mathbf{W}_p(Q_T)$ the space of vectors

$$\mathbf{U}(x, t) = \begin{pmatrix} \mathbf{w} \\ \mathbf{h} \end{pmatrix} \in \mathbf{W}_p^{2,1}(Q_T) \times \mathbf{W}_p^{2,1}(Q_T),$$

which for almost all t belong to $\mathbf{D}_p(\Omega)$ (thus they are solenoidal and satisfy (3.5)). We introduce a norm in $\mathbf{W}_p(Q_T)$ by the formula

$$\|\mathbf{U}\|_{\mathbf{W}_p(Q_T)} = \|\mathbf{w}\|_{\mathbf{W}_p^{2,1}(Q_T)} + \|\mathbf{h}\|_{\mathbf{W}_p^{2,1}(Q_T)}. \tag{3.6}$$

In studying problem (3.2), (3.4) in this space and also the linearized problem

$$\frac{d\mathbf{V}}{dt} + A_0\mathbf{V} + 2K(\mathbf{u}', \mathbf{V}) = \mathbf{G}, \tag{3.7}$$

$$\mathbf{V}|_{t=0} = 0 \tag{3.8}$$

we shall use the following estimates for the operator K.

Lemma 3'. *Let $\mathbf{u}' \in \mathbf{W}_r^{2,1}(Q_T)$ with $r > 5/3$. Then for any $\mathbf{V} \in \mathbf{W}_p^{2,1}(Q_T)$, $p > 1$, satisfying the initial condition* (3.8) *we have*

$$\|K(\mathbf{u}', \mathbf{V})\|_{\mathbf{L}_p(Q_T)} \leqslant \delta \|\mathbf{u}'\|_{\mathbf{W}_r^{2,1}(Q_T)} \|\mathbf{V}\|_{\mathbf{W}_p^{2,1}(Q_T)} + C_\delta \|\mathbf{u}'\|_{\mathbf{W}_r^{2,1}(Q_T)} \|\mathbf{V}\|_{\mathbf{L}_p(Q_T)} \tag{3.9}$$

with $\delta > 0$ arbitrarily small. Moreover, if $p \geq 5/3$ there exists an $\omega(p) \geq 0$ (with $\omega(p) > 0$ for $p > 5/3$) such that

$$\|K(\mathbf{V}, \mathbf{V})\|_{\mathbf{L}_p(Q_T)} \leqslant C \min(1, T^{\omega}) \|\mathbf{V}\|^2_{\mathbf{W}_p^{2,1}(Q_T)}. \tag{3.10}$$

This lemma is a corollary of Theorem 4.2 and Lemma 10.1 of [9]. The following lemmas are well known.

Lemma 4. *If the functions $y_0(t)$ and $y(t)$ are such that $y(t) \leq y_0(t) + C\int_0^t y(\tau)d\tau$, then*

$$y(t) \leqslant e^{Ct} \int_0^t y_0(\tau)\, d\tau. \tag{3.11}$$

Lemma 5. *Suppose that the numbers* $\alpha_n \geq 0$, $n = 0, 1, \ldots,$ *satisfy the recurrence relation* $\alpha_{n+1} \leq a\alpha_n^2 + b$. *If* $ab < 1/4$ *and* $\alpha_0 \leq 2b/(1 + \sqrt{1 - ab})$, *then*

$$\alpha_n \leqslant \frac{2b}{1 + \sqrt{1 - 4ab}}.$$

Theorem 7. *Let* $S \in C^3$ *and* $\mathbf{u}' \in W_r^{2,1}(Q_T)$, $r > 5/3$. *Then for any* $\mathbf{G} \in \mathring{\mathbf{J}}_p(Q_T) \times \mathring{\mathbf{J}}_p(Q_T)$, $p > 1$, *problem* (3.7), (3.8) *has a unique solution* $\mathbf{V} \in \mathbf{W}_p(Q_T)$ *which satisfies the inequality*

$$\| \mathbf{V} \|_{\mathbf{W}_p(Q_T)} \leqslant C(T) \| \mathbf{G} \|_{L_p(Q_T)}. \tag{3.12}$$

Proof. We note first of all that if $\mathbf{u}' = 0$ the theorem to be proved follows from Theorems 2 and 5 (the vector $\mathbf{j}$ in (2.2) can be found from the problem $\operatorname{curl} \mathbf{j} = \mathbf{w}$, $\operatorname{div} \mathbf{j} = 0$, $\mathbf{j}_\tau|_S = 0$ which is solvable for all $\mathbf{w} \in \mathring{\mathbf{J}}_p(Q_T)$ in the class of vectors with generalized first derivatives in $\mathbf{L}_p(Q_T)$). In the case $\mathbf{u}' \neq 0$ we shall solve problem (3.7), (3.8) by the method of successive approximations, setting $\mathbf{V}^{(0)} = 0$ and defining $\mathbf{V}^{(n+1)}$, $n \geq 0$, from the problems

$$\frac{d\mathbf{V}^{(n+1)}}{dt} + A_0 \mathbf{V}^{(n+1)} = -2K(\mathbf{u}', \mathbf{V}^{(n)}) + \mathbf{G},$$
$$\mathbf{V}^{(n+1)}|_{t=0} = \mathbf{0}.$$

We prove the convergence of $\mathbf{V}^{(n)}$. The differences $\mathbf{Z}^{(n+1)} = \mathbf{V}^{(n+1)} - \mathbf{V}^{(n)}$ are solutions of the problems

$$\frac{d\mathbf{Z}^{(n+1)}}{dt} + A_0 \mathbf{Z}^{(n+1)} = -2K(\mathbf{u}', \mathbf{Z}^{(n)}), \quad \mathbf{Z}^{(n+1)}|_{t=0} = \mathbf{0},$$

and from (2.4) and (2.16) it follows that for all $t \leq T$ and $\epsilon > 0$

$$\| \mathbf{Z}^{(n+1)} \|_{\mathbf{W}_p(Q_t)} \leqslant \varepsilon \| \mathbf{Z}^{(n)} \|_{\mathbf{W}_p(Q_t)} + C_\varepsilon \| \mathbf{Z}^{(n)} \|_{\mathbf{L}_p(Q_t)}. \tag{3.13}$$

Since

$$\| \mathbf{Z}^{(n)} \|^p_{\mathbf{L}_p(\Omega)} \leqslant p \| \mathbf{Z}^{(n)} \|^{p-1}_{\mathbf{L}_p(Q_t)} \cdot \| \mathbf{Z}_t^{(n)} \|_{\mathbf{L}_p(Q_t)}$$

and hence

$$\| \mathbf{Z}^{(n)} \|_{\mathbf{L}_p(Q_t)} \leqslant p \int_0^t \| \mathbf{Z}_\tau^{(n)} \|_{\mathbf{L}_p(Q_\tau)} d\tau,$$

it follows that

$$\|\mathbf{Z}^{(n+1)}\|_{\mathbf{W}_p(Q_t)} \leqslant \varepsilon \|\mathbf{Z}^{(n)}\|_{\mathbf{W}_p(Q_t)} + C_\varepsilon p \int_0^t \|\mathbf{Z}^{(n)}\|_{\mathbf{W}_p(Q_\tau)} d\tau.$$

We sum this inequality over n. For the function

$$\xi_N(t) = \sum_{n=1}^{N} \|\mathbf{Z}^{(n)}\|_{\mathbf{W}_p(Q_t)} \tag{3.14}$$

we obtain the estimate

$$\xi_{N+1}(t) \leqslant \varepsilon \xi_N(t) + \xi_1(t) + C_\varepsilon p \int_0^t \xi_N(\tau)\, d\tau$$

and thus for $\epsilon < 1$

$$\xi_{N+1}(t) \leqslant \frac{1}{1-\varepsilon} \xi_1(t) + \frac{C_\varepsilon p}{1-\varepsilon} \int_0^t \xi_{N+1}(\tau)\, d\tau.$$

By Lemma 4, $\xi_N(t)$ is thus bounded on any interval $[0, T]$ by a constant independent of N; the series (3.14) and hence also the sequence $\mathbf{V}^{(n+1)}$ converge.

From (2.4) and (2.16) it also follows that any solution of problem (3.7), (3.8) satisfies

$$\|\mathbf{V}\|_{\mathbf{W}_p(Q_t)} \leqslant C \left(\|\mathbf{G}\|_{\mathbf{L}_p(Q_t)} + \varepsilon \|\mathbf{V}\|_{\mathbf{W}_p(Q_t)} + C_\varepsilon \int_0^t \|\mathbf{V}\|_{\mathbf{W}_p(Q_\tau)} d\tau \right),$$

from which (3.12) is obtained for sufficiently small ϵ; in the case $\mathbf{G} = 0$ we have $\mathbf{V} = 0$. The proof of the theorem is complete.

Remark. The theorem remains in force if the operator $K(\mathbf{u}', \cdot)$ is replaced by any linear operator $A_1(t)\cdot$ which satisfies (3.9) with arbitrary small δ (without the norm of $\mathbf{u}'$).

Theorem 8. *Let* $S \in C^3$ *and* $p \geq 5/3$. *If the condition*

$$C_1^2 C_2 \min(1, T^\omega) \|\mathbf{G}\|_{\mathbf{L}_p(Q_T)} < \frac{1}{4} \tag{3.15}$$

is satisfied, where C_1 *is the constant of* (3.12) *and* C_2 *and* ω *are those of* (3.10), *then problem* (3.2), (3.4) *is uniquely solvable for* $0 \leq t \leq T$ *in the class* $\mathbf{W}_p(Q_T)$, *and for the solution there is the estimate*

$$\|\mathbf{U}\|_{\mathbf{W}_p(Q_T)} \leqslant \frac{2C_1 \|\mathbf{G}\|_{\mathbf{L}_p(Q_T)}}{1 + \sqrt{1 - 4C_1^2 C_2 \min(1, T^\omega) \|\mathbf{G}\|_{\mathbf{L}_p(Q_T)}}}. \tag{3.16}$$

Proof. We shall solve problem (3.2), (3.4) by the method of successive approximations, setting $\mathbf{U}^{(0)} = 0$ and

$$\frac{d\mathbf{U}^{(n+1)}}{dt}+A_0\mathbf{U}^{(n+1)}+2K(\mathbf{u}',\mathbf{U}^{(n+1)})=\mathbf{G}-K(\mathbf{U}^{(n)},\mathbf{U}^{(n)}),$$
$$\mathbf{U}^{(n+1)}|_{t=0}=\mathbf{0}. \tag{3.17}$$

We prove the boundedness of the norms $\|\mathbf{U}^{(n+1)}\|_{\mathbf{W}_p(Q_T)}$ under condition (3.15). It follows from (3.10) and (3.12) that

$$\|\mathbf{U}^{(n+1)}\|_{\mathbf{W}_p(Q_T)}\leqslant C_1\left(\|\mathbf{G}\|_{\mathbf{L}_p(Q_T)}+C_2\min(1,T^{\omega})\|\mathbf{U}^{(n)}\|^2_{\mathbf{W}_p(Q_T)}\right).$$

By Lemma 5, $\|\mathbf{U}^{(n+1)}\|_{\mathbf{W}_p(Q_T)}$ is bounded by the right side of (3.16). Since the sequence $\mathbf{U}^{(n)}$ converges in $\mathbf{W}_p(Q_T)$, it follows that (3.16) is also valid for the limit vector $\mathbf{U}$ which is a solution of the problem (3.2), (3.4). In order to prove the convergence of $\mathbf{U}^{(n)}$ we note that $\mathbf{Z}^{(n+1)}=\mathbf{U}^{(n+1)}-\mathbf{U}^{(n)}$ is a solution of the problem

$$\frac{d\mathbf{Z}^{(n+1)}}{dt}+A_0\mathbf{Z}^{(n+1)}+2K(\mathbf{u}',\mathbf{Z}^{(n+1)})=-K(\mathbf{U}^{(n)}+\mathbf{U}^{(n-1)},\mathbf{Z}^{(n)}),$$
$$\mathbf{Z}^{(n+1)}|_{t=0}=\mathbf{0}.$$

Since the norms $\|\mathbf{U}^{(n)}\|_{\mathbf{W}_p(Q_T)}$ have already been estimated, we can use (3.9) with $r=p$ and a constant independent of n to estimate $K(\mathbf{U}^{(n)}+\mathbf{U}^{(n-1)},\mathbf{Z}^{(n)})$. Therefore for any $t\leq T$ we have an inequality

$$\|\mathbf{Z}^{(n+1)}\|_{\mathbf{W}_p(Q_t)}\leqslant\varepsilon\|\mathbf{Z}^{(n)}\|_{\mathbf{W}_p(Q_t)}+C_\varepsilon\|\mathbf{Z}^{(n)}\|_{\mathbf{L}_p(Q_t)},$$

analogous to (3.13). We have already seen that this inequality implies the convergence of the series $\sum_1^\infty\|\mathbf{Z}^{(n)}\|$, i.e. the convergence of $\mathbf{U}^{(n)}$.

The solvability of the problem and the estimate (3.16) have been established. We shall prove uniqueness. If $\mathbf{U}$ and $\mathbf{U}'$ are two solutions in $\mathbf{W}_p(Q_T)$, then their difference $\mathbf{Z}$ is a solution of the linear problem

$$\frac{d\mathbf{Z}}{dt}+A_0\mathbf{Z}+2K(\mathbf{u}',\mathbf{Z})+K(\mathbf{U},\mathbf{Z})+K(\mathbf{U}',\mathbf{Z})=\mathbf{0},$$
$$\mathbf{Z}|_{t=0}=\mathbf{0}.$$

Hence $\mathbf{Z}=0$. The proof of the theorem is complete.

The unique solvability of problem (1.4)–(1.7) in the class $\mathbf{W}_p^{2,1}(Q_T)$ follows from the theorem just proved. The number T depends on the norms of the vectors $\mathbf{a}$, $\mathbf{v}_0$, $\mathbf{H}_0$ and $\mathbf{j}$, which must satisfy the consistency conditions indicated in Theorems 2 and 5.

We proceed to consider problem (3.2), (3.4) in the class $\mathbf{C}^{2+\alpha,1+\alpha/2}(Q_T)$, assuming that $S\in C^{3+\alpha}$. For the existence of a solution in this class it is necessary that the vectors $\mathbf{v}_0$, $\mathbf{H}_0\in\mathbf{C}^{2+\alpha}(\Omega)$, $\mathbf{a}\in\mathbf{C}^{2+\alpha,1+\alpha/2}(S_T)$, $\mathbf{f}\in\mathring{\mathbf{J}}^{\alpha}(Q_T)$ and $\mathbf{j}_\tau\in\mathbf{C}^{1+\alpha,1/2+\alpha/2}(S_T)$ satisfy the following consistency conditions:

$$\mathbf{v}_0|_S = \mathbf{a}|_{t=0},$$
$$\frac{\partial \mathbf{a}}{\partial t}\Big|_{t=0} = \mathbf{P}\nu\Delta\mathbf{v}_0 - \mathbf{P}\sum_{k=1}^{3}\Big(v_{0k}\mathbf{v}_{0x_k} + \frac{\mu}{\rho}H_{0k}\mathbf{H}_{0x_k}\Big) + \mathbf{f}(x, 0)\Big|_S,$$
$$H_{0n}|_S = 0, \quad \operatorname{curl}_\tau \mathbf{H}_0|_S = \mathbf{j}_\tau|_{\substack{t=0\\ x\in S}}.$$

We shall construct vectors $\mathbf{v}'$, $\mathbf{H}' \in C^{2+\alpha,1+\alpha/2}(Q_T)$ satisfying (1.7). To this end we solve the problems

$$\frac{\partial \mathbf{v}'}{\partial t} - \nu\Delta\mathbf{v}' + \operatorname{grad} p' = \mathbf{f} - \mathbf{P}\sum_{k=1}^{3}\Big(v_{0k}\mathbf{v}_{0x_k} + \frac{\mu}{\rho}H_{0k}\mathbf{H}_{0x_k}\Big),$$
$$\operatorname{div}\mathbf{v}' = 0, \quad \mathbf{v}'|_{t=0} = \mathbf{v}_0(x), \qquad \mathbf{v}'|_S = \mathbf{a},$$
$$\frac{\partial \mathbf{H}'}{\partial t} - \frac{1}{\sigma\mu}\Delta\mathbf{H}' = \frac{1}{\sigma\mu}\operatorname{curl}\mathbf{j} - \operatorname{curl}[\mathbf{v}_0(x)\times\mathbf{H}_0(x)],$$
$$\operatorname{div}\mathbf{H}' = 0, \quad \mathbf{H}'|_{t=0} = \mathbf{H}_0(x), \quad H'_n|_S = 0, \quad \operatorname{curl}_\tau \mathbf{H}'|_S = \mathbf{j}_\tau|_S.$$

It is not difficult to verify that all the conditions of Theorems 3 and 6 are satisfied and the vectors $\mathbf{v}'$, $\mathbf{H}' \in C^{2+\alpha,1+\alpha/2}(Q_T)$ can be found; these satisfy

$$|\mathbf{v}'|_{Q_T}^{(2+\alpha)} \leqslant C\Big[|\mathbf{f}|_{Q_T}^{(\alpha)} + |\mathbf{v}_0|_\Omega^{(2+\alpha)} + |\mathbf{a}|_{S_T}^{(2+\alpha)} + (|\mathbf{v}_0|_\Omega^{(1+\alpha)})^2 + (|\mathbf{H}_0|_\Omega^{(1+\alpha)})^2\Big],$$
$$|\mathbf{H}'|_{Q_T}^{(2+\alpha)} \leqslant C\Big[|\operatorname{rot}\mathbf{j}|_{Q_T}^{(\alpha)} + |\mathbf{H}_0|_\Omega^{(2+\alpha)} + |\mathbf{j}_\tau|_{S_T}^{(1+\alpha)} + |\mathbf{v}_0|_\Omega^{(1+\alpha)}|\mathbf{H}_0|_\Omega^{(1+\alpha)}\Big].$$

The vector $\mathbf{U} = \mathbf{u} - \mathbf{u}'$ must be a solution of problem (3.2), (3.4), in which

$$\mathbf{G} = \mathbf{F} - \frac{d\mathbf{u}'}{dt} - A_0\mathbf{u}' - K(\mathbf{u}', \mathbf{u}') = K(\mathbf{u}', \mathbf{u}')|_{t=0} - K(\mathbf{u}', \mathbf{u}') \tag{3.18}$$

belongs to the class $\mathring{\mathbf{J}}^\alpha(Q_T)\times\mathring{\mathbf{J}}^\alpha(Q_T)$ and satisfies

$$\mathbf{G}|_{t=0} = \mathbf{0}. \tag{3.19}$$

We consider problems (3.2), (3.4) and (3.7), (3.8) in which the vector $\mathbf{G}$ satisfies these conditions. The operators A_0 and $K(\mathbf{u}', \cdot)$ we define on the set $\mathring{D}^{2+\alpha}(Q_T)$ of solenoidal vectors in

$$\mathbf{C}^{2+\alpha,\,1+\alpha/2}(Q_T)\times\mathbf{C}^{2+\alpha,\,1+\alpha/2}(Q_T),$$

which satisfy zero boundary conditions (3.5) and the initial conditions $\mathbf{u}|_{t=0} = 0$ and $\partial\mathbf{u}/\partial t|_{t=0} = 0$. In place of (3.9) and (3.10) we use estimates which follow from Lemma 10.3 of [9]:

$$\begin{gathered}|K(\mathbf{u}', \mathbf{V})|_{Q_T}^{(\alpha)} \leqslant \delta\sup_{Q_T}|\mathbf{u}'||\mathbf{V}|_{Q_T}^{(2+\alpha)} + C_\delta|\mathbf{u}'|_{Q_T}^{(1+\alpha)}\sup_{Q_T}|\mathbf{V}|,\\ |K(\mathbf{U}, \mathbf{U})|_{Q_T}^{(\alpha)} \leqslant C\min(1, T)\big(|\mathbf{U}|_{Q_T}^{(2+\alpha)}\big)^2,\end{gathered} \tag{3.20}$$

valid for all $\mathbf{V}, \mathbf{U} \in \overset{\circ}{\mathbf{D}}{}^{2+\alpha}(Q_T)$ and $\mathbf{u}' \in \mathbf{C}^{1+\alpha, 1/2+\alpha/2}(Q_T) \cap \mathbf{J}^{\alpha}(Q_T)$. They make it possible to prove the following theorems.

Theorem 9. *For any* $\mathbf{G} \in \overset{\circ}{\mathbf{J}}{}^{\alpha}(Q_T) \times \overset{\circ}{\mathbf{J}}{}^{\alpha}(Q_T)$ *satisfying the condition* $\mathbf{G}|_{t=0} = 0$, *problem* (3.7), (3.8) *has a unique solution in the class* $\overset{\circ}{\mathbf{D}}{}^{2+\alpha}(Q_T)$ *which satisfies the inequality*

$$|\mathbf{V}|^{(2+\alpha)}_{Q_T} \leqslant C(T)\,|\mathbf{G}|^{(\alpha)}_{Q_T}. \tag{3.21}$$

Theorem 10. *If the condition* $C_1^2 C_2 \min(1, T)|\mathbf{G}|^{(\alpha)}_{Q_T} < 1/4$ *is satisfied, where* C_1 *is the constant of* (3.21) *and* C_2 *that of* (3.20), *then problem* (3.2), (3.4) *is uniquely solvable for* $0 \leq t \leq T$ *in the class* $\overset{\circ}{\mathbf{D}}{}^{2+\alpha}(Q_T)$, *and*

$$|\mathbf{U}|^{(2+\alpha)}_{Q_T} \leqslant \frac{2C_1\,|\mathbf{G}|^{(\alpha)}_{Q_T}}{1+\sqrt{1-C_1^2 C_2 \min(1, T)\,|\mathbf{G}|^{(\alpha)}_{Q_T}}}. \tag{3.22}$$

The proofs of these theorems repeat the proofs of Theorems 7 and 8 almost word for word.

Theorems 7–10 make it possible to carry over to the spaces $\mathbf{W}_p^{2,1}(Q_T)$ and $\mathbf{C}^{2+\alpha,1+\alpha/2}(Q_T)$ the results on the stability of solutions of problem (1.4)–(1.7) which were obtained in [3] for the classes $\mathbf{W}_2^{2,1}(Q_T)$, but to do this it is necessary to trace the dependence of the constants in (3.12) and (3.21) on T and obtain estimates in norms with an exponential weight $e^{-\gamma t}$ (see [3] and [9]).

BIBLIOGRAPHY

1. O. A. Ladyženskaja and V. A. Solonnikov, *Solution of some nonstationary problems of magnetohydrodynamics for a viscous incompressible fluid*, Trudy Mat. Inst. Steklov. 59 (1960), 115–173. (Russian) MR 30 #370.

2. O. A. Ladyženskaja, *Mathematical problems in the dynamics of a viscous incompressible fluid*, 2nd rev. aug. ed., "Nauka", Moscow, 1970; English transl. of 1st ed., *The mathematical theory of viscous incompressible flow*, Gordon and Breach, New York, 1963; rev. 1969. MR 42 #6442; 27 #5034b; 40 #7610.

3. O. A. Ladyženskaja and V. A. Solonnikov, *On the linearization principle and invariant manifolds for problems of magnetohydrodynamics*, Zap. Naučn. Sem. Leningrad. Otdel. Mat. Inst. Steklov. (LOMI) 38 (1973), 46–93 = J. Soviet Math. (to appear).

4. S. L. Sobolev, *Applications of functional analysis in mathematical physics*, Izdat. Leningrad. Univ., Leningrad, 1950; English transl., Amer. Math. Soc., Providence, R. I., 1963. MR 14, 565; 29 #2624.

5. L. N. Slobodeckiĭ, *Estimates of solutions of elliptic and parabolic systems*, Dokl. Akad. Nauk SSSR 120 (1958), 468–471. (Russian) MR 21 #5060.

6. V. A. Solonnikov, *A priori estimates for second-order parabolic equations*, Trudy Mat. Inst. Steklov. 70 (1964), 133–212; English transl., Amer. Math. Soc. Transl. (2) 65 (1967), 51–137. MR 28 #5267.

7. ———, *Estimates of the solutions of a nonstationary linearized system of Navier-Stokes equations*, Trudy Mat. Inst. Steklov. 70 (1964), 213–317; English transl., Amer. Math. Soc. Transl. (2) 75 (1968), 1–116. MR 30 #1325.

8. V. A. Solonnikov, *On the differential properties of the solution of the first boundary-value problem for the nonstationary system of Navier-Stokes equations*, Trudy. Mat. Inst. Steklov. 73 (1964), 222–291. (Russian) MR 30 #2240.

9. ———, *Estimates of the solutions of the nonstationary Navier-Stokes system*, Zap. Naučn. Sem. Leningrad. Otdel. Mat. Inst. Steklov.(LOMI) 38 (1973),153–231 = J. Soviet Math. (to appear).

10. Š. Sahaev, *Solution of the first initial-boundary-value problem for the nonstationary system of Maxwell equations*, Akad. Nauk Kazah. SSR Trudy Inst. Mat. Meh. 2 (1972), 69–77. (Russian)

11. ———, *Estimation of the solution of an overdetermined parabolic initial-boundary-value problem*, Trudy Mat. Inst. Steklov. 127 (1975), 58–75 = Proc. Steklov Inst. Math. 127 (1975), 67–86.

12. V. A. Solonnikov, *Investigation of overdetermined elliptic boundary value problems in the fractional spaces of K. K. Golovkin*, Trudy Mat. Inst. Steklov. 127 (1975), 93–114 = Proc. Steklov Inst. Math. 127 (1975), 109–134.

13. L. N. Slobodeckiĭ, *Generalized Sobolev spaces and their application to boundary value problems for partial differential equations*, Leningrad. Gos. Ped. Inst. Učen. Zap. 197 (1958), 54–112; English transl., Amer. Math. Soc. Transl. (2) 57 (1966), 207–276. MR 34 #3075.

14. S. Agmon, A. Douglis and L. Nirenberg, *Estimates near the boundary for solutions of elliptic partial differential equations satisfying general boundary conditions*. I, Comm. Pure Appl. Math. 12 (1959), 623–727. MR 23 #A2610.

15. V. A. Solonnikov, *On boundary value problems for linear parabolic systems of differential equations of general form*, Trudy Mat. Inst. Steklov. 83 (1965) = Proc. Steklov Inst. Math. 83 (1965). MR 35 #1965.

Translated by J. R. SCHULENBERGER

Trudy Mat. Inst. Steklov.
127 (1975)

Proc. Steklov Inst. Math.
127 (1975)

INVESTIGATION OF OVERDETERMINED ELLIPTIC BOUNDARY VALUE PROBLEMS IN THE FRACTIONAL SPACES OF K. K. GOLOVKIN

UDC 517.946

V. A. SOLONNIKOV

Abstract. We prove sharp coercive estimates for the solutions of overdetermined elliptic boundary value problems, and establish their normal solvability in the fractional spaces of functions introduced by K. K. Golovkin; these spaces include, in particular, the spaces of functions with Hölder continuous derivatives.

Bibliography: 10 titles.

§1. Main definitions and subsidiary propositions

In our article [1] we consider overdetermined elliptic boundary value problems in a bounded domain $\Omega \subset R^n$ with a smooth boundary S

$$\begin{aligned} \mathcal{L}\left(x, \frac{\partial}{\partial x}\right)u(x) &= f(x), \\ \mathcal{B}\left(x, \frac{\partial}{\partial x}\right)u(x)|_S &= \Phi(x). \end{aligned} \tag{1.1}$$

Here $\mathfrak{L}(x, \partial/\partial x)$ is an $M \times m$ matrix $(M \geq m)$ whose elements are the differential operators $L_{kj}(x, \partial/\partial x)$ of order not greater than $s_k + t_j$, where s_k and t_j are integers, $s_k \leq 0$, $\max s_k = 0$, $t_j > 0$, $k = 1, \ldots, M$, $j = 1, \ldots, m$, and $\mathfrak{B}(x, \partial/\partial x)$ is a matrix with elements $B_{qj}(x, \partial/\partial x)$ of order $\leq \sigma_q + t_j$, where $q = 1, \ldots, Q$, and the σ_q are also integers. By the principal parts of the operators L_{kj} and B_{qj} we mean the sum of their terms of orders $s_k + t_j$ and $\sigma_q + t_j$. We denote these principal parts by L^0_{kj} and B^0_{qj}, and their corresponding matrices by $\mathfrak{L}_0(x, \partial/\partial x)$ and $\mathfrak{B}_0(x, \partial/\partial x)$; we assume that the following conditions hold:

1) *Condition of uniform ellipticity.* The rank of the matrix $\mathfrak{L}_0(x, \xi)$ is m for all $x \in \overline{\Omega} = \Omega \cup S$ and for all nonzero $\xi = (\xi_1, \ldots, \xi_n) \in R^n$.

2) *The matrix* $\mathfrak{B}$ *covers* $\mathfrak{L}$. The problem

$$\begin{aligned} \mathcal{L}_0\left(x, i\zeta + \nu(x)\frac{d}{dz}\right)\hat{u}(z) &= 0, \quad z > 0, \\ \mathcal{B}_0\left(x, i\zeta + \nu(x)\frac{d}{dz}\right)\hat{u}(z)|_{z=0} &= 0, \\ \hat{u} \to 0,\ z &\to +\infty \end{aligned} \tag{1.2}$$

AMS (MOS) subject classifications (1970). Primary 35N10, 46E15, 35J40, 35B45; Secondary 46E30, 41A30.

has only the zero solution for all $x \in S$ and for all $\zeta \in R^n$ satisfying the condition $\zeta \cdot \nu(x) = 0$, where $\nu(x)$ is the inward unit normal to S at x.

In [1] we constructed a left regularizer for the problem (1.1), and proved a sharp coercive estimate for the solution in the norms of the Sobolev spaces $W_p^l(\Omega)$, where l is an integer and $p > 1$. The aim of the present paper is to extend these results to the wide class of "fractional" or (J, r)-norms introduced by Golovkin [2]–[5]. We give their definition.

Let $\|\cdot\|$ be an arbitrary norm which is defined on functions defined on the whole space R^n and is translation invariant and monotonic:

$$\|u(x+\xi)\| = \|u(x)\|, \quad \forall \xi \in R^n,$$
$$\|u(x)\| \leqslant \|v(x)\| \quad \text{for} \quad |u(x)| \leqslant |v(x)|.$$

Let J be the "maximizing type functional" [2] defined by the formula $J[\psi(h)] = N[\psi(e^x)]$, where N is a monotonic translation invariant norm of functions of a single variable. The functional J is zero-dimensional; that is,

$$J[\psi(\lambda h)] = N[\psi(e^{x+\log\lambda})] = J[\psi(h)], \quad \forall \lambda > 0. \tag{1.3}$$

The (J, r)-norm of functions defined on R^n is defined by

$$\|u\|_{(J,r)} = [u]_{(J,r)} + \|u\|, \tag{1.4}$$

where

$$[u]_{(J,r)} = \sum_{i=1}^{n} J[h^{-r} \|\Delta_i^l(h)u\|], \tag{1.5}$$

and

$$\Delta_i^l(h)u(x) = \sum_{j=1}^{n} (-1)^{l-j} C_l^j u(x_1, \ldots, x_i + jh, \ldots, x_n)$$

is the finite difference of order $l > r > 0$ and with step h of the function u with respect to x_i.

We assume that the set $E(R^n)$ of functions for which the norm $\|u\|$ is finite is a complete Banach space containing bounded functions with compact support. It follows from the inequality

$$\left\| \int_B u(x+\xi)\,d\xi \right\| \leqslant \int_B \|u(x+\xi)\|\,d\xi = \operatorname{meas} B\, \|u(x)\| \tag{1.6}$$

that the elements of $E(R^n)$ are locally integrable functions: if K_ρ denotes the ball $|x| \le \rho$, and $\chi(x; K_\rho)$ is its characteristic function, then

$$\begin{aligned} \int_{K_\rho} |u(y)|\,dy &= \inf_{x \in K_\rho} \int_{|x-y| \leqslant 2\rho} |u(y)|\,\chi(y; K_\rho)\,dy \\ &\leqslant \|\chi(x, K_\rho)\|^{-1} \left\| \chi(x, K_\rho) \int_{|z| \leqslant 2\rho} |u(x-z)|\,\chi(x-z; K_\rho)\,dz \right\| \\ &\leqslant \operatorname{meas} K_{2\rho} \|\chi(x, K_\rho)\|^{-1} \|u(x)\,\chi(x, K_\rho)\|. \end{aligned} \tag{1.7}$$

It follows (see [3]) from the completeness of $E(R^n)$ that the set of functions with a finite norm (1.4) is a complete Banach space, which we denote by (J, r, R^n). By replacing the norm $\|\cdot\|$ in (1.4) and (1.5) by $\|v\|_R = \|v(x)\chi(x; R)\|$, where R is the domain $x_{i_1} \geq 0, \dots, x_{i_k} \geq 0$, we obtain the space (J, r, R). We will omit the letter R in this notation when it does not lead to misunderstanding. Below we consider, in the main, the cases $R = R^n$ and $R = R^n_+(x_n \geq 0)$.

If

$$\|u\| = \|u\|_p = \left(\int |u(x)|^p \, dx\right)^{1/p}, \qquad J[\psi] = J_\theta[\psi] = \left(\int_0^\infty |\psi(h)|^\theta \frac{dh}{h}\right)^{1/\theta},$$

$1 \leq p, \theta \leq \infty$, then the space (J, r, R) is the same as the Besov space $B^r_{p,\theta}(R)$ [6]; for $p = \theta = \infty$, $\|u\| = \|u\|_\infty = \sup|u(x)|$ and $J[\psi] = J_\infty[\psi] = \sup_{h>0}|\psi(h)|$ we obtain the space H^r of functions that are Hölder continuous together with their derivatives of order less than r.

We list the properties of the (J, r)-norms that we need.

Lemma 1. *For functions satisfying the condition*

$$J[h^{-r}\theta(h-1)\|\Delta_1^{l_1}(h)\dots\Delta_n^{l_n}(h)u\|] < \infty \tag{1.8}$$

for $\sum_1^n l_i > r > 0$, $\theta(t) = 1$ *for* $t \geq 0$, $\theta(t) = 0$ *for* $t < 0$ *(and, furthermore, for* $u \in (J, r)$*), all the seminorms of the form*

$$\sum_{|\alpha|=k} \sum_{i=1}^n J[h^{-r+k}\|\Delta_i^l(h) D^\alpha u\|]$$

with r *fixed and* $l > r - k > 0$ *are equivalent. The derivatives*

$$D^\alpha u = \frac{\partial^{|\alpha|} u}{\partial x_1^{\alpha_1} \dots \partial x_n^{\alpha_n}}$$

are to be understood as the limits of the corresponding difference quotients in the norm $\|\cdot\|$.

Lemma 2. *Under the condition* (1.8) *we have*

$$J[h^{-r+|\alpha|}\|\Delta_j^l(h) D^\alpha u\|] \leq c \prod_{k=1}^n \{J[h^{-r}\|\Delta_k^l(h)u\|]\}^{\alpha_k/r}\{J[h^{-r}\|\Delta_j^l(h)u\|\}^{1-|\alpha|/r}, \quad |\alpha| < r, \tag{1.9}$$

and hence, for every $\epsilon > 0$,

$$J[h^{-r+|\alpha|}\|\Delta_j^l(h) D^\alpha u\|] \leq c\,(\varepsilon^{|\alpha|-\alpha_j} J[h^{-r}\|\Delta_j^l(h)u\|] + \varepsilon^{-r+|\alpha|-\alpha_j}\sum_{i\neq j} J[h^{-r}\|\Delta_i^l(h)u\|]. \tag{1.9a}$$

Lemma 3. *The following inequalities hold for every* $\epsilon > 0$:

$$[u]_{(J,r)} \leq \varepsilon^{r_1-r}[u]_{(J,r_1)} + c\varepsilon^{r_2-r}[u]_{(J,r_2)}, \quad r_2 < r < r_1, \tag{1.10}$$

$$[u]_{(J,r)} \leq \varepsilon^{r_1-r}[u]_{(J,r_1)} + c\varepsilon^{-r}\|u\|, \quad r_1 > r, \tag{1.11}$$

$$[u]_{(J,r)} \leqslant \varepsilon^{l-r} \sum_{i=1}^{n} \| D_i^l u \| + c\varepsilon^{-r} \| u \|, \quad l > r, \tag{1.12}$$

$$\| D^\alpha u \| \leqslant \varepsilon^{r-|\alpha|} [u]_{(J,r)} + c\varepsilon^{-|\alpha|} \| u \|, \quad |\alpha| < r. \tag{1.13}$$

Lemma 1 is a theorem of Golovkin [2] about equivalent normalizations; Lemma 2 is stated in [4] and is proved in [7] for the spaces $H^r(R^n)$. Lemma 3 is well known for many concrete spaces, and the general case of it is stated without proof in [5]. We prove the estimates (1.9)–(1.13). For the proof of (1.10)–(1.12) we use the inequality

$$J[h^{-r} \| \Delta_i^l (h) u \|] \leqslant J\left[\chi\left(\frac{h}{\varepsilon}\right) h^{-r} \| \Delta_i^l (h) u \|\right] + J\left[\theta\left(\frac{h}{\varepsilon} - 1\right) h^{-r} \| \Delta_i^l (h) u \|\right],$$

in which $\chi(t) = \theta(t) - \theta(t-1)$. The norms on the right-hand side are estimated as follows:

$$J\left[h^{-r}\chi\left(\frac{h}{\varepsilon}\right) \| \Delta_i^l (h) u \|\right] \leqslant \varepsilon^{r_1 - r} J\left[h^{-r_1}\chi\left(\frac{h}{\varepsilon}\right) \| \Delta_i^l (h) u \|\right],$$

$$J\left[h^{-r}\chi\left(\frac{h}{\varepsilon}\right) \| \Delta_i^l (h) u \|\right] \leqslant \| D_i^l u \| J\left[h^{l-r}\chi\left(\frac{h}{\varepsilon}\right)\right] = c_{l-r}\varepsilon^{l-r} \| D_i^l u \|,$$

$$J\left[h^{-r}\theta\left(\frac{h}{\varepsilon} - 1\right) \| \Delta_i^l (h) u \|\right] \leqslant \varepsilon^{r_2 - r} J\left[h^{-r_2}\theta\left(\frac{h}{\varepsilon} - 1\right) \| \Delta_i^l (h) u \|\right],$$

$$J\left[h^{-r}\theta\left(\frac{h}{\varepsilon} - 1\right) \| \Delta_i^l (h) u \|\right] \leqslant 2^l \| u \| J\left[h^{-r}\theta\left(\frac{h}{\varepsilon} - 1\right)\right] = 2^l d_r \varepsilon^{-r} \| u \|,$$

where $c_k = J[h^k \chi(h)] < \infty$ and $d_k = J[h^{-k}\theta(h-1)] < \infty$ for $k > 0$ [5]. The estimates (1.10)–(1.12) follow from these inequalities. It is sufficient to obtain (1.9) and (1.13) for $|\alpha| = 1$, because then it is not difficult to prove them in the general case by induction on $|\alpha|$. We use the identity

$$D_i u(x) = \int_0^1 (-1)^m \Delta_i^m (h\varepsilon t) D_i u(x)\, dt$$

$$- \frac{1}{h\varepsilon} \sum_{k=1}^{m} (-1)^k C_m^k \frac{1}{k} [u(x_1, \ldots, x_i + k\varepsilon h, \ldots, x_n) - u(x)], \tag{1.14}$$

which is valid for every $\epsilon > 0$. First of all it follows from this identity that

$$J[h^{-r+1} \| \Delta_j^m (h) D_i u \|] \leqslant 2^m \int_0^1 J[h^{-r+1} \| \Delta_i^m (h\varepsilon t) D_i u \|]\, dt$$

$$+ \frac{2}{\varepsilon} \sum_{k=1}^{m} C_m^k \frac{1}{k} J[h^{-r} \| \Delta_j^m (h) u \|]$$

$$\leqslant 2^m \varepsilon^{r-1} J[h^{-r+1} \| \Delta_i^m (h) D_i u \|] \int_0^1 t^{r-1} dt + \frac{2}{\varepsilon} \sum_{k=1}^{m} C_m^k k^{-1} J[h^{-r} \| \Delta_j^m (h) u \|],$$

which is equivalent to (1.9) with $|\alpha| = 1$. Furthermore,

$$J\left[h^{-r+1}(\theta(h-1)-\theta(h-2))\right]\|D_i u\|$$
$$\leqslant \int_0^1 J\left[h^{-r+1}\|\Delta_i^m(h\varepsilon t)D_i u\|\right]dt + c\varepsilon^{-1}J\left[h^{-r}(\theta(h-1)-\theta(h-2))\right]\|u\|,$$

whence follows (1.13). Thus the inequalities (1.9)–(1.13) are proved.

We denote by (J, r, p), $1 \leq p \leq \infty$, the space of functions with a finite norm (1.4) if $\|u\| = \|u\|_p$ (as usual, $\|u\|_\infty = \sup|u(x)|$). Note that $(J, r, \infty) \subset H^r = (J_\infty, r, \infty)$.

Lemma 4. *If $u(x) \in (J, r)$ and $v(x) \in (J, r, \infty)$, then $uv \in (J, r)$ and*

$$\|uv\| \leqslant \|u\|\|v\|_\infty, \tag{1.15}$$

$$[uv]_{(J,r)} \leqslant \|v\|_\infty [u]_{(J,r)} + c\sum_{i=1}^n \sum_{0 \leqslant l < r} \|D_i^l u\| \cdot J\left[h^{-r+l}\|\Delta_i^s(h)v\|_\infty\right]. \tag{1.16}$$

The estimate (1.15) is obvious, and (1.16) follows from Leibniz' identity

$$\Delta_i^s(h)uv = \sum_{l=0}^s C_s^l \Delta_i^l(h)u(x) \cdot \Delta_i^{s-l}(h)v(x_1, \ldots, x_i + lh, \ldots, x_n). \tag{1.17}$$

If, in addition, $\operatorname{supp} v(x) \in Q_\rho(z) \cap R$ ($Q_\rho(z)$ always denotes the cube $-\rho/2 \leq x_i - z_i \leq \rho/2$), then the right-hand side of the identity is zero for $x \notin Q_{\rho+sh}(z)$. Therefore for every $d > 0$ we have

$$\sum_{i=1}^n J\left[h^{-r}\chi\left(\frac{h}{d}\right)\|\Delta_i^s(h)uv\|\right] \leqslant \|v\|_\infty \sum_{i=1}^n J\left[h^{-r}\chi\left(\frac{h}{d}\right)\|\Delta_i^s(h)u\|_{Q_{\rho+sd}}\right]$$
$$+ c\sum_{i=1}^n \sum_{0 \leqslant l < r} \|D_i^l u\|_{Q_{\rho+sd}} J\left[h^{-r+l}\chi\left(\frac{h}{d}\right)\|\Delta_i^s(h)v\|_\infty\right], \tag{1.18}$$

where, as usual, $\|u\|_{Q_\rho} = \|u(x)\chi(x;\, Q_\rho)\|$.

Lemma 5. *If $\sum_{|\alpha|<r}\|D^\alpha u\|_{Q_\rho} < \infty$, then for every $t \leq \rho/2$*

$$\sum_{|\alpha|<r} \|D^\alpha u\|_{Q_{\rho-t}} t^{-r+|\alpha|} \leqslant c\left\{\sum_{i=1}^n J\left[h^{-r}\chi\left(\frac{h}{t}\right)\|\Delta_i^s(h)u\|_{Q_\rho} + t^{-r}\|u\|_{Q_\rho}\right]\right\}. \tag{1.19}$$

Proof. Suppose that $\psi_t(x) \in C_0^\infty(Q_{\rho-t/2})$, $\psi_t(x) = 1$ for $x \in Q_{\rho-t}$, $0 \leq \psi_t \leq 1$, and $|D^\alpha\psi(x)| \leq c_\alpha t^{-|\alpha|}$. It follows from (1.12) with $\epsilon = t$ that for every $l < r$

$$J\left[h^{-r+l}\|\Delta_i^s(h)\psi_t(x)\|_\infty\right] \leqslant c\left(t^{s-r+l}\sum_{|\alpha|=s}\|D^\alpha\psi_t\|_\infty + t^{-r+l}\|\psi_t\|_\infty\right) \leqslant ct^{-r+l}. \tag{1.20}$$

We estimate $f(t) = \sum_{|\alpha|<r} t^{|\alpha|}\|D^\alpha u\|_{Q_{\rho-t}}$ by using (1.13) with $\epsilon = t\delta$, $0 \leq \delta \leq 1$. We have

$$f(t) \leqslant \sum_{|\alpha|<r} t^{|\alpha|}\|D^\alpha u\psi_t\| \leqslant t^r\delta_1[u\psi_t]_{(J,r)} + c_{\delta_1}\|u\psi_t\|,$$

where $\delta_1 = \sum_{|\alpha|<r}\delta^{r-|\alpha|}$ is an arbitrary number from the interval $(0, \delta_0]$, $\delta_0 > 0$. Then by (1.18) and (1.20) we have

$$[u\psi_t]_{(J,r)} \leqslant \sum_{i=1}^{n} J\left[h^{-r}\chi\left(\frac{4sh}{t}\right)\|\Delta_i^s(h)\,u\|_{Q_\rho}\right] + c\sum_{i=1}^{}\sum_{0\leqslant l<r} t^{-r+l}\|D_i^l u\|_{Q_{\rho-t/4}},$$

so that

$$f(t) \leqslant c_0\delta_1 f\left(\frac{t}{4}\right) + K(t, \delta_1), \tag{1.21}$$

where

$$K(t, \delta_1) = \delta_1\sum_{i=1}^{n} J\left[h^{-r}\chi\left(\frac{h}{t}\right)\|\Delta_i^s(h)\,u\|_{Q_\rho}\right]t^r + c'_{\delta_1}\|u\|_{Q_\rho}.$$

In (1.21) we take $\delta_1 = \min(1/2c_0, \delta_0)$. Since $K(t, \delta_1)$ is an increasing function of t, it follows from (1.21) that

$$f(t) \leqslant K(t, \delta)\sum_{i=0}^{\infty}(c_0\delta_1)^i \leqslant 2K(t, \delta_1),$$

which is equivalent to (1.19). This proves the lemma.

Lemma 6. *Every function $u \in (J, r, R)$ can be approximated in the following sense by infinitely differentiable functions: $u_n \to u$ in the norms $\|\cdot\|$ and of (J, r') with any $r' < r$, and $\|u_n\|_{(J,r)} \leq \|u\|_{(J,r)}$.*

Proof. We consider the average

$$u_\rho(x) = \int_R \omega_\rho(z)\,u(x+z)\,dz = \int_R \omega_\rho(y-x)\,u(y)\,dy \tag{1.22}$$

with the kernel $\omega_\rho(z) = \rho^{-n}\omega(z\rho^{-1})$, where $\omega(z) \in C_0^\infty(R)$, $\omega \geq 0$ and $\int\omega(z)dz = 1$. Clearly,

$$\|u_\rho(x)\|_{(J,r)} \leqslant \|u(x)\|\int_R \omega_\rho(z)\,dz = \|u(x)\|.$$

Furthermore,

$$\|u_\rho - u\| \leqslant \sup_{\substack{|z|\leqslant\rho\\ z\in R}}\|u(x+z) - u(x)\| \leqslant \sum_{i=1}^{n}\sup_{t\leqslant\rho}\|\Delta_i(t)\,u(x)\|$$

$$\leqslant \frac{\sum_{i=1}^{n} J\left[h^{-\beta}\left(\theta\left(\frac{h}{\rho}-1\right) - \theta\left(\frac{h}{\rho}-2\right)\right)\sup_{t\leqslant h}\|\Delta_i(t)\,u\|\right]}{J\left[h^{-\beta}\left(\theta\left(\frac{h}{\rho}-1\right) - \theta\left(\frac{h}{\rho}-2\right)\right)\right]}$$

$$\leqslant \frac{\sum_{i=1}^{n} J\,[h^{-\beta}\sup_{t\leqslant h}\|\Delta_i(t)\,u\|]}{J\left[h^{-\beta}\left(\theta\left(\frac{h}{\rho}-1\right) - \theta\left(\frac{h}{\rho}-2\right)\right)\right]} \leqslant c\rho^\beta[u]_{(J,\beta)} \xrightarrow[\rho\to 0]{} 0, \tag{1.23}$$

if $\beta < 1$, $\beta \leq r$ (regarding the last estimate, see [2]). It follows from (1.11) that $\|u_\rho - u\|_{(J,r')} \to 0$ as $\rho \to 0$ for every $r' \leq r$. It is clear from the example of the spaces H^r that we cannot always approximate a function $u \in (J, r)$ in the (J, r)-norm by functions from C^∞.

Suppose that $R = R^n_+$. We denote by H_0 the closure (in the sense of Lemma 6) of the set of infinitely differentiable functions that vanish for $x_n = 0$. It may coincide with (J, r, R^n_+), since this holds for the space $B^r_{p,p}(R^n_+)$ with $r < 1/p$. If it is not the same space, then we can speak about the traces of functions from (J, r, R^n_+) on R^{n-1} $(x_n = 0)$, and assume that $u \in H_0$ satisfies the condition $u|_{x_n=0} = 0$. We denote the space of traces by $(J, r, R^n_+)_{R^{n-1}}$; we can define it as $(J, r, R^n_+)/H_0$, and in it we can define the norm

$$\inf_{v \in H_0} \|u - v\|_{(J, r, R^n_+)} = \inf_{u^* - u \in H_0} \|u^*\|_{(J, r, R^n_+)}.$$

This space is nonempty if $r > 1$. For, if otherwise, by (1.7) the following inequality would hold for any smooth function $u(x)$ and for every $\rho > 0$:

$$\inf_{\substack{v|_{x_n=0}=0 \\ v \in C^\infty}} \sum_{i=1}^{n} \int_{\substack{|x| \leq \rho \\ x_n > 0}} |D_i u(x) - D_i v(x)|\, dx \leqslant C(\rho) \inf_{\substack{v|_{x_n=0}=0 \\ v \in C^\infty}} \|u - v\|_{(J, r)} = 0;$$

but this is impossible.

It follows from the results in [8] that the space $(J, R, p, R^n_+)_{R^{n-1}}$ with $\|\cdot\| = \|\cdot\|_p$ and $r > 1/p$ is isomorphic to $(J, r - 1/p, p, R^{n-1})$. (This result is established in [6] for the spaces $B^r_{p,\theta}$.) The spaces $(J, r, R^n)_{R^m}$ for any $R^m \subset R^n$ are defined similarly.

§2. Estimates for the solutions of elliptic boundary value problems in the half-space R^n_+

We consider in the half-space R^n_+ the problem

$$\mathscr{L}_0\left(\frac{\partial}{\partial x}\right) u(x) = f(x), \quad x \in R^n_+,$$

$$\mathscr{B}_0\left(\frac{\partial}{\partial x}\right) u|_{x_n=0} = \Phi(x'), \quad x' = (x_1, \ldots, x_{n-1}) \in R^{n-1}, \tag{2.1}$$

where $\mathfrak{L}_0$ and $\mathfrak{B}_0$ are operators with constant coefficients satisfying the conditions stated at the beginning of §1, and not containing lower order terms. We derived in [1] a representation of the following form for the solution of this problem:

$$u_j(x) = \sum_{k=1}^{M} \sum_{|\gamma| = b+1-s_k} \int_{R^n_+} [Z^\gamma_{jk}(x - y) + \tilde{Z}^\gamma_{jk}(x' - y', x_n, y_n)] D^\gamma_y f_k(y)\, dy$$

$$+ \sum_{q=1}^{Q} \sum_{|\gamma| = b_1 - \sigma_q} \int_{R^n_+} U^\gamma_{jq}(x' - y', x_n, y_n) D^\gamma_y \Phi^*_q(y)\, dy, \tag{2.2}$$

where Φ_q^* is an arbitrary extension of the function Φ_q from R^{n-1} to R_+^n, and b and b_1 are numbers satisfying the conditions

$$b \geqslant \max(-1, \sigma_1, \dots, \sigma_Q), \quad b_1 \geqslant \max(0, 1+\sigma_1, \dots, 1+\sigma_Q).$$

The functions Z_{jk}^γ have the form

$$Z_{jk}^\gamma(x) = M_{jk}^\gamma\left(\frac{\partial}{\partial x}\right)\Gamma(x), \tag{2.3}$$

where M_{jk}^γ is a differential operator of order $2R - t_j - b - 1$, $R = \Sigma_{j=1}^m t_j + m(b+1)$, $\Gamma(x)$ is the fundamental solution of the equation $L(\partial/\partial x)\Gamma = \delta(x)$, and $L(\partial/\partial x)$ is a homogeneous elliptic operator of order $2R$. $\Gamma(x)$ satisfies the following estimate:

$$|D_x^\mu \Gamma(x)| \leqslant c\varphi(|x|; -n + 2R - |\mu|), \tag{2.4}$$

where

$$\varphi(r; k) = \begin{cases} r^k, & k < 0, \\ r^k(1 + |\log r|), & k \geqslant 0. \end{cases}$$

Hence Z_{jk}^γ is subject to an analogous inequality, but with the function $\phi(|x|; -n + t_j + b + 1)$ on the right-hand side.

The functions $\tilde{Z}_{jk}^\gamma$ and U_{jq}^γ can be represented in the form

$$\tilde{Z}_{jk}^\gamma(x' - y', x_n, y_n) = \Delta'^p \tilde{Z}_{jk}^{\gamma p}, \quad U_{jq}^\gamma = \Delta'^p U_{jq}^{\gamma p}, \tag{2.5}$$

where $\Delta' = \Sigma_1^{n-1} D_i^2$, and $\tilde{Z}_{jk}^{\gamma p}$ and $U_{jq}^{\gamma p}$ satisfy the equation $L(\partial/\partial x)w = 0$ and the inequalities

$$\begin{aligned} |D_x^\mu \tilde{Z}_{jk}^{\gamma p}(x' - y', x_n, y_n)| &\leqslant c\varphi(|x - y^*|; -n + t_j + b + 1 + 2p - |\mu|), \\ |D_x^\mu U_{jq}^{\gamma p}(x' - y', x_n, y_n)| &\leqslant c\varphi(|x - y^*|; -n + t_j + b_1 + 2p - |\mu|), \end{aligned} \tag{2.6}$$

in which $y^* = (y_1, \dots, -y_n)$, so that $|x - y^*|^2 = |x' - y'|^2 + (x_n + y_n)^2$.

We estimate the functions (2.2).

Theorem 1. *The functions u_j in (2.2) are subject to the inequality*

$$\sum_{j=1}^m [u_j]_{(J,\, l+t_j,\, R_+^n)} \leqslant c\left(\sum_{j=1}^M [f_j]_{(J,\, l-s_j,\, R_+^n)} + \sum_{q=1}^Q [\Phi_q^*]_{(J,\, l-\sigma_q,\, R_+^n)}\right), \tag{2.7}$$

provided that the number l satisfies the condition

$$l > \max(b+1, b_1), \tag{2.8}$$

and f_k and Φ_q^* are functions with finite $(J, l - s_k, R_+^n)$- and $(J, l - \sigma_q, R_+^n)$-norms.

Proof. The formulas (2.3) and (2.5) allow us to write (2.2) in the form

$$u_j(x)=\sum_{k=1}^{M}\sum_{|\gamma|=b+1-s_k}\left[M^{\gamma}_{jk}\left(\frac{\partial}{\partial x}\right)u^{\gamma}_k(x)+\sum_{|\alpha|=p_j}c_\alpha D^\alpha v^{\gamma\alpha}_{jk}(x)\right]$$
$$+\sum_{q=1}^{Q}\sum_{|\gamma|=b_1-\sigma_q}\sum_{|\beta|=p^j}c_\beta D^\beta w^{\gamma\beta}_{jq}(x), \tag{2.9}$$

where $p_j = 2R - t_j - b - 1$, $p^j = 2R - t_j - b_1$,

$$u^{\gamma}_k(x)=\int_{R^n_+}\Gamma(x-y)\,D^\gamma f_k(y)\,dy,\quad |\gamma|=b+1-s_k; \tag{2.10}$$

$$v^{\gamma\alpha}_{jk}(x)=\int_{R^n_+}V^{\gamma\alpha}_{jk}(x'-y',\ x_n,\ y_n)\,D^\gamma f_k(y)\,dy,\quad |\gamma|=b+1-s_k; \tag{2.11}$$

$$w^{\gamma\beta}_{jq}(x)=\int_{R^n_+}W^{\gamma\beta}_{jq}(x'-y',\ x_n,\ y_n)\,D^\gamma\Phi^*_q(y)\,dy,\quad |\gamma|=b_1-\sigma_q, \tag{2.12}$$

and the kernels $V^{\gamma\alpha}_{jk} = D^\alpha \tilde{Z}^{\gamma p_j}_{jk}$ and $W^{\gamma\beta}_{jq} = D^\beta U^{\gamma p^j}_{jq}$ satisfy the inequality

$$|D^\mu_x V^{\gamma\alpha}_{jk}|+|D^\mu_x W^{\gamma\beta}_{jq}|\leqslant c\varphi(|x-y^*|;\ -n+2R). \tag{2.13}$$

Furthermore, the following equations hold:

$$L\left(\frac{\partial}{\partial x}\right)u^{\gamma}_k(x)=D^\gamma f_k(x), \tag{2.14}$$

$$L\left(\frac{\partial}{\partial x}\right)v^{\gamma\alpha}_{jk}(x)=0,\quad L\left(\frac{\partial}{\partial x}\right)w^{\gamma\beta}_{jq}(x)=0. \tag{2.15}$$

The inequality (2.7) reduces to an estimate of the potentials (2.10)–(2.12) in the norms of $(J,\ l+2R-b-1)$, $(J,\ l+2R-b-1)$ and $(J,\ l+2R-b_1)$, respectively.

We estimate $u^{\gamma}_k(x)$. We form the finite difference $\Delta^{s+t}_i(h)u^{\gamma}_k(x)$ with $s > l + 2R$, $t > l$ and $i < n$. By using the identity

$$\Delta^k_i(h)\,F(x)=\int_0^h\cdots\int_0^h D^k_i F(x+e_i(\xi_1+\cdots+\xi_k))\,d\xi_1,\ldots,d\xi_k$$

where e_i is the point of R^n with the coordinates δ_{ij}, we obtain

$$\|\Delta^{s+t}_i(h)\,u^{\gamma}_k(x)\|\leqslant h^{2R-1}\|\Delta^{s-2R+1+t}_i(h)\,D^{2R-1}_i u^{\gamma}_k(x)\|. \tag{2.16}$$

But since

$$\Delta^{s-2R+1+t}_i(h)\,D^{2R-1}_i u^{\gamma}_k=\int_{R^n_+}\Delta^{s-2R+1}_i(h)\,D^{2R-1}_i\Gamma(x-y)\,\Delta^t_i(h)\,D^\gamma f_k(y)\,dy,$$

and

$$|\Delta_i^{s-2R+1}(h)\,D_i^{2R-1}\Gamma(x-y)|\leqslant\psi(|x-y|,\ h)$$
$$\equiv c\begin{cases}\sum\limits_{j=0}^{s-2R+1}|x-y+e_ijh|^{-n+1}, & |x-y|\leqslant 2sh,\\ h^{s-2R+1}|x-y|^{-n-s+2R}, & |x-y|\geqslant 2sh,\end{cases} \tag{2.17}$$

we have

$$|\Delta_i^{s+t-2R+1}(h)\,D_i^{2R-1}u_k^\gamma|\leqslant\int\limits_{R^n}\psi(|z|;\ h)\,\theta(x_n-z_n)\,|\Delta_i^t(h)\,D^\gamma f_k(x-z)|\,dz$$

and hence

$$\|\Delta_i^{s+t-2R+1}(h)\,D_i^{2R-1}u_k^\gamma\|\leqslant\|\Delta_i^t(h)\,D^\gamma f_k(x)\|\int\limits_{R^n}\psi(|z|;\ h)\,dz$$
$$\leqslant ch\,\|\Delta_i^t(h)\,D^\gamma f_k\|, \tag{2.18}$$

$$J\,[h^{-l-2R+b+1}\,\|\Delta_i^{s+t}(h)\,u_k^\gamma(x)\|]\leqslant c\,[f_k]_{(J,\ l-s_k)}. \tag{2.19}$$

To estimate this norm for u_k^γ with $i=n$ we use (2.14) and Lemma 2. For every $N\leq\sigma$ we have

$$\|\Delta_n^\sigma(h)\,u_k^\gamma(x)\|\leqslant h^N\,\|\Delta_n^{\sigma-N}(h)\,D_n^Nu_k^\gamma\|.$$

If $N\geq 2R$, then by using (2.14) we can express $D_n^Nu_k^\gamma$ in terms of $D_n^{N-2R}D^\gamma f_k$ and a linear combination of the derivatives $D_x^\alpha u_k^\gamma$, $|\alpha|=N$, $\alpha_k\leq N-1$, and also obtain the estimate

$$\|\Delta_n^\sigma(h)\,u_k^\gamma(x)\|\leqslant ch^N\Bigl(\sum_{\substack{|\alpha|=N\\ \alpha_n\leqslant N-1}}\|\Delta_n^{\sigma-N}(h)\,D^\alpha u_n^\gamma\|+\|\Delta_n^{\sigma-N}(h)\,D_n^{N-2R}D^\gamma f_k\|\Bigr). \tag{2.20}$$

We fix an N for which $0<l+2R-b-1-N\leq 1$. Then $N\geq 2R$, and by Lemma 2

$$J\,[h^{-l-2R+b+1}\,\|\Delta_n^\sigma(h)\,u_k^\gamma(x)\|]\leqslant cJ\,[h^{-l-2R+b+1+N}\,\|\Delta_n^{\sigma-N}(h)\,D_n^{N-2R}D^\gamma f_k\|]$$
$$+\,c\varepsilon J\,[h^{-l-2R+b+1}\,\|\Delta_n^\sigma(h)\,u_k^\gamma(x)\|]+c_\varepsilon\sum_{i=1}^{n-1}J\,[h^{-l-2R+b+1}\,\|\Delta_i^\sigma(h)\,u_k^\gamma(x)\|]. \tag{2.21}$$

By taking $\epsilon=1/2c$ and using (2.19) we obtain the required inequality

$$[u_k^\gamma]_{(J,\ l+2R-b-1)}\leqslant c\,[f_k]_{(J,\ l-s_k)}. \tag{2.22}$$

But, in applying Lemma 2, we need to verify that (1.8) holds. We establish that if $\Sigma_1^n m_i>l+2R-b-1$, then

$$J\,[h^{-l-2R+b+1}\theta(h-1)\,\|\Delta_1^{m_1}(h),\ \ldots,\ \Delta_n^{m_n}u_k^\gamma(x)\|]$$
$$\leqslant J\,[h^{-l+b}\theta(h-1)\,\|\Delta_1^{r_1}(h)\ldots\Delta_n^{r_n}(h)\,D_x^\alpha u_k^\gamma(x)\|]<\infty,$$

where $\alpha_i=m_i-r_i$ and $|\alpha|=2R-1$. By analogy with (2.17),

$$|\Delta_1^{r_1}(h)\ldots\Delta_n^{r_n}(h)\,D_x^\alpha\,\Gamma(x-y)|\leqslant\psi'(|x-y|;\ h),$$

where $\int_{R^n}\psi'(|x-y|;\ h)dx\leq ch$. Therefore

$$\|\Delta_1^{r_1}(h)\ldots\Delta_n^{r_n}(h)\,D_x^\alpha u_k^\gamma\|\leqslant ch\|D^\gamma f_k\|$$

and

$$J\left[h^{-l+b}\theta(h-1)\|\Delta_1^{r_1}(h)\ldots\Delta_n^{r_n}(h)\,D_x^\alpha u_k^\gamma\|\right]\leqslant c\|D^\gamma f_k\|\,J\left[h^{-l+b+1}\theta(h-1)\right]<\infty.$$

This proves (2.22).

The potentials (2.11) and (2.12) are estimated in a similar way, since their kernels depend on the difference $x'-y'$ and satisfy inequalities of the type (2.17), for example,

$$|\Delta_i^{s-2R+1}(h)\,D_i^{2R-1}V_{jk}^{\gamma\alpha}(x'-y',\ x_n,\ y_n)|\leqslant c\psi(|x-y^*|;\ h)\leqslant c\psi(|x-y|;\ h),$$

and the homogeneous equation (2.15). The estimates for these potentials are as follows:

$$[v_{jk}^{\gamma\alpha}]_{(J,\ l+2R-b-1)}\leqslant c[f_k]_{(J,\ l-s_k)},$$
$$[w_{jq}^{\gamma\alpha}]_{(J,\ l+2R-b_1)}\leqslant c[\Phi_q^*]_{(J,\ l-\sigma_q)}.$$

From these inequalities and (2.22) follows (2.7), and the theorem is proved.

From Theorem 1 it follows, in particular, that any smooth finite solution of the problem (2.1) satisfies (2.7) with any

$$l>\max(0,\ 1+\sigma_1,\ \ldots,\ 1+\sigma_Q) \tag{2.23}$$

(because we can then take $b<l-1$ and $b_1<l$ in (2.2)). In the general case it seems impossible to obtain (2.7) for smaller l, although this may be possible under certain conditions on $\|\cdot\|$. We state the relevant theorem but do not prove it in detail.

Theorem 2. *Suppose that* $\|\cdot\|=\|\cdot\|_p$, $p\in[1,\infty]$. *Then for any*

$$l>\max\left(0,\ \sigma_1+\frac{1}{p},\ \ldots,\ \sigma_Q+\frac{1}{p}\right) \tag{2.24}$$

we have

$$\sum_{j=1}^m[u_j]_{(I,\ l+t_j,\ p,\ R_+^n)}\leqslant c\left(\sum_{j=1}^M[f_j]_{(I,\ l-s_j,\ p,\ R_+^n)}+\sum_{q=1}^Q[\Phi_q]_{(I,\ l-\sigma_q-1/p,\ p,\ R^{n-1})}\right).$$

As we show in [1], the third term u_j' in (2.2) can be reduced to the form

$$u_j'(x)=\sum_{q=1}^Q\ \sum_{|\gamma|=b'-\sigma_q}\ \int_{R^{n-1}}P_{jq}^\gamma(x'-y',\ x_n)\,D_y^\gamma\Phi_q(y')\,dy',$$

where $b'\geq\max\sigma_q$ and P_{jq}^γ satisfies the conditions

$$P_{jq}^\gamma=\Delta'^sP_{jq}^{\gamma s}(x'-y',\ x_n,\ y_n);\quad L\left(\frac{\partial}{\partial x}\right)P_{jq}^{\gamma s}=0,$$
$$|D_x^\mu P_{jq}^{\gamma s}|\leqslant c\varphi(|x-y'|;\ -n+1+t_j+b'+2s-|\mu|).$$

Potentials with these kernels are estimated in [4] (but under slightly more general assumptions about $\|\cdot\|$). Now we estimate $D^{\alpha}u_j$, $|\alpha| < l + t_j$.

Theorem 3. *Suppose that* $\operatorname{supp} f$, $\operatorname{supp}\Phi^* \subset Q^+_\lambda(Z) = Q_\lambda(Z) \cap R^n_+$, $\lambda \leq 1$. *Then, for* $|\alpha| < \min(b+1, b_1) + t_j$,

$$\|\chi(x;\ Q^+_\lambda)\,D^\alpha u_j\| \leqslant c\Bigl(\sum_{k=1}^{M}\ \sum_{|\gamma|=b+1-s_k}\varphi(\lambda;\ t_j+b+1-|\alpha|)\,\|D^\gamma f_k\| + \sum_{q=1}^{Q}\ \sum_{|\gamma|=b_1-\sigma_q}\varphi(\lambda;\ t_j+b_1-|\alpha|)\,\|D^\gamma\Phi^*_q\|\Bigr). \tag{2.25}$$

Proof. It follows from (2.2) that

$$|D^\alpha u_j(x)| \leqslant c\sum_{k=1}^{M}\ \sum_{|\gamma|=b+1-s_k}\int_{R^n}[\varphi(|x-y|;\ -n+t_j+b+1-|\alpha|) + \varphi(|x-y^*|;\ -n+t_j+b+1-|\alpha|)]\,|D^\gamma_y f_k(y)|\,dy + \sum_{q=1}^{Q}\ \sum_{|\gamma|=b_1-\sigma_q}\int_{R^n_+}\varphi(|x-y^*|;\ -n+t_j+b_1-|\alpha|)\,|D^\gamma_y\Phi^*_q(y)|\,dy.$$

For $x,\ y\in Q^+_\lambda$ we have

$$\varphi(|x-y^*|;\ d) \leqslant c\begin{cases}\varphi(|x-y|;\ d), & d\leqslant 0,\\ \varphi|x-y|+2\lambda;\ d), & d>0\end{cases} \equiv \psi_\lambda(|x-y|;\ d),$$

and therefore

$$|D^\alpha u_j(x)| \leqslant c\Bigl[\sum_k\sum_\gamma\int_{R^n}\psi_\lambda(|y|;\ -n+t_j+b+1-|\alpha|) \times |D^\gamma f_k(x-y)|\,\chi(x-y;\ Q^+_\lambda)\,dy + \sum_q\sum_\gamma\int_{R^n}\psi_\lambda(|y|;\ -n+t_j+b_1-|\alpha|)\,|D^\gamma\Phi^*_q(x-y)|\,\chi(x-y;\ Q^+_\lambda)\,dy\Bigr].$$

We multiply both sides of this inequality by $\chi(x;\ Q^+_\lambda)$ and take into account that

$$\chi(x;\ Q^+_\lambda)\,\chi(x-y;\ Q^+_\lambda) \leqslant \chi(y;\ Q^+_{2\lambda}(0))\,\chi(x-y;\ Q^+_\lambda).$$

By forming the norm $\|\cdot\|$ we obtain

$$\|\chi(x;\ Q^+_\lambda)\,D^\alpha u_j(x)\| \leqslant c\Bigl[\sum_k\sum_\gamma\|D^\gamma f_k\|\int_{R^n}\chi(z;\ Q_{2\lambda}(0))\,\psi_\lambda(|z|;\ -n+t_j+b+1-|\alpha|)\,dz + \sum_q\sum_\gamma\|D^\gamma\Phi^*_q\|\int_{R^n}\chi(z;\ Q_{2\lambda}(0))\,\psi_\lambda(|z|;\ -n+t_j+b_1-|\alpha|)\,dz,$$

whence follows (2.25), and the theorem is proved.

Corollary. *Suppose that the conditions of Theorems* 1 *and* 3 *hold. Then*

$$\sum_{j=1}^{m}\Big([u_j]_{(J,\,l+t_j)}+\sum_{|\alpha|<l+t_j}\|D^\alpha u_j\|_{Q_\lambda^+}\lambda^{-(l+t_j)+|\alpha|}\Big)$$
$$\leqslant c\Big\{\sum_{k=1}^{M}\Big([f_k]_{(J,\,l-s_k)}+\sum_{|\alpha|<l-s_k}\lambda^{-l+s_k+|\alpha|}(1+|\log\lambda|)\|D^\alpha f_k\|_{Q_\lambda^+}\Big)$$
$$+\sum_{q=1}^{Q}\Big([\Phi_q]_{(J,\,l-\sigma_q)}+\sum_{|\alpha|<l-\sigma_q}\lambda^{-l+\sigma_q+|\alpha|}(1+|\log\lambda|)\|D^\alpha\Phi_q^*\|_{Q_\lambda^+}\Big)\Big\}. \tag{2.26}$$

For by Lemma 5 we have

$$\sum_{|\alpha|<l+t_j}\|D^\alpha u_j\|_{Q_\lambda^+}\lambda^{-(l+t_j)+|\alpha|}\leqslant c\Big([u_j]_{(J,\,l+t_j)}+\lambda^{-(l+t_j)}\|u_j\|_{Q_{2\lambda}^+}\Big).$$

We estimate the first term on the right-hand side by using Theorem 1, and the second term by using Theorem 3; we then obtain (2.26).

The condition $\Sigma_{|\alpha|<l+t_j}\|D^\alpha u_j\|_{Q_{2\lambda}}<\infty$ in Lemma 5 follows from Theorem 2; if f and Φ^* have a finite norm

$$\sum_{j=1}^{M}\sum_{|\alpha|<l+1-s_k}\|D^\alpha f_j\|+\sum_{q=1}^{Q}\sum_{|\alpha|<l-\sigma_q}\|D^\alpha\Phi_q^*\|$$

(this holds for smooth finite vectors), then (2.25) ensures that $\Sigma_{|\alpha|<l+t_j}\|D^\alpha u_j\|_{Q_{2\lambda}^+}$ is finite. We can dispense with the unnecessary smoothness condition by approximating the compactly-supported functions f and Φ^* by their averages as in Lemma 6.

Similarly we can establish the following theorem.

Theorem 4. *Under the conditions of Theorem* 2,

$$\sum_{j=1}^{M}\Big([u_j]_{(J,\,l+t_j,\,p)}+\sum_{|\alpha|<l+t_j}\|D^\alpha u_j\|_{Q_\lambda^+}\lambda^{-(l+t_j)+|\alpha|}\Big)$$
$$\leqslant c\Big\{\sum_{k=1}^{M}\Big([f_k]_{(J,\,l-s_k,\,p)}+\sum_{|\alpha|<l-s_k}\lambda^{-l+s_k+|\alpha|}(1+|\log\lambda|)\|D^\alpha f_k\|_{Q_\lambda^+}\Big)$$
$$+\sum_{q=1}^{Q}\Big([\Phi_q]_{(J,\,l-\sigma_q-1/p,\,p,\,R^{n-1})}+\sum_{|\alpha|<l-\sigma_q}\lambda^{-l+\sigma_q+1/p+|\alpha|}$$
$$\times(1+|\log\lambda|)\|D^\alpha\Phi_q\|_{\Sigma_\lambda}\Big)\Big\}, \tag{2.26a}$$

where $Q_\lambda^+=Q_\lambda(Z)\cap R_+^n$, $Z\in R^{n-1}$ $(Z_n=0)$ *and* $\Sigma_\lambda=Q_\lambda^+\cap R^{n-1}$.

In addition to the problem (2.4) we need to consider the system

$$\mathscr{L}_0\Big(\frac{\partial}{\partial x}\Big)u(x)=f(x) \tag{2.27}$$

on the whole of R^n. Every solution of this system is given by

$$u_j(x)=\sum_{k=1}^{m}\sum_{|\gamma|=b+1-s_k}\int_{R_+^n}Z_{jk}^\gamma(x-y)\,D^\gamma f_k(y)\,dy \tag{2.28}$$

(the Z_{jk}^γ are the same as in (2.2)), and hence the following theorem holds.

Theorem 5. *If* $f_k \in (J, l - s_k, R^n)$, *then*

$$\sum_{j=1}^{m} [u_j]_{(J,\, l+t_j,\, R^n)} \leqslant c \sum_{k=1}^{M} [f_k]_{(J,\, l-s_k,\, R^n)}$$

for every $l > 0$. *Furthermore,*

$$\sum_{j=1} \Big([u_j]_{(J,\, l+t_j)} + \sum_{|\alpha|<l+t_j} \| D^\alpha u_j \|_{Q_\lambda} \lambda^{-(l+t_j)+|\alpha|} \Big)$$
$$\leqslant c \sum_{k=1}^{M} \Big([f_k]_{(J,\, l-s_k)} + \sum_{|\alpha|<l-s_k} \| D^\alpha f_k \|_{Q_\lambda} (1 + |\log \lambda|) \lambda^{-l+s_k+|\alpha|} \Big), \qquad (2.29)$$

provided that $\operatorname{supp} f \in Q_\lambda(Z)$, $Z \in R^n$.

§3. The spaces (J, r, Ω)

By following the ideas in [5] we define the (J, r)-norms of functions defined in a bounded domain Ω with a smooth (as a minimum, Ljapunov) boundary S. For every $z \in \bar{\Omega} = \Omega \cup S$ there is a closed neighborhood (or, if $z \in S$, a semi-neighborhood) $U_z \subset \bar{\Omega}$ which can be mapped onto the domain $Q_d^+(Z) = Q_d(Z) \cap R^n_+$ ($Q_d(Z)$ denotes the cube with center Z and edges of length d parallel to the coordinate axes) by the smooth transformation

$$y = T_z(x): \quad y_i = T_i(x;\, z), \quad i = 1, \ldots, n, \qquad (3.1)$$

such that if the set $\sigma_z = U_z \cap S$ is nonempty, then its image is the face Σ_d ($y_n = 0$) of $Q_d^+(Z)$, and Z has the coordinates $(0, \ldots, 0, \eta)$, where $\eta \in [0, d/2]$ and $\eta < d/2$ only if $\sigma_z \neq 0$. We assume that the absolute value of the Jacobian of T_z is bounded both above and below by a number independent of z. We may assume that d is arbitrarily small, because it can always be decreased by contracting U_z (without changing T_z). Finally, we may assume that the transformations T_z and T_z^{-1} are defined on the whole of R^n, since this can always be achieved in the following way. Let $\zeta(y) \in C_0^\infty(Q_{3/4}(0))$, $\zeta = 1$ for $y \in Q_{1/2}(0)$, and $\zeta_\rho(y) = \zeta(y/\rho)$. Suppose that the transformation $T_z^{-1}(y) = x$ is given by the functions $T^{(i)}(y;\, z) = x_i$, $i = 1, \ldots, n$, defined in Q_d^+. We consider the transformation $x_i = \tilde{T}^{(i)}(y;\, z)$, where

$$\tilde{T}^{(i)}(y;\, z) = T^{(i)}(Z;\, z) + \sum_{k=1}^{n} \frac{\partial T^{(i)}(y;\, z)}{\partial y_k}\Big|_{y=Z} (y_k - Z_k) + R^{(i)}(y,\, z),$$

$$R^{(i)}(y,\, z) = \zeta_d(y - Z) \Big[T^{(i)}(y;\, z) - T^{(i)}(Z;\, z) - \sum_{k=1}^{n} \frac{\partial T^{(i)}(y;\, z)}{\partial y_k}\Big|_{y=Z} (y_k - Z_k) \Big].$$

It is defined at least in R^n_+, coincides with T_z^{-1} in $Q_{d/2}^+(Z)$, is linear outside $Q_d^+(Z)$, and is invertible for sufficiently small d since the expression in the square brackets tends to zero faster than d. Finally, we extend the functions $R^{(i)}$ from R^n_+ into R^n by using the Hestenes-Whitney formula; that is, by putting for $y_n < 0$

$$R^{(i)}(y;\ z)=\sum_{j=1}^{N}\lambda_j R^{(i)}(y_1,\ldots,y_{n-1},\ -jy_n;\ z), \tag{3.2}$$

where the λ_j are determined from the system $\sum_{j=1}^{N}\lambda_j(-j)^s = 1$, $s = 0,\ldots, N-1$. It is easy to see that if $d \leq d_0(N)$, then the transformation obtained as a result of this extension is invertible and linear outside some cube.

We assume that this procedure has been carried out, and that d is a fixed sufficiently small number. We say that $S \in (J, \rho, \infty)$, $\rho > 1$, if $\zeta_d(y-Z)T^{(i)}(y;\ z) \in (J, \rho, \infty, R^n_+)$ for every $z \in \Omega$, and the norms of all of these functions are bounded by a common constant. By taking $N > \rho$ in (3.2) we can establish that

$$\zeta_d(y-Z)\,T^{(i)}(y;\ z) \in (J,\ \rho,\ \infty,\ R^n)$$

(the proof is the same as that of Theorem 2 in [9]).

For a bounded domain Ω with the boundary $S \in (J, \rho, \infty)$, $\rho > 1$, we define the norm for (J, r, Ω), $\rho \geq r$, by using a sufficiently fine and regular partition of unity. For all $z \in \Omega$ and $\lambda \in (0, d]$ the domains $\omega(\lambda;\ z) \equiv T_z^{-1}(Q_\lambda^+(Z))$ cover Ω. We form a finite covering of Ω by taking the domains $\omega_k^{(\lambda)} \equiv \omega(\lambda/2;\ z)$ of multiplicity $N \leq N_0$, where N_0 does not depend on λ. Every ball $K_\lambda(z)$: $|y - z| \leq \lambda$, $z \in \Omega$, contains not more than cN points z_k. Therefore the domains $\Omega_k^{(\lambda)} = \omega(\lambda;\ z_k)$ also form a covering of Ω of multiplicity N_1 (c and N_1 are independent of λ).

Now we introduce the functions

$$\psi_k^{(\lambda)}(x)=\zeta_\lambda(T_k x - Z_k), \quad \varphi_k^{(\lambda)}(x)=\frac{\psi_k^{(\lambda/2)}(x)}{\sum_j \psi_j^{(\lambda/2)}(x)},$$

where $T_k x = T(x;\ z_k)$. They are such that

$$\begin{gathered}\psi_k^{(\lambda)}(x)\,\varphi_k^{(\lambda)}(x)=\varphi_k^{(\lambda)}(x),\\ \sum_k \varphi_k^{(\lambda)}(x)=1.\end{gathered} \tag{3.3}$$

We define the norm in (J, r, Ω) by the relation

$$\begin{aligned}\|u\|_{(J,\,r,\,\Omega)} &= \max_k \|\zeta_\lambda(y-Z_k)\,u(T_k^{-1}y)\|^{(\lambda)}_{(J,\,r,\,R^n_+)}\\ &= \max_k \|T_k \circ \psi_k^{(\lambda)}(x)\,u(x)\|^{(\lambda)}_{(J,\,r,\,R^n_+)},\end{aligned} \tag{3.4}$$

where $T_k \circ v(x) = v(T_k^{-1}y)$, and

$$\|v\|^{(\lambda)}_{(J,\,r,\,R^n_+)} = [v]_{(J,\,r,\,R^n_+)} + \sum_{|\alpha|<r}\lambda^{-r+|\alpha|}\|D^\alpha v\|_{R^n_+}. \tag{3.5}$$

The multipliers $\lambda^{-r+|\alpha|}$ and the derivatives $D^\alpha v$ are introduced into the norm to make our subsequent calculations easier. It follows from Lemma 7 below that for different $\lambda > 0$ and different partitions of unity (3.3), the norms (3.4) are equivalent to one another, and also to the norm

$$\sup_{z\in\Omega}\|\zeta_\rho(x-z)\,T_z\circ u(x)\|_{(J,\,r)},$$

introduced by Golovkin [5]. (Regarding the equivalence, see also [5], Lemma 1).

We will always assume that on functions $v(x)$ of compact support the norm $\|\cdot\|$ satisfies

$$\|T^{-1}\circ v\|\leqslant c\|v\|, \tag{3.6}$$

where $T(x)=y$ is any invertible transformation of R^n of class (J,ρ,∞) with $\rho>1$. We establish that this inequality also holds for the (J,r)-norms of functions of compact support.

Lemma 7. *Suppose that* (3.6) *holds with a constant* $c(\lambda_0)$ *for all functions* $v(x)$ *with* $\operatorname{supp}v(x)\in Q_{\lambda_0}(z)\in R^n$, *and that* $T(x)$ *is such that*

$$\|T_i(x)\,\zeta_{\lambda_0}(x-z)\|_{(J,\,\rho,\,\infty)}\leqslant A<\infty,\quad \rho>1. \tag{3.7}$$

Then for every $u(x)\in(J,r)$, $r\le\rho$, *such that* $\operatorname{supp}u\in Q_{\lambda_0/2}(z)$,

$$\|T^{-1}\circ u\|^{(\lambda)}_{(J,\,r)}\leqslant c\|u\|^{(\lambda)}_{(J,\,r)},\quad \lambda\leqslant\frac{\lambda_0}{2}, \tag{3.8}$$

where the constant c *is independent of* u *and* λ.

Proof. Let $v(x)=u(Tx)=T^{-1}\circ u$; then

$$D^\alpha v(x)=\sum_{\beta\leqslant\alpha}A_{\alpha\beta}(x)\,D^\beta_y u(y)\Big|_{y=T(x)}=\sum_{\beta\leqslant\alpha}A_{\alpha\beta}(x)\,\zeta_{\lambda_0}(x-z)\,D^\beta u(y)\Big|_{y=T(x)},$$

where the $A_{\alpha\beta}(x)$ are polynomials in the derivatives $D^\gamma T_i(x)$, $\gamma\le\alpha-\beta$. By (3.6) and (3.7) we have

$$\|D^\alpha v(x)\|\leqslant\sum_{\beta\leqslant\alpha}\sup|A_{\alpha\beta}(x)\,\zeta_{\lambda_0}(x-z)|\cdot\|T^{-1}\circ D^\beta u(y)\|\leqslant c\sum_{\beta\leqslant\alpha}\|D^\beta u(y)\|, \tag{3.9}$$

and hence

$$\sum_{|\alpha|<r}\|D^\alpha v(x)\|\cdot\lambda^{-r+|\alpha|}\leqslant c\sum_{\beta<r}\|D^\beta u(y)\|^{-r+|\beta|}.$$

Furthermore,

$$J[h^{-r+|\alpha|}\|\Delta_i^s(h)\,D^\alpha v(x)\|]\leqslant c\sum_{\beta\leqslant\alpha}J[h^{-r+|\alpha|}\|\Delta_i^k(h)\,u_\beta(x)\|], \tag{3.10}$$

where $u_\beta(x)=T\circ D^\beta u(y)$.

We first estimate the norm $[v]_{(J,r)}$ for $r<1$. It follows from the identity

$$v(x+\xi)-v(x)=\frac{1}{\varkappa|\xi|^n}\int_{|x-z|\leqslant|\xi|}(v(z)-v(x))\,dz+\frac{1}{\varkappa|\xi|^n}\int_{|x-z|\leqslant|\xi|}(v(x+\xi)-v(z))\,dz, \tag{3.11}$$

in which κ denotes the volume of the unit ball in R^n, that

$$\|v(x+\xi)-v(x)\| \leqslant \|f(x, |\xi|)\| + 2^n \|f(x+\xi; 2|\xi|\| \leqslant 2^{n+1}\|f(x, 2|\xi|)\|,$$

where

$$f(x, t) = \frac{1}{\varkappa t^n} \int\limits_{|x-z|\leqslant t} |v(z)-v(x)|\, dz.$$

Since T is invertible, we can find a $k>0$ such that $|T(x)-T(z)| \leq kt$ for $|x-z| \leq t$. Therefore

$$f(x; t) \leqslant \frac{1}{\varkappa t^n} \int\limits_{|T(x)-T(z)|\leqslant kt} |v(z)-v(x)|\, dz$$
$$\leqslant \frac{I}{\varkappa t^n} \int\limits_{|T(x)-z|\leqslant kt} |u(z)-u(Tx)|\, dz \equiv g(Tx; t),$$

where I is the maximum value of the modulus of the Jacobian of the transformation T^{-1}.

If $|\xi| \leq \lambda_0/4k$, then $\operatorname{supp} g(x; 2|\xi|) \subset Q_{\lambda_0}(z)$, and by (3.6)

$$\|g(Tx; 2|\xi|)\| \leqslant c(\lambda_0)\|g(x; 2|\xi|)\| \leqslant c(\lambda_0) I k^n \sup_{|\eta|\leqslant 2k|\xi|} \|u(x+\eta)-u(x)\|.$$

Hence

$$\|v(x+\xi)-v(x)\| \leqslant 2^{n+1} c(\lambda_0) I k^n \sup_{|\eta|\leqslant 2k|\xi|} \|u(x+\eta)-u(x)\|.$$

It follows from this and (3.9) with $\alpha = 0$ that

$$[v]_{(J, r)} \leqslant n\left\{J\left[h^{-r}\chi\left(\frac{4kh}{\lambda_0}\right)\sup_{|\xi|\leqslant h}\|v(x+\xi)-v(x)\|\right] + 2d_r\|v\|\left(\frac{4k}{\lambda_0}\right)^r\right\}$$
$$\leqslant c\{J[h^{-r}\sup_{|\eta|\leqslant h}\|u(x+\eta)-u(x)\|] + \lambda_0^{-r}\|u\|\} \leqslant c\|u\|^{(\lambda)}_{(J, r)}$$

(regarding this last inequality, see [2]).

We turn to the case when $r = 1$. By integrating both sides of the identity

$$v(x+2\xi) - 2v(x+\xi) + v(x) = [v(x+2\eta) - 2v(x+\eta) + v(x)]$$
$$-[v(x+\xi) - 2v(x+\eta) + v(x+2\eta-\xi)] - \left[v(x+\xi) - 2v\left(x+\eta+\frac{\xi}{2}\right)\right.$$
$$\left. + v(x+2\eta)\right] + \left[v(x+2\xi) - 2v\left(x+\frac{\xi}{2}+\eta\right) + v(x+2\eta-\xi)\right]$$

over the ball $|\eta| \leq |\xi|$, and then forming the norm $\|\cdot\|$, we obtain

$$\|v(x+2\xi) - 2v(x+\xi) + v(x)\| \leqslant c\left\|f_1\left(x; \frac{5}{2}|\xi|\right)\right\|,$$

where

$$f_1(x; d) = \frac{1}{d^n} \int\limits_{|\eta|\leqslant d} |v(x) - 2v(x+\eta) + 2v(x+2\eta)|\, d\eta \leqslant f_2(x, d) + f_3(x, d),$$

$$f_2(x, d) = \frac{1}{(2d)^n} \int\limits_{|x-z|\leqslant 2d} \left|v(x) - 2v\left(T^{-1}\left(\frac{T(x)+T(z)}{2}\right)\right) + v(z)\right|\, dz,$$

$$f_3(x, d) = \frac{1}{(2d)^n} \int\limits_{|x-z|\leqslant 2d} \left| v\left(\frac{x+z}{2}\right) - v\left(T^{-1}\frac{T(x)+T(z)}{2}\right)\right| dz$$
$$= (2d)^{-n} \int\limits_{|\eta|\leqslant 2d} \left| v\left(x+\frac{\eta}{2}\right) - v\left(T^{-1}\frac{T(x)+T(x+\eta)}{2}\right)\right| d\eta.$$

By arguing as in the case $r < 1$, it is not difficult to show that for $d \leq \lambda_0/4k$

$$\|f_2(x;\ d)\| \leqslant c_0(\lambda_0) I\varkappa k^n \sup_{|\eta|\leqslant dk} \|u(x) - 2u(x+\eta) + u(x+2\eta)\|. \quad (3.12)$$

We estimate the norm of f_3. First of all we have

$$\|f_3(x;\ d)\| \leqslant (2d)^{-n} \int\limits_{|\eta|\leqslant 2d} \left\| v\left(x+\frac{\eta}{2}\right) - v\left(T^{-1}\frac{T(x)+T(x+\eta)}{2}\right)\right\| d\eta$$
$$\leqslant (2d)^{-n} \int\limits_{|\eta|\leqslant 2d} \|v(x) - v(x+\tau(x,\ \eta))\| d\eta, \quad (3.13)$$

where

$$\tau(x,\ \eta) = T^{-1}\frac{T\left(x-\frac{\eta}{2}\right)+T\left(x+\frac{\eta}{2}\right)}{2} - x.$$

Since T is invertible, and since by (3.7) we have $T_i \in (J,\rho) \subset H^{1+\epsilon}$, $\epsilon > 0$, in any bounded domain, it follows that

$$|\tau_k(x,\ \eta)| \leqslant c\left| \frac{T\left(x-\frac{\eta}{2}\right)+T\left(x+\frac{\eta}{2}\right)}{2} - T(x)\right| \leqslant b|\eta|^{1+\varepsilon}, \quad (3.14)$$
$$\left|\frac{\partial \tau_k(x,\ \eta)}{\partial x_i}\right| \leqslant b_1|\eta|^{\varepsilon}.$$

To estimate the right-hand side of (3.13) we use a formula analogous to (3.11):

$$v(x+\tau(x,\ \eta)) - v(x) = \frac{1}{\varkappa b^n|\eta|^{(1+\varepsilon)n}} \int\limits_{|\tau+x-z|\leqslant b|\eta|^{1+\varepsilon}} [v(x+\tau(x,\ \eta)) - v(z)]\, dz$$
$$- \frac{1}{\varkappa b^n|\eta|^{(1+\varepsilon)n}} \int\limits_{|\tau+x-z|\leqslant b|\eta|^{1+\varepsilon}} [v(x) - v(z)]\, dz \equiv V_1(x,\ \eta) + V_2(x,\ \eta). \quad (3.15)$$

We consider the function V_1. Note that by (3.14) the transformation

$$y = T_\eta x = x + \tau(x,\ \eta) = T^{-1}\frac{T\left(x-\frac{\eta}{2}\right)+T\left(x+\frac{\eta}{2}\right)}{2}$$

is invertible for small values of η. In addition, if $|T_\eta x - z| \leq b|\eta|^{1+\epsilon}$, then $|TT_\eta x - Tz| \leq bk|\eta|^{1+\epsilon}$. Therefore

$$|V_1(x,\ \eta)| \leqslant \frac{1}{\varkappa b^n|\eta|^{(1+\varepsilon)n}} \int\limits_{|TT_\eta x - Tz|\leqslant bk|\eta|^{1+\varepsilon}} |u(TT_\eta x) - u(Tz)|\, dz$$

and for $bk|\eta| \leq \lambda_0/2$

$$\| V_1(x, \eta)\| \leqslant c(\lambda_0) Ik^n \sup_{|\sigma| \leqslant bk|\eta|^{1+\varepsilon}} \|u(x+\sigma) - u(x)\|.$$

Since $|T_\eta x - z| \leq b|\eta|^{1+\epsilon}$ implies that $|x - z| \leq |x - T_\eta x| + |T_\eta x - z| \leq 2b|\eta|^{1+\epsilon}$, for $2b|\eta| \leq \lambda_0/2$ we have

$$\| V_2(x, \eta)\| \leqslant c(\lambda_0) Ik^n \sup_{|\sigma| \leqslant 2bk|\eta|^{1+\varepsilon}} \|u(x+\sigma) - u(x)\|.$$

It follows from the above estimates and from (3.13) and (3.15) that

$$\|f_3(x, d)\| \leqslant 2c(\lambda_0) Ik^n (2d)^{-n} \int_{|\eta| \leqslant 2d} \sup_{|\sigma| \leqslant 2bk|\eta|^{1+\varepsilon}} \|u(x+\sigma) - u(x)\| \, d\eta,$$

and hence for $|\xi| \leq \theta\lambda_0$, $\theta > 0$,

$$\begin{aligned} &\|v(x+2\xi) - 2v(x+\xi) + v(x)\| \\ &\qquad \leqslant c\Big\{ \sup_{|\eta| \leqslant \frac{5}{2} k|\xi|} \|u(x+2\eta) - 2u(x+\eta) + u(x)\| \\ &\qquad\qquad + \sup_{|\sigma| \leqslant 2bk(5|\xi|)^{1+\varepsilon}} \|u(x+\sigma) - u(x)\| \Big\}. \end{aligned}$$

Therefore

$$\begin{aligned} &J\Big[h^{-1}\chi\Big(\frac{h}{\theta\lambda_0}\Big) \sup_{|\xi| \leqslant h} \|v(x+2\xi) - 2v(x+\xi) + v(x)\|\Big] \\ &\qquad \leqslant c\Big\{ J\Big[h^{-1}\chi\Big(\frac{h}{\theta\lambda_0}\Big) \sup_{|\eta| \leqslant \frac{5}{2} hk} \|u(x+2\eta) - 2u(x+\eta) + u(x)\|\Big] \\ &\qquad\qquad + J\Big[h^{-1}\chi\Big(\frac{h}{\theta\lambda_0}\Big) \sup_{|\eta| \leqslant 2bk(5h)^{1+\varepsilon}} \|u(x+\eta) - u(x)\|\Big\}. \end{aligned} \tag{3.16}$$

The last term is estimated by

$$cJ\Big[h^{\varepsilon}\chi\Big(\frac{h}{\theta\lambda_0}\Big)\Big] J_\infty \Big[h^{-\frac{1+\varepsilon_1}{1+\varepsilon}} \sup_{|\eta| \leqslant h} \|u(x+\eta) - u(x)\|\Big] \leqslant c_1 [u]_{\left(J, \frac{1+\varepsilon_1}{1+\varepsilon}, R^n\right)}, \tag{3.17}$$

where $\epsilon_1 < \epsilon$. This inequality is valid because J_∞ is the weakest of the functionals J (see, for example, [3]). Thus

$$\begin{aligned} [v]_{(J, 1)} \leqslant c\Big\{ J\Big[h^{-1}\chi\Big(\frac{h}{\theta\lambda_0}\Big) \sup_{|\eta| \leqslant \frac{5}{2} hk} \|u(x+2\eta) - 2u(x+\eta) + u(x)\|\Big] \\ + [u]_{\left(J, \frac{1+\varepsilon_1}{1+\varepsilon}\right)} + \lambda_0^{-1}\|u\|\Big\}. \end{aligned}$$

When we estimate the norm $[u]_{(J,(1+\epsilon_1)/(1+\epsilon))}$ by using (1.11) with $\epsilon = \lambda_0$, we obtain (3.8).

In the case when $r > 1$ it is sufficient to estimate the highest derivatives $D^{\alpha}v$ for $r-1 \leq |\alpha| < r$ by using (3.9). The norm of the function u_{β} on the right-hand side of (3.9) is estimated in the same way as the norm of $u(x)$ when $r \leq 1$. This proves the lemma.

Lemma 8. *Suppose that the transformation T satisfies the conditions of Lemma 7, and that the function $\Phi^{(\lambda)}(x)$ satisfies the inequalities*

$$|D^{\alpha}\Phi^{(\lambda)}(x)| \leqslant c_{\alpha}|\lambda|^{-|\alpha|+b}, \quad |\alpha| \leqslant r+1. \tag{3.18}$$

Then

$$\|T^{-1} \circ \Phi^{(\lambda)}u\|^{(\lambda)}_{(J, r, R^n_+)} \leqslant c\lambda^b \|u\|^{(\lambda)}_{(J, r, R^n_+)} \tag{3.19}$$

with a constant independent of $\lambda \leq \lambda_0/2$.

The estimate (3.19) follows from (3.8) and (1.18) if we note that, by (1.12), $[\Phi^{(\lambda)}(x)]_{(J,\rho,\infty)} \leq c|\lambda|^{-\rho+b}$, $\rho \leq r$.

Lemma 9. *Every set of functions with a uniformly bounded norm $\|u\|_{(J,r,\Omega)}$ is compact in the space $E(\Omega)$ with the norm $\|u\|_{\Omega} = \|\chi(\Omega)u\|$, and in (J, r', Ω) with any $r' < r$.*

Proof. Any function $u(x) \in (J, r, \Omega)$ can be approximated (in the same sense as in Lemma 6) by functions

$$u_{\rho}(x) = \sum_k \varphi^{(\lambda)}_k(x)\, T_k^{-1} \circ u_{k\rho},$$

where $u_{k\rho}$ is the average (1.22) of $u_k(y) = \zeta_\lambda(y - Z_k)u(T_k^{-1}y)$. Since

$$u_{\rho} - u = \sum_k \varphi^{(\lambda)}_k(x)\, T_k^{-1} \circ [u_{k\rho}(y) - \zeta_\lambda(y - Z_k)\, u(T_k^{-1}y)],$$

as a consequence of (3.6) and (1.23) we have

$$\begin{aligned}\|u(x) - u_{\rho}(x)\|_{\Omega} &\leqslant c\sum_k \|u_{k\rho}(y) - u_k(y)\| \\ &\leqslant c\rho^{\beta}\sum_k \|u_k(y)\|_{(J, \beta, R^n_+)} \leqslant c\rho^{\beta}\sum_k \|T_k \circ \psi^{(\lambda)}_k u\|_{(J, r, R^n_+)} \leqslant c(\lambda)\,\rho^{\beta}\|u\|_{(J, r, \Omega)}\end{aligned} \tag{3.20}$$

for every $\beta \in (0, \min(1, r))$.

Let M be the set of functions with a uniformly bounded (J, r, Ω)-norm. By (1.7), for every $\rho > 0$ the functions $u_{k\rho}$, and hence the u_{ρ}, form a compact set in $C(\Omega)$. Since Ω is a bounded domain, it follows that this set is compact in $E(\Omega)$. It follows from this and (3.20) that M is also compact in $E(\Omega)$. The compactness of M in (J, r', Ω) for $r' < r$ follows from the compactness of M in $E(\Omega)$ and the inequality

$$\|u\|_{(J, r', \Omega)} \leqslant \varepsilon\|u\|_{(J, r, \Omega)} + c(\varepsilon, \lambda)\|u\|_{\Omega}, \quad \forall \varepsilon \leqslant \varepsilon_0,$$

which is easily deduced from (1.10) and (1.13).

In the same way as is done in §1 for the case $\Omega = R^n_+$, we can define the space $(J, r, \Omega)_S = (J, r, \Omega)/H_0$, where H_0 is the closure in (J, r, Ω) of the set of smooth functions that are zero on S. This space is nonempty in any case for $r > 1$; the norm in it is defined as

$$\inf_{v \in H_0} \|u - v\|_{(J, r, \Omega)} = \inf_{u^* - u \in H_0} \|u^*\|_{(J, r, R^n_+)}.$$

For $r > 1/p$ the space $(J, r, p, \Omega)_S$ is isomorphic to the space $(J, r - 1/p, p, S)$ with the norm

$$\|u\|_{(J, r-1/p, p, S)} = \max_k \|T_k \circ \psi_k^{(\lambda)} u)|_{z_n=0}\|^{(\lambda)}_{(J, r-1/p, p, R^{n-1})}.$$

§4. Investigation of the problem (1.1)

We write the problem (1.1) as

$$Au = h, \text{ where } h = \begin{pmatrix} f \\ \Phi \end{pmatrix}, \quad Au = \begin{pmatrix} \mathscr{L}u \\ \mathscr{B}u|_S \end{pmatrix}.$$

We claim that under our conditions on the coefficients and the surface S, A is a bounded operator from the space

$$\mathscr{D}_l = \prod_{j=1}^m (J, l + t_j, \Omega)$$

into

$$\mathscr{R}_l = \prod_{k=1}^M (J, l - s_k, \Omega) \times \prod_{q=1}^Q (J, l - \sigma_q, \Omega)_S.$$

As usual, the norms in these spaces are defined as the sums of the norms of all the components of the vectors that are their elements.

Theorem 6. *Suppose that the number l satisfies the condition* (2.23). *If $S \in (J, l + \max t_j, \infty)$ and the coefficients of the operators L_{kj} and B_{qj} belong to the classes $(J, l - s_k, \infty, \Omega)$ and $(J, l - \sigma_q, \infty, \Omega)_S$, respectively, then $\|Au\|_{\mathfrak{R}_l} \le c\|u\|_{\mathfrak{D}_l}$.*

Proof. We consider one of the terms in $\|Au\|_{\mathfrak{R}_l}$, for example,

$$\left\| \sum_{j=1}^m B_{qj}\left(x, \frac{\partial}{\partial x}\right) u_j \right\|_{(J, l-\sigma_q, \Omega)_S} \leqslant \sum_{j=1}^m \left\| B_{qj}\left(x, \frac{\partial}{\partial x}\right) u_j \right\|_{(J, l-\sigma_q, \Omega)}.$$

Let

$$B_{qj}\left(x, \frac{\partial}{\partial x}\right) u_j = \sum_{|\alpha| \leqslant \sigma_q + t_j} b^\alpha_{qj}(x) D^\alpha u_j, \quad b^\alpha_{qj} \in (J, l - \sigma_q, \infty, \Omega).$$

We estimate the norm

$$\|b^\alpha_{qj} D^\alpha u_j\|_{(J, l-\sigma_q, \Omega)} = \max_k \|T_k \circ \psi_k^{(\lambda)} b^\alpha_{qj} D^\alpha u_j\|^{(\lambda)}_{(J, l-\sigma_q, R^n_+)}.$$

By (3.3) we have $u_j = \Sigma_l u_j \phi_l^{(\lambda)} \psi_l^{(\lambda)}$, and the number of terms for which $\psi_k^{(\lambda)}\phi_l^{(\lambda)} \neq 0$ is independent of λ. Hence

$$\| b_{qj}^{\alpha} D^{\alpha} u_j \|_{(J,\, l-\sigma_q,\, \Omega)} \leqslant \sum_{\beta \leqslant \alpha} c_{\alpha\beta} \max_{k,\, l} \| T_k \circ \psi_k^{(\lambda)} b_{qj}^{\alpha} D^{\alpha-\beta} \varphi_l^{(\lambda)} \cdot D^{\beta} (u_j \psi_l^{(\lambda)}) \|_{(J,\, l-\sigma_q)}^{(\lambda)}.$$

Then, since the transformation $T_k T_l^{-1}$ and the functions $\Phi_{kl}^{(\lambda)} = \psi_k^{(\lambda)} D^{\alpha-\beta} \phi_l^{(\lambda)}$ satisfy the condition of Lemma 8 with $b = -|\alpha| + |\beta|$, the norms on the right-hand side are estimated by

$$c \max_l \| T_l \circ b_{qj}^{\alpha} D^{\beta} (u_j \psi_l^{(\lambda)}) \|_{(J,\, l-\sigma_q,\, R_+^n)}^{(\lambda)} \lambda^{-|\alpha|+|\beta|}.$$

Finally, by Lemma 4 we have

$$\| b_{qj}^{\alpha} D^{\alpha} u_j \|_{(J,\, l-\sigma_q,\, \Omega)} \leqslant c \sum_{\beta \leqslant \alpha} \lambda^{-|\alpha|+|\beta|} \max_l \| T_l \circ D^{\beta} (u_j \psi_l^{(\lambda)}) \|_{(J,\, l-\sigma_q)}^{(\lambda)}$$
$$\leqslant c \max_l \| T_l \circ (u_j \psi_l^{(\lambda)}) \|_{(J,\, l+t_j,\, R_+^n)}^{(\lambda)},$$

as required. The norm $\|L_{kj}(x, \partial/\partial x) u_j\|_{(J, l-s_k, \Omega)}$ can be estimated similarly.

The main result of this article is the proof of the following theorem.

Theorem 7. *Under the conditions of Theorem 6 the following assertions are equivalent*:

1. *The operator* $\mathfrak{L}$ *is elliptic in* $\Omega \cap S$, *and* $\mathfrak{B}$ *covers it.*
2. *The operator* A *has a left regularizer* $R\colon \mathfrak{R}_l \to \mathfrak{D}_l$ *satisfying*

$$RA = I - T, \tag{4.1}$$

where T *is a completely continuous operator in* $\mathfrak{D}_l$.

3. *The null subspace of* A *is finite-dimensional, and the range of* A *is closed.*
4. *The following inequality holds*:

$$\|u\|_{\mathscr{D}_l} \leqslant c_1 \|Au\|_{\mathscr{R}_l} + c_2 \sum_j \|u_j\|_{\Omega}. \tag{4.2}$$

We only construct the regularizer, since the other assertions are proved by a standard scheme (see [1]), and the inequality (4.2) also follows from the results in [5]. We follow the treatment in [1] and introduce the following notation: $\mathfrak{L}^{(k)}$ and $\mathfrak{B}^{(k)}$ are the operators $\mathfrak{L}$ and $\mathfrak{B}$ in the variables $y^{(k)} = T_k x$; $\mathfrak{L}_0^{(k)}$ and $\mathfrak{B}_0^{(k)}$ denote their principal parts; $\mathfrak{L}^{(k)}$ and $\mathfrak{B}^{(k)}$ are the operators $\mathfrak{L}_0^{(k)}$ and $\mathfrak{B}_0^{(k)}$ in the variables x; $l^{(k)} = \mathfrak{L} - \mathfrak{L}^{(k)}$ and $b^{(k)} = \mathfrak{B} - \mathfrak{B}^{(k)}$. Obviously, $l^{(k)}$ and $b^{(k)}$ consist of lower order terms; that is, of operators whose orders are less than $s_k + t_j$ and $\sigma_q + t_j$. Suppose also that

$$a_k^0 v = \begin{pmatrix} \left[\mathfrak{L}^{(k)}\left(x, \frac{\partial}{\partial x}\right) - \mathfrak{L}^{(k)}\left(z_k, \frac{\partial}{\partial x}\right)\right] v, \\ \left[\mathfrak{B}^{(k)}\left(x, \frac{\partial}{\partial x}\right) - \mathfrak{B}^{(k)}\left(z_k, \frac{\partial}{\partial x}\right)\right] v \Big|_{\sigma_{z_k}} \end{pmatrix}, \tag{4.3}$$

$$a_k' v = \begin{pmatrix} l^{(k)} v, \\ b^{(k)} v \big|_{\sigma_{z_k}} \end{pmatrix};$$

where the R'_k are the integral operators in (2.28) and (2.2) that give a solution of the system $\mathfrak{L}_0^{(k)}(Z_k, \partial/\partial y)v = f^{(k)}$ in the whole space R^n, or if $\sigma_{z_k} \neq 0$, of the problem

$$\mathcal{L}_0^{(k)}\left(Z_k, \frac{\partial}{\partial y}\right)v = f^{(k)}, \quad \mathcal{B}_0^{(k)}\left(Z_k, \frac{\partial}{\partial y}\right)v\bigg|_{y_n=0} = \Phi^{(k)};$$

finally, $R_k h = T_k^{-1} \circ R'_k(T_k \circ h)$. We consider the operator

$$R'h = \sum_k \varphi_k^{(\lambda)} R_k (\psi_k^{(\lambda)s} h), \tag{4.4}$$

where $s > 0$ is an integer that will be chosen later (in [1], $s = 1$).

It follows from Theorems 1–4 and Lemma 8 that R' is a bounded linear operator from $\mathfrak{R}_l$ into $\mathfrak{D}_l$. As we have shown in [1],

$$R'Au = u + tu + T'u, \tag{4.5}$$

where t and T' are bounded operators in $\mathfrak{D}_l$:

$$tu = \sum_k \varphi_k^{(\lambda)} R_k a_k^{(0)} (u\psi_k^{(\lambda)s}), \tag{4.6}$$

$$T'u = \sum_k \varphi_k^{(\lambda)} R_k [\psi_k^{(\lambda)s}, A]u + \sum_k \varphi_k R_k a'_k (u\psi_k^{(\lambda)s}), \quad [\psi, A]u = \psi Au - A\psi u. \tag{4.7}$$

We establish that $I + t$ is invertible for small λ, and that T' is completely continuous. We estimate $w = tu$, paying attention to the dependence of the constants on λ. Since for every i the number of nonzero terms in

$$\sum_k \psi_i^{(\lambda)} \varphi_k^{(\lambda)} R_k a_k^{(0)} (u\psi_k^{(\lambda)s})$$

is determined by the multiplicity of the covering of Ω and by the domains $\Omega_k^{(\lambda)}$, but it does not depend on λ, we have

$$\|w_j\|_{(J, l+t_j, \Omega)} \leqslant c \max_{i,k} \|T_i \circ \psi_i^{(\lambda)} \varphi_k^{(\lambda)} T_k^{-1} \circ W_{kj}^{(\lambda)}\|_{(J, l+t_j, R_+^n)}^{(\lambda)},$$

and

$$W_k^{(\lambda)} = R'_k T_k^{-1} \circ a_k^{(0)} (u\psi_k^{(\lambda)s}). \tag{4.8}$$

Since the transformation $T_i T_k^{-1}$ and the functions $\Phi_{ik}^{(\lambda)} = T'_k \circ \psi_i^{(\lambda)} \phi_k^{(\lambda)}$ satisfy the conditions of Lemma 8, where $b = 0$, we have

$$\|w_j\|_{(J, l+t_j, \Omega)} \leqslant c \max_k \|W_{kj}^{(\lambda)}\|_{(J, l+t_j, R_+^n)}^{(\lambda)}$$

where c is a constant independent of λ.

To estimate the norm of $W_{kj}^{(\lambda)}$ we use (2.26) and (2.29), and obtain

$$\sum_{j=1}^m \|w_j\|_{(J, l+t_j, \Omega)} \leqslant c(1 + |\log\lambda|)\left[\sum_{j=1}^M \max_k \|g_{kj}^{(\lambda)}(y)\|_{(J, l-s_j, R_+^n)}^{\lambda} + \sum_{q=1}^Q \max_{k \in \mathfrak{N}} \|Y_{kq}^{(\lambda)}(y)\|_{(J, l-\sigma_q, R_+^n)}\right], \tag{4.9}$$

where $\mathfrak{N}$ is the set of the indices of those $\Omega_k^{(\lambda)}$ whose boundaries intersect S, and the $g_{kj}^{(\lambda)}$ and the $Y_{kq}^{(\lambda)}$ are the components of the vector $T_k^{-1} \circ a_k(u\psi_k^{(\lambda)s})$; that is, by (4.3),

$$g_k^{(\lambda)}(y) = \left[\mathscr{L}_0^{(k)}\left(y, \frac{\partial}{\partial y}\right) - \mathscr{L}_0^{(k)}\left(Z_k, \frac{\partial}{\partial y}\right)\right] u^{(k)}(y)\,\zeta_\lambda^s(y - Z_k),$$

$$Y_k^{(\lambda)}(y) = \left[\mathscr{B}_0^{(k)}\left(y, \frac{\partial}{\partial y}\right) - \mathscr{B}_0^{(k)}\left(Z_k, \frac{\partial}{\partial y}\right)\right] u^{(k)}(y)\,\zeta_\lambda^s(y - Z_k), \quad k \in \mathfrak{N},$$

$$u^{(k)} = T_k \circ u.$$

We estimate the right-hand side of (4.9) by using (1.18), bearing in mind that the coefficients of the operators

$$\mathscr{L}_0^{(k)}\left(y, \frac{\partial}{\partial y}\right) - \mathscr{L}_0^{(k)}\left(Z_k, \frac{\partial}{\partial y}\right) \quad \text{and} \quad \mathscr{B}_0^{(k)}\left(y, \frac{\partial}{\partial y}\right) - \mathscr{B}_0^{(k)}\left(Z_k, \frac{\partial}{\partial y}\right)$$

are estimated by $c|y - Z_k|^{l-[l]}$, and hence, for $y \subset Q_\lambda(Z_k)$, by $c\lambda^{l-[l]}$. As a result we obtain

$$\sum_{j=1}^{M} \| g_{kj}^{(\lambda)}(y) \|_{(J,\, l-s_j,\, R_+^n)}^{(\lambda)} \leqslant c\lambda^{l-[l]} \sum_{i=1}^{m} \| u_i^{(k)} \zeta_\lambda^s (y - Z_k) \|_{(J,\, l+t_j,\, R_+^n)}^{(\lambda)},$$

$$\sum_{q=1}^{Q} \| Y_{kq}^{(\lambda)}(y) \|_{(J,\, l-\sigma_q,\, R_+^n)}^{(\lambda)} \leqslant c\lambda^{l-[l]} \sum_{i=1}^{m} \| u_i^{(k)}(y)\, \zeta_\lambda^s (y - Z_k) \|_{(J,\, l+t_j,\, R_+^n)}^{(\lambda)}.$$

It follows from these inequalities and from (4.9) that

$$\sum_{j=1}^{m} \| w_j \|_{(J,\, l+t_j,\, \Omega)} \leqslant c\,(1 + |\log \lambda|)\, \lambda^{l-[l]} \sum_{i=1}^{m} \max_k \| u_i^{(k)}(y)\, \zeta_\lambda^s (y - Z_k) \|_{(J,\, l+t_j,\, R_+^n)}^{(\lambda)}.$$

By applying Lemma 8 again we can estimate the norm on the right-hand side by the same norm with $s = 1$, and rewrite the resulting inequality in the form

$$\| tu \|_{D_l} \leqslant c\,(1 + |\log \lambda|)\, \lambda^{l-[l]} \| u \|_{D_l}. \tag{4.10}$$

Since c does not depend on λ, for small λ the operator t is a contraction operator in $\mathfrak{D}_l$, and hence the operator $I + t$ is invertible.

We consider the operator $T' = T_1 + T_2$ (4.7). The vector $T_2 u$ is estimated in exactly the same way as tu, but since a_k' is an operator of lower order in comparison with A, instead of (4.10) we have $\|T_2 u\|_{\mathfrak{D}_l} \leq c(\lambda) \|u\|_{\mathfrak{D}_{l-1}}$, where

$$\mathscr{D}_{l-1} = \prod_{j=1}^{m} (J,\, l + t_j - 1,\, \Omega).$$

In the estimation of $T_1 u$ we choose s so that

$$[\psi_k^{(\lambda)s}, A]u = \psi_k^{(\lambda)s} Au - Au\psi_n^{(\lambda)s} = a_k''(u\psi_k^{(\lambda)}),$$

where a_k'' is an operator of lower order in comparison with A (we can do this by choosing s sufficiently large). Then we obtain for $T_1 u$ the same estimate as for $T_2 u$, and hence

$$\| T'u \|_{\mathfrak{D}_l} \leqslant c(\lambda) \| u \|_{\mathfrak{D}_{l-1}}. \tag{4.11}$$

Since by Lemma 9 a set bounded in $\mathfrak{D}_l$ is compact in $\mathfrak{D}_{l-1}$, it follows that T' is a completely continuous operator in $\mathfrak{D}_l$. It follows from (4.5) that the operator $R = (I + t)^{-1}R'$ satisfies (4.1) with the completely continuous operator $T = -(I + t)^{-1}T'$, and hence it is a left regularizer.

We continue to investigate the differential properties of the solutions of the problem (1.1).

Theorem 8. *Suppose that the conditions of Theorem 7 hold with some* $l > 0$. *If the coefficients of the operators* L_{kj} *and* B_{qj} *belong to the classes* $(J, l' - s_k, \infty, \Omega')$ *and* $(J, l' - \sigma_q, \infty, \Omega')$, *respectively, where* $l' > l$, $\Omega' \subset \Omega$, *and* $S' \equiv \partial\Omega' \cap S \in (J, l' + \max t_j, \infty)$, *and if* $f_j \in (J, l' - s_j, \Omega')$ *and* $\Phi_q \in (J, l' - \sigma_q, \Omega')_{S'}$, *then every solution of the problem* (1.1) *of the class* $\mathfrak{D}_l$ *belongs to* $\mathfrak{D}_{l'}(\Omega'') = \Pi_j(J, l' + t_j, \Omega'')$, *where* Ω'' *is any strictly interior subdomain of* Ω'.

Proof. Let $\zeta(x) \in C^\infty(\Omega)$, $\zeta(x) = 1$ for $x \in \tilde{\Omega}$ and $\zeta(x) = 0$ for $x \in \Omega \backslash \Omega'$, where $\Omega' \supset \tilde{\Omega} \supset \Omega''$. From (4.1) we have $u\zeta = R'Au\zeta - tu\zeta - T'u\zeta = \tilde{R}'Au\zeta - \tilde{t}u\zeta - \tilde{T}'u\zeta$ ($\tilde{R}'$, $\tilde{t}$ and $\tilde{T}'$ are the operators defined by (4.4), (4.6) and (4.7), respectively, where the summation is over all those k for which $\psi_k^{(\lambda)}\zeta \neq 0$. If λ is sufficiently small, then $\tilde{R}'$ acts from $\mathfrak{R}_{\tilde{l}}(\Omega)$ into $\mathfrak{D}_{\tilde{l}}(\Omega)$ for all $\tilde{l} \in [l, l']$, $\tilde{t}$ is a contraction operator in $\mathfrak{D}_{\tilde{l}}$, and $\tilde{T}'$ acts from $\mathfrak{D}_{\tilde{l}-1}$ into $\mathfrak{D}_{\tilde{l}}$. Therefore, if $u \in \mathfrak{D}_l(\Omega)$, then $v = \tilde{R}'Au\zeta - \tilde{T}'u\zeta \in \mathfrak{D}_{l_1}(\Omega)$, where $l_1 = \min(l', l + 1)$; hence $u\zeta = (I + \tilde{t})^{-1}v \in \mathfrak{D}_{l_1}(\Omega)$ and $u \in \mathfrak{D}_{l_1}(\Omega)$. Since $\tilde{\Omega}$ is any interior subdomain of Ω', by repeating this argument if necessary, we prove that $u \in \mathfrak{D}_{l'}(\Omega'')$.

The theorems we have proved can be slightly sharpened if $\|\cdot\| = \|\cdot\|_p$, $1 \leq p \leq \infty$. In this case, the operator A acts from the space

$$\mathfrak{D}_{l,p} = \prod_{j=1}^{m} (J, l + t_j, p, \Omega)$$

into

$$\mathfrak{R}_{l,p} = \prod_{k=1}^{M} (J, l - s_k, p, \Omega) \times \prod_{q=1}^{Q} \left(J, l - \sigma_q - \frac{1}{p}, p, S\right),$$

where the number l must satisfy (2.24). The conditions on the coefficients of the operators L_{kj} and B_{qj} can be weakened and stated in terms of the classes $(J, l - s_k, r, \Omega)$ and $(J, l - \sigma_q - 1/p, \tilde{r}, S)$, where r and $\tilde{r}$ depend on the order of the derivative associated with a given coefficient (conditions of this kind are given in [10]).

Finally, in the case in question, Theorem 8 can be formulated thus: *Every solution of problem* (1.1) *in the class* $\mathfrak{D}_{l,p}$ *belongs to* $\mathfrak{D}_{l',q}(\Omega'')$ *with* $l' > l$ *and* $q > p$, *provided that* $h \in R_{l',q}(\Omega)$, $S' \in (J, l' + \max t_j, \infty)$ *and* $\Omega'' \subset \Omega'$. The proof is the same as for Theorem 8; one uses the imbedding theorem from [8].

BIBLIOGRAPHY

1. V. A. Solonnikov, *Overdetermined elliptic boundary value problems*, Zap. Naučn. Sem. Leningrad. Otdel. Mat. Inst. Steklov. (LOMI) 21 (1971), 112–158 = J. Soviet Math. 1 (1973), 477–512. MR 44 #7121.

2. K. K. Golovkin, *On equivalent norms for fractional spaces*, Trudy Mat. Inst. Steklov. 66 (1962), 364–383; English transl., Amer. Math. Soc. Transl. (2) 81 (1969), 257–280. MR 27 #3988.

3. ———, *Parametric-normed spaces and normed massives*, Trudy Mat. Inst. Steklov. 106 (1969) = Proc. Steklov Inst. Math. 106 (1969)(1972).

4. K. K. Golovkin and V. A. Solonnikov, *Estimates for integral operators in translation-invariant norms*. II, Trudy Mat. Inst. Steklov. 92 (1966), 5–30 = Proc. Steklov Inst. Math. 92 (1966), 3–32(1968). MR 34 #4840.

5. K. K. Golovkin, *The principle of localness in the theory of boundary value problems for partial differential equations*. I, Izv. Akad. Nauk SSSR Ser. Mat. 31 (1967), 1027–1056 = Math. USSR Izv. 1 (1967), 983–1010. MR 36 #1819.

6. O. V. Besov, *Investigation of a family of function spaces in connection with imbedding and extension theorems*, Trudy Mat. Inst. Steklov. 60 (1961), 42–81; English transl., Amer. Math. Soc. Transl. (2) 40 (1964), 85–126. MR 24 #A3501.

7. V. A. Solonnikov, *Estimates of the solutions of a nonstationary linearized system of Navier-Stokes equations*, Trudy Mat. Inst. Steklov 70 (1964), 213–317; English transl., Amer. Math. Soc. Transl. (2) 75 (1968), 1–116. MR 30 #1325.

8. K. K. Golovkin, *Imbedding theorems for fractional spaces*, Trudy Mat. Inst. Steklov. 70 (1964), 38–46; English transl., Amer. Math. Soc. Transl. (2) 91 (1970), 57–67. MR 28 #5282.

9. V. P. Il′in and V. A. Solonnikov, *On some properties of differentiable functions of several variables*, Trudy Mat. Inst. Steklov. 66 (1962), 205–226; English transl., Amer. Math. Soc. Transl. (2) 81 (1969), 67–90. MR 27 #2768.

10. V. A. Solonnikov, *Estimates in L_p of solutions of elliptic and parabolic systems*, Trudy Mat. Inst. Steklov. 102 (1967), 137–160 = Proc. Steklov Inst. Math. 102 (1967), 157–185 (1970). MR 37 #4388.

Translated by L. W. LONGDON

Trudy Mat. Inst. Steklov.
127 (1975)

Proc. Steklov Inst. Math.
127 (1975)

INITIAL-BOUNDARY VALUE PROBLEMS FOR EQUATIONS OF MIXED TYPE

UDC 517.946

L. STUPJALIS

Abstract. Initial-boundary value problems for equations that abruptly change type on passing from one part of the domain to another are considered. Appropriate junction conditions are given on the interface between the two parts. Unique solvability of the initial-boundary value problems for these equations is proved.
Bibliography: 10 titles.

In [1]–[4], initial-boundary value problems have been considered in which one must find functions satisfying equations of different type in different parts of a space and satisfying diffraction-type junction conditions on the interface. In the present article, we consider the same initial-boundary value problems with different junction conditions, namely, with those with which they are considered in problems arising in biology (see [5]). The results given below show that these junction conditions are "more natural". It is precisely when these conditions are present that problems in which an elliptic equation is given in one of the domains and a hyperbolic equation is given in the other are uniquely solvable in the class of functions which have a "finite energy integral" for any square integrable right side. This situation does not occur when combining equations under diffraction-type junction conditions (see [3]). In addition, in order to combine equations under such conditions, it was required that the elliptic operator in the elliptic equation be formally selfadjoint (see [4]). In the case here, this restriction is not imposed.

§1. The formulation of the problems

Assume that the bounded domain Ω in n-dimensional Euclidean space is the sum of two domains, Ω_1 and Ω_2, and the $(n-1)$-dimensional surface Γ separating them, i.e. $\Omega = \Omega_1 \cup \Gamma \cup \Omega_2$. Denote the boundary of Ω by S and the boundaries of Ω_1 and Ω_2 by S_1 and S_2. The following two problems will be considered.

Problem I. Find the function $u(x, t)$ which satisfies an elliptic equation of the form

$$\mathscr{L}_1^{(1)}u^{(1)} \equiv -\frac{\partial}{\partial x_i}\left[a_{ij}^{(1)}(x, t)\frac{\partial u^{(1)}}{\partial x_j}\right] + a_i^{(1)}(x, t)\frac{\partial u^{(1)}}{\partial x_i} + a^{(1)}(x, t)\,u^{(1)} = f_1^{(1)}(x, t) \qquad (1)$$

AMS (MOS) subject classifications (1970). Primary 35M05; Secondary 92A05.

in one of the domains $Q_T^{(k)} = \Omega_k \times [0, T]$, $k = 1, 2$ (to be definite, let it be $Q_T^{(1)}$); a hyperbolic equation of the form

$$\mathcal{L}_3^{(2)} u^{(2)} \equiv u_{tt}^{(2)} + \mathcal{L}_1^{(2)} u^{(2)} = f_3^{(2)}(x, t) \tag{2}$$

in the domain $Q_T^{(2)}$, where

$$\mathcal{L}_1^{(2)} = -\frac{\partial}{\partial x_i}\left[a_{ij}^{(2)}(x, t)\frac{\partial}{\partial x_j}\right] + a_i^{(2)}(x, t)\frac{\partial}{\partial x_i} + a^{(2)}(x, t);$$

the initial conditions

$$u^{(2)}|_{t=0} = \varphi_0^{(2)}(x), \quad u_t^{(2)}|_{t=0} = \varphi_1^{(2)}(x) \tag{3}$$

in the domain Ω_2 at $t = 0$; one of the classical boundary conditions, say, for example, the first condition,

$$u^{(k)}|_{S_k \setminus \Gamma} = \psi(s, t), \quad t \in [0, T], \quad k = 1, 2, \tag{4}$$

on the boundary S; and the junction conditions

$$\left.\begin{aligned} u^{(1)}|_\Gamma - \left.\frac{\partial u^{(2)}}{\partial t}\right|_\Gamma &= \psi_1(s, t), \\ b^{(1)}(x, t)\left.\frac{\partial u^{(1)}}{\partial N^{(1)}}\right|_\Gamma - b^{(2)}(x, t)\left.\frac{\partial u^{(2)}}{\partial N^{(2)}}\right|_\Gamma &= \psi_2(s, t), \quad t \in [0, T], \end{aligned}\right\} \tag{5}$$

on the boundary Γ, where

$$\frac{\partial}{\partial N^{(k)}} = a_{ij}^{(k)} \frac{\partial}{\partial x_j} \cos(\mathbf{n}, x_i);$$

$\mathbf{n}$ being the normal to Γ in the outward direction with respect to Ω_1. Here the $b^{(k)}(x, t) \geq \beta > 0$, $k = 1, 2$, are known functions; $u(x, t) = u^{(1)}(x, t)$ for $x \in \Omega_1$, $t \geq 0$; and $u(x, t) = u^{(2)}(x, t)$ for $x \in \Omega_2$, $t \geq 0$.

Problem II. Determine the function $u(x, t)$ which satisfies a parabolic equation of the form

$$\mathcal{L}_2^{(1)} u^{(1)} \equiv u_t + \mathcal{L}_1^{(1)} u^{(1)} = f_2^{(1)}(x, t) \tag{6}$$

in the domain $Q_T^{(1)}$, a hyperbolic equation of form (2) in $Q_T^{(2)}$, the single initial condition

$$u^{(1)}|_{t=0} = \varphi_0^{(1)}(x) \tag{7}$$

in Ω_1 at $t = 0$, the two initial conditions (3) in Ω_2 at $t = 0$, condition (4) on the boundary S for $t \in [0, T]$, and conditions (5) on Γ for $t \in [0, T]$.

The boundary condition (4) and conditions (5) can be reduced to homogeneous conditions by introducing a function $v(x, t) = u(x, t) - w(x, t)$ to replace $u(x, t)$, where $w(x, t)$ coincides with $u(x, t)$ on S and undergoes the jumps prescribed by conditions (5) on Γ. Assume that this change has already been made; that is, that $u(x, t)$ satisfies the conditions

$$\left.\begin{aligned} &u^{(k)}|_{S_k\setminus\Gamma}=0, \quad k=1,\ 2, \\ &u^{(1)}|_{\Gamma}-\frac{\partial u^{(2)}}{\partial t}\Big|_{\Gamma}=0, \quad b^{(1)}(x,\ t)\frac{\partial u^{(1)}}{\partial N^{(1)}}\Big|_{\Gamma}-b^{(2)}(x,\ t)\frac{\partial u^{(2)}}{\partial N^{(2)}}\Big|_{\Gamma}=0, \quad t\in[0,\ T],\end{aligned}\right\} \tag{8}$$

on the surfaces S and Γ.

Without explicitly stating so below, we will assume everywhere in this article that the following conditions are fulfilled.

1. The coefficients of $\mathfrak{L}_1^{(k)}$, $k = 1, 2$, satisfy the ellipticity conditions

$$a_{ij}^{(k)}=a_{ji}^{(k)}, \quad \sum_{i,\ j=1}^{n} a_{ij}^{(k)}\xi_i\xi_j \geqslant \alpha \sum_{i=1}^{n} \xi_i^2, \quad \alpha=\text{const}>0,$$

in $\overline{Q}_T^{(k)}$, where $\xi_1, \ldots, \xi_n$ are any real numbers.

2. The coefficients of the operator $\mathfrak{L}_1^{(1)}$ of equation (1) (but not of equation (6)) satisfy the condition

$$\int_{\Omega_1}\left[b^{(1)}a_{ij}^{(1)}\frac{\partial v}{\partial x_i}\frac{\partial v}{\partial x_j}+\left(\frac{\partial b^{(1)}}{\partial x_j}a_{ij}^{(1)}+b^{(1)}a_i^{(1)}\right)\frac{\partial v}{\partial x_i}v+b^{(1)}a^{(1)}v^2\right]dx \geqslant \nu\|v\|^2_{W_2^1(\Omega_1)}, \tag{9}$$

$\nu = \text{const} > 0$, for any function $v \in W_2^1(\Omega_1)$ such that $v|_{S_1\setminus\Gamma}=0$ for all $t \in [0, T]$.

Let us introduce the following notation. Let $W_2^{1,0}(Q_T^{(k)})$ be the Hilbert space with the scalar product

$$(u,\ v)_{W_2^{1,0}(Q_T^{(k)})}=\int_{Q_T^{(k)}} (uv+u_{x_i}v_{x_i})\, dxdt$$

and $W_2^{1,1}(Q_T^{(k)})$ the Hilbert space with the scalar product

$$(u,\ v)_{W_2^{1,1}(Q_T^{(k)})}=\int_{Q_T^{(k)}} (uv+u_{x_i}v_{x_i}+u_tv_t)\, dxdt, \qquad k = 1, 2.$$

Let W_T^1 be the Hilbert space consisting of all elements $w(x, t) = \{v^{(1)}(x, t), v^{(2)}(x, t)\} \in W_2^{1,0}(Q_T^{(1)}) \times W_2^{1,1}(Q_T^{(2)})$ which satisfy the conditions $v^{(2)}|_{t=T} = 0$, $x \in \Omega_2$, $v^{(k)}|_{S_T^{(k)}\setminus\Gamma_T} = 0$, $k = 1, 2$, and $v^{(1)}|_{\Gamma_T} = v^{(2)}|_{\Gamma_T}$, where $S_T^{(k)} = S_k \times [0, T]$ and $\Gamma_T = \Gamma \times [0, T]$. Let $\widehat{W}_T^1$ be the Hilbert space consisting of all elements $w(x, t) = \{v^{(1)}(x, t), v^{(2)}(x, t)\} \in W_2^{1,1}(Q_T^{(1)}) \times W_2^{1,1}(Q_T^{(2)})$ that satisfy the conditions $v^{(k)}|_{t=T} = 0$, $x \in \Omega_k$, $v^{(k)}|_{S_T^{(k)}\setminus\Gamma_T} = 0$, $k = 1, 2$, and $v^{(1)}|_{\Gamma_T} = v^{(2)}|_{\Gamma_T}$. Let V^1 be the Hilbert space consisting of all elements of the space $W_2^{1,0}(Q_T^{(1)}) \times W_2^{1,1}(Q_T^{(2)})$ that satisfy the conditions $v^{(k)}|_{S_T^{(k)}\setminus\Gamma_T} = 0$, $k = 1, 2$, and $\int_0^t v^{(1)}(x, \tau)\, d\tau|_{\Gamma_T} = v^{(2)}(x, t)|_{\Gamma_T}$.

Let us prove one particular property of the elements of the space V^1. Let $\{v^{(1)}, v^{(2)}\}$ be an arbitrary element of V^1. For almost all t in $[0, T]$, the inequality $\|v^{(2)}\|_{W_2^1(\Omega_2)} \leq C'\|v^{(2)}\|_{W_2^{1/2}(\Gamma)}$ holds for it, where the constant C' does not depend on $v^{(2)}$ (see [6]). Furthermore,

$$\|v^{(2)}(x, t)\|_{W_2^{1/2}(\Gamma)} = \left\| \int_0^t v^{(1)}(x, \tau)\, d\tau \right\|_{W_2^{1/2}(\Gamma)}$$

$$\leqslant C'' \left\| \int_0^t v^{(1)}(x, \tau)\, d\tau \right\|_{W_2^1(\Omega_1)} \leqslant \tilde{C} \|v^{(1)}\|_{W_2^{1,0}(Q_T^{(1)})}.$$

This implies the inequality

$$\operatorname*{ess\,sup}_{0\leqslant t\leqslant T} \|v^{(2)}(x, t)\|_{W_2^1(\Omega_2)} \leqslant \hat{C} \| v \|_{V^1} \equiv \hat{C} \left(\|v^{(1)}\|^2_{W_2^{1,0}(Q_T^{(1)})} + \|v^{(2)}\|^2_{W_2^{1,1}(Q_T^{(2)})} \right)^{1/2}.$$

An element $u(x, t) = \{u^{(1)}(x, t), u^{(2)}(x, t)\}$ of V^1 that satisfies the condition

$$u^{(2)}|_{t=0} = \varphi_0^{(2)}(x), \quad x \in \Omega_2 \tag{10}$$

and the integral identity

$$-\int_{Q_T^{(2)}} \frac{\partial u^{(2)}}{\partial t} \cdot \frac{\partial}{\partial t}(b^{(2)}\psi^{(2)})\, dxdt + \sum_{k=1}^{2} \int_0^T \mathscr{L}_1^{(k)}(u^{(k)}, \psi^{(k)})\, dt = \int_{Q_T^{(1)}} b^{(1)} f_1^{(1)} \psi^{(1)} dxdt$$
$$+ \int_{Q_T^{(2)}} b^{(2)} f_3^{(2)} \psi^{(2)} dxdt + \int_{\Omega_2} b^{(2)} \varphi_1^{(2)} \psi^{(2)} dx\,|_{t=0} \tag{11}$$

for any $\psi(x, t) = \{\psi^{(1)}(x, t), \psi^{(2)}(x, t)\}$ in W_T^1 will be called a generalized solution of Problem I. Here,

$$\mathscr{L}_1^{(k)}(u^{(k)}, \psi^{(k)}) = \int_{\Omega_k} \Big[b^{(k)} a_{ij}^{(k)} \frac{\partial u^{(k)}}{\partial x_i} \cdot \frac{\partial \psi^{(k)}}{\partial x_j} + \Big(\frac{\partial b^{(k)}}{\partial x_j} a_{ij}^{(k)} + b^{(k)} a_i^{(k)} \Big) \frac{\partial u^{(k)}}{\partial x_i} \psi^{(k)}$$
$$+ b^{(k)} a^{(k)} u^{(k)} \psi^{(k)} \Big]\, dx.$$

If the function $\phi_0^{(2)}(x)$ is in $\overset{\circ}{W}{}_2^1(\Omega_2)$, then the initial condition (10) can be reduced to a homogeneous condition. Indeed, if we introduce the function $v^{(2)}(x, t) = u^{(2)}(x, t) - \phi_0^{(2)}(x)$ to replace $u^{(2)}(x, t)$, then from (11) we obtain for it

$$-\int_{Q_T^{(2)}} \frac{\partial v^{(2)}}{\partial t} \frac{\partial}{\partial t}(b^{(2)}\psi^{(2)})\, dxdt + \int_0^T [\mathscr{L}_1^{(1)}(u^{(1)}, \psi^{(1)}) + \mathscr{L}_1^{(2)}(v^{(2)}, \psi^{(2)})]\, dt$$
$$= \int_{Q_T^{(1)}} b^{(1)} f_1^{(1)} \psi^{(1)} dxdt + \int_{Q_T^{(2)}} b^{(2)} f_3^{(2)} \psi^{(2)} dxdt + \int_{\Omega_2} b^{(2)} \varphi_1^{(2)} \psi^{(2)} dx\,|_{t=0}$$
$$-\int_0^T \mathscr{L}_1^{(2)}(\varphi_0^{(2)}, \psi^{(2)})\, dt, \tag{12}$$

and from (10)

$$v^{(2)}|_{t=0} = 0, \quad x \in \Omega_2. \tag{13}$$

Instead of seeking $u(x, t) = \{u^{(1)}(x, t), u^{(2)}(x, t)\}$, we will seek $w(x, t) = \{u^{(1)}(x, t), v^{(2)}(x, t)\}$ as an element of V^1 that satisfies (12) and (13). By finding it, we automatically find $u(x, t) = \{u^{(1)}(x, t), v^{(2)}(x, t) + \phi_0^{(2)}(x)\}$, the generalized solution of Problem I.

An element $u(x, t) = \{u^{(1)}(x, t), u^{(2)}(x, t)\}$ of V^1 that satisfies (10) and the integral identity

$$-\int_{Q_T^{(1)}} u^{(1)}\frac{\partial}{\partial t}(b^{(1)}\psi^{(1)})\,dxdt - \int_{Q_T^{(2)}} u_t^{(2)}\frac{\partial}{\partial t}(b^{(2)}\psi^{(2)})\,dxdt + \sum_{k=1}^{2}\int_0^T \mathscr{L}_1^{(k)}(u^{(k)}, \psi^{(k)})\,dt$$
$$= \int_{Q_T^{(1)}} b^{(1)}f_2^{(1)}\psi^{(1)}dxdt + \int_{Q_T^{(2)}} b^{(2)}f_3^{(2)}\psi^{(2)}dxdt + \int_{\Omega_1} b^{(1)}\varphi_0^{(1)}\psi^{(1)}dx\,|_{t=0} \tag{14}$$
$$+ \int_{\Omega_2} b^{(2)}\varphi_1^{(2)}(x)\,\psi^{(2)}dx\,|_{t=0}$$

for any $\psi(x, t) = \{\psi^{(1)}(x, t), \psi^{(2)}(x, t)\}$ in $\hat{W}_T^1$ will be called a generalized solution of Problem II.

If the function $\phi_0^{(2)}(x)$ is in $\overset{\circ}{W}{}_2^1(\Omega_2)$, then, in a manner analogous to that used for Problem I, the initial condition (10) can be reduced to a homogeneous condition. Indeed, if we introduce a function $v^{(2)}(x, t) = u^{(2)}(x, t) - \phi_0^{(2)}(x)$ to replace $u^{(2)}(x, t)$, then from (14) we obtain

$$-\int_{Q_T^{(1)}} u^{(1)}\frac{\partial}{\partial t}(b^{(1)}\psi^{(1)})\,dxdt - \int_{Q_T^{(2)}} \frac{\partial v^{(2)}}{\partial t}\cdot\frac{\partial}{\partial t}(b^{(2)}\psi^{(2)})\,dxdt + \int_0^T [\mathscr{L}_1^{(1)}(u^{(1)}, \psi^{(1)})$$
$$+ \mathscr{L}_1^{(2)}(v^{(2)}, \psi^{(2)})]\,dt = \int_{Q_T^{(1)}} b^{(1)}f_2^{(1)}\psi^{(1)}dxdt + \int_{Q_T^{(2)}} b^{(2)}f_3^{(2)}\psi^{(2)}dxdt \tag{15}$$
$$+ \int_{\Omega_1} b^{(1)}\varphi_0^{(1)}\psi^{(1)}dx\,|_{t=0} + \int_{\Omega_2} b^{(2)}\varphi_1^{(2)}\psi^{(2)}dx\,|_{t=0} - \int_0^T \mathscr{L}_1^{(2)}(\varphi_0^{(2)}, \psi^{(2)})\,dt,$$

and from condition (10) we obtain (13).

It is easy to verify that if the functions $u^{(k)}(x, t)$, $k = 1, 2$, are generalized solutions of Problem I (Problem II) and the generalized derivatives of these functions that enter into equations (1) and (2) ((6) and (2)) are square integrable, then they satisfy (1) and (2) ((6) and (2)) almost everywhere in $Q_T^{(k)}$ and satisfy the corresponding conditions in the sense of the imbedding theorems.

§2. Uniqueness theorems

Let us show that the generalizations of the concepts of the solutions of Problems I and II that we have introduced are permissible in the sense that these problems cannot have more than one generalized solution.

Theorem 1. *Let the following conditions be satisfied*: a) *the functions*

$$a_{ij}^{(k)},\ \frac{\partial a_{ij}^{(k)}}{\partial t},\ \frac{\partial a_{ij}^{(2)}}{\partial x_i},\ a_i^{(k)},\ \frac{\partial a_i^{(1)}}{\partial t},\ \frac{\partial a_i^{(2)}}{\partial x_j},\ a^{(k)},\ \frac{\partial a^{(1)}}{\partial t},$$
$$b^{(k)},\ \frac{\partial b^{(k)}}{\partial t},\ \frac{\partial b^{(k)}}{\partial x_j},\ \frac{\partial^2 b^{(1)}}{\partial t\partial x_j},\ \frac{\partial^2 b^{(2)}}{\partial x_i\partial x_j},\ \frac{\partial^2 b^{(2)}}{\partial t^2}$$

are bounded and measurable in $\overline{Q}_T^{(k)}$, $k = 1, 2$; b) *the surface* Γ *is piecewise smooth; and* c) *for any function* $v \in W_2^1(\Omega_1)$ *such that* $v|_{S_1\setminus\Gamma} = 0$, *and for all* $t \in [0, T]$, *the inequality*

$$\int_{\Omega_1}\left[b^{(1)}a_{ij}^{(1)}\frac{\partial v}{\partial x_i}\cdot\frac{\partial v}{\partial x_j}+\left(\frac{\partial b^{(1)}}{\partial x_j}a_{ij}^{(1)}+b^{(1)}a_i^{(1)}\right)\frac{\partial v}{\partial x_i}v+b^{(1)}a^{(1)}v^2\right]dx \geqslant \nu\|v\|^2_{W_2^1(\Omega_1)}$$

holds, where $\nu = \text{const} > 0$. *Then Problem* I *has no more than one generalized solution.*

Proof. Let $u(x, t) = \{u^{(1)}(x, t), u^{(2)}(x, t)\}$ and $u_1(x, t) = \{u_1^{(1)}(x, t), u_1^{(2)}(x, t)\}$ be two generalized solutions of Problem I. Subtract the integral identity (11) corresponding to the second of these solutions from the identity corresponding to the first and, in the difference, set

$$\psi^{(1)}(x, t)=\begin{cases}\displaystyle\int_h^t d\tau\int_0^\tau w^{(1)}(x, \xi)\,d\xi, & 0\leqslant t\leqslant h,\\ 0, & h\leqslant t\leqslant T,\end{cases}$$

$$\psi^{(2)}(x, t)=\begin{cases}\displaystyle\int_h^t w^{(2)}(x, \tau)\,d\tau, & 0\leqslant t\leqslant h,\\ 0, & h\leqslant t\leqslant T,\end{cases}$$

where $w^{(1)}(x, t) = u^{(1)}(x, t) - u_1^{(1)}(x, t)$ and $w^{(2)}(x, t) = u^{(2)}(x, t) - u_1^{(2)}(x, t)$.

We then have

$$-\int_0^h\int_{\Omega_2}\frac{\partial w^{(2)}}{\partial t}\cdot\frac{\partial}{\partial t}(b^{(2)}\psi^{(2)})\,dxdt+\sum_{k=1}^{2}\int_0^h\mathscr{L}_1^{(k)}(w^{(k)}, \psi^{(k)})\,dt=0.$$

Express all terms under the integral signs using the functions $\psi^{(k)}$, $k = 1, 2$, and their derivatives. Doing this, one obtains

$$\frac{1}{2}\int_{\Omega_2}b^{(2)}\left(\frac{\partial\psi^{(2)}}{\partial t}\right)^2dx\Big|_{t=h}+\int_0^h\mathscr{L}_1^{(1)}\left(\frac{\partial\psi^{(1)}}{\partial t}, \frac{\partial\psi^{(1)}}{\partial t}\right)dt+\frac{1}{2}\int_{\Omega_2}b^{(2)}a_{ij}^{(2)}\frac{\partial\psi^{(2)}}{\partial x_i}\cdot\frac{\partial\psi^{(2)}}{\partial x_j}dx\Big|_{t=0}$$
$$=\frac{3}{2}\int_0^h\int_{\Omega_2}\frac{\partial b^{(2)}}{\partial t}\left(\frac{\partial\psi^{(2)}}{\partial t}\right)^2dxdt+\int_0^h\int_{\Omega_2}\frac{\partial^2 b^{(2)}}{\partial t^2}\cdot\frac{\partial\psi^{(2)}}{\partial t}\cdot\psi^{(2)}dxdt$$
$$-\int_0^h\int_{\Omega_1}\left[\frac{\partial(b^{(1)}a_{ij}^{(1)})}{\partial t}\cdot\frac{\partial^2\psi^{(1)}}{\partial x_i\partial t}\cdot\frac{\partial\psi^{(1)}}{\partial x_j}+\frac{\partial}{\partial t}\left(\frac{\partial b^{(1)}}{\partial x_j}a_{ij}^{(1)}+b^{(1)}a_i^{(1)}\right)\cdot\frac{\partial^2\psi^{(1)}}{\partial x_i\partial t}\psi^{(1)}\right. \tag{16}$$
$$\left.+\frac{\partial(b^{(1)}a^{(1)})}{\partial t}\cdot\frac{\partial\psi^{(1)}}{\partial t}\psi^{(1)}\right]dxdt+\int_0^h\int_{\Omega_2}\left[-\frac{1}{2}\frac{\partial(b^{(2)}a_{ij}^{(2)})}{\partial t}\cdot\frac{\partial\psi^{(2)}}{\partial x_i}\cdot\frac{\partial\psi^{(2)}}{\partial x_j}\right.$$
$$\left.+\left(\frac{\partial b^{(2)}}{\partial x_j}a_{ij}^{(2)}+b^{(2)}a_i^{(2)}\right)\frac{\partial^2\psi^{(2)}}{\partial x_i\partial t}\psi^{(2)}+b^{(2)}a^{(2)}\frac{\partial\psi^{(2)}}{\partial t}\psi^{(2)}\right]dxdt.$$

The absolute value of the first three integrals on the right side of this equality does not exceed the quantity

$$M\left(\frac{3}{2}+h\right)\int_0^h\int_{\Omega_2}\left(\frac{\partial\psi^{(2)}}{\partial t}\right)^2dxdt+M^2[(n+1)\sqrt{n}+2n]h\cdot\int_0^h\left\|\frac{\partial\psi^{(1)}}{\partial t}\right\|^2_{W_2^1(\Omega_1)}dt,$$

where M is an upper bound for the functions

$$|a_{ij}^{(k)}|,\ \left|\frac{\partial a_{ij}^{(k)}}{\partial t}\right|,\ \left|\frac{\partial a_{ij}^{(2)}}{\partial x_i}\right|,\ |a_i^{(k)}|,\ \left|\frac{\partial a_i^{(1)}}{\partial t}\right|,\ \left|\frac{\partial a_i^{(2)}}{\partial x_j}\right|,\ |a^{(k)}|,\ \left|\frac{\partial a^{(1)}}{\partial t}\right|,$$
$$b^{(k)},\ \left|\frac{\partial b^{(k)}}{\partial t}\right|,\ \left|\frac{\partial b^{(k)}}{\partial x_j}\right|,\ \left|\frac{\partial^2 b^{(1)}}{\partial t\partial x_j}\right|,\ \left|\frac{\partial^2 b^{(2)}}{\partial x_i\partial x_j}\right|,\ \left|\frac{\partial^2 b^{(2)}}{\partial t^2}\right|,$$
$$i,\ j=1,\ 2,\ \ldots,\ n,$$

in $\overline{Q}_T^{(k)}$, $k=1,2$.

In order to bound the last integral on the right side of equality (16), we transform the integral

$$I=\int_0^h\int_{\Omega_2}\left(\frac{\partial b^{(2)}}{\partial x_j}a_{ij}^{(2)}+b^{(2)}a_i^{(2)}\right)\frac{\partial^2\psi^{(2)}}{\partial x_i\partial t}\psi^{(2)}dxdt$$

in the following manner:

$$I=-\int_0^h\int_{\Omega_2}\left(\frac{\partial b^{(2)}}{\partial x_j}a_{ij}^{(2)}+b^{(2)}a_i^{(2)}\right)\frac{\partial\psi^{(2)}}{\partial x_i}\cdot\frac{\partial\psi^{(2)}}{\partial t}dxdt-\int_0^h\int_{\Omega_2}\frac{\partial}{\partial x_i}\left(\frac{\partial b^{(2)}}{\partial x_j}a_{ij}^{(2)}\right.$$
$$\left.+b^{(2)}a_i^{(2)}\right)\frac{\partial\psi^{(2)}}{\partial t}\psi^{(2)}dxdt+\int_0^h\int_\Gamma\left(\frac{\partial b^{(2)}}{\partial x_j}a_{ij}^{(2)}+b^{(2)}a_i^{(2)}\right)\cdot\frac{\partial\psi^{(1)}}{\partial t}\psi^{(1)}\cos(\mathbf{n},\ x_i)\,d\Gamma dt,$$

where $\mathbf{n}$ is the normal to the boundary Γ in the outward direction with respect to Ω_2. Here we have made use of the facts that on Γ_T the function $\psi^{(2)}$ satisfies the conditions $\psi^{(2)}|_{\Gamma_T}=\psi^{(1)}|_{\Gamma_T}$ and

$$\left.\frac{\partial\psi^{(2)}}{\partial t}\right|_{\Gamma_T}=\left.\frac{\partial\psi^{(1)}}{\partial t}\right|_{\Gamma_T},$$

and that on $S_T^{(2)}\setminus\Gamma_T$ it satisfies the condition $\psi^{(2)}|_{S_T^{(2)}\setminus\Gamma_T}=0$.

Making use of the inequality $\|v\|_{L_2(\Gamma)}\le C_1\|v\|_{W_2^1(\Omega_1)}$, which holds for any function in $W_2^1(\Omega_1)$ with a constant that does not depend on v, we can bound the contour integral by the quantity

$$C_1^2M^2(n+1)nh\int_0^h\left\|\frac{\partial\psi^{(1)}}{\partial t}\right\|^2_{W_2^1(\Omega_1)}dt.$$

The rest of the expression I can be bounded by the quantity

$$\frac{M^2(n+1)\sqrt{n}}{2}\int_0^h\int_{\Omega_2}\sum_{i=1}^n\left(\frac{\partial\psi^{(2)}}{\partial x_i}\right)^2dxdt+M^2(n+1)\left(2nh+\frac{\sqrt{n}}{2}\right)\int_0^h\int_{\Omega_2}\left(\frac{\partial\psi^{(2)}}{\partial t}\right)^2dxdt.$$

For

$$I_1=\int_0^h\int_{\Omega_2}\left[-\frac{1}{2}\frac{\partial\,(b^{(2)}a_{ij}^{(2)})}{\partial t}\cdot\frac{\partial\psi^{(2)}}{\partial x_i}\cdot\frac{\partial\psi^{(2)}}{\partial x_j}+b^{(2)}a^{(2)}\frac{\partial\psi^{(2)}}{\partial t}\psi^{(2)}\right]dxdt$$

we have the bound

$$|I_1|\leqslant M^2n\int_0^h\int_{\Omega_2}\sum_{i=1}^n\left(\frac{\partial\psi^{(2)}}{\partial x_i}\right)^2dxdt+M^2h\int_0^h\int_{\Omega_2}\left(\frac{\partial\psi^{(2)}}{\partial t}\right)^2dxdt.$$

By virtue of condition (9), the left side of (16) is not less than

$$\frac{\beta}{2}\int\limits_{\Omega_2}\left(\frac{\partial\psi^{(2)}}{\partial t}\right)^2 dx\Big|_{t=h}+\nu\int\limits_0^h\left\|\frac{\partial\psi^{(1)}}{\partial t}\right\|^2_{W_2^1(\Omega_1)}dt+\frac{\alpha\beta}{2}\int\limits_{\Omega_2}\sum_{i=1}^n\left(\frac{\partial\psi^{(2)}}{\partial x_i}\right)^2 dx\Big|_{t=0}.$$

Combining these bounds, we obtain

$$\frac{\beta}{2}\int\limits_{\Omega_2}\left(\frac{\partial\psi^{(2)}}{\partial t}\right)^2 dx\Big|_{t=h}+\nu\int\limits_0^h\left\|\frac{\partial\psi^{(1)}}{\partial t}\right\|^2_{W_2^1(\Omega_1)}dt+\frac{\alpha\beta}{2}\int\limits_{\Omega_2}\sum_{i=1}^n\left(\frac{\partial\psi^{(2)}}{\partial x_i}\right)^2 dx\Big|_{t=0}$$
$$\leqslant C_2\int\limits_0^h\int\limits_{\Omega_2}\left(\frac{\partial\psi^{(2)}}{\partial t}\right)^2 dx\,dt+C_3h\int\limits_0^h\left\|\frac{\partial\psi^{(1)}}{\partial t}\right\|^2_{W_2^1(\Omega_1)}dt+C_4\int\limits_0^h\int\limits_{\Omega_2}\sum_{i=1}^n\left(\frac{\partial\psi^{(2)}}{\partial x_i}\right)^2 dx\,dt,$$

where

$$C_2=M^2\left[h+(n+1)\left(2nh+\frac{\sqrt{n}}{2}\right)\right]+M\left(h+\frac{3}{2}\right),$$

$$C_3=M^2\left[C_1^2(n+1)n+(n+1)\sqrt{n}+2n\right],\qquad C_4=\frac{M^2}{2}\left[(n+1)\sqrt{n}+2n\right].$$

Now choose $h_0=\nu/2C_3$. Then from this last equality we obtain

$$\int\limits_{\Omega_2}\left(\frac{\partial\psi^{(2)}}{\partial t}\right)^2 dx\Big|_{t=h}+\int\limits_0^h\left\|\frac{\partial\psi^{(1)}}{\partial t}\right\|^2_{W_2^1(\Omega_1)}dt+\int\limits_{\Omega_2}\sum_{i=1}^n\left(\frac{\partial\psi^{(2)}}{\partial x_i}\right)^2 dx\Big|_{t=0}$$
$$\leqslant\frac{2C_2}{\gamma}\int\limits_0^h\int\limits_{\Omega_2}\left(\frac{\partial\psi^{(2)}}{\partial t}\right)^2 dx\,dt+\frac{2C_4}{\gamma}\int\limits_0^h\int\limits_{\Omega_2}\sum_{i=1}^n\left(\frac{\partial\psi^{(2)}}{\partial x_i}\right)^2 dx\,dt,\tag{17}$$

which holds for all $h\in[0,h_0]$, where $\gamma=\min(\beta,\nu,\alpha\beta)$.

Let us introduce the new functions

$$\Phi_i^{(2)}(x,t)=\int\limits_t^0\frac{\partial w^{(2)}(x,\tau)}{\partial x_i}d\tau,$$

$i=1,\dots,n$, into consideration. Obviously

$$\frac{\partial\psi^{(2)}(x,t)}{\partial x_i}=\Phi_i^{(2)}(x,h)-\Phi_i^{(2)}(x,t).$$

We substitute these expressions into (17):

$$\int\limits_{\Omega_2}(w^{(2)}(x,h))^2dx+\int\limits_0^h\left\|\frac{\partial\psi^{(1)}}{\partial t}\right\|^2_{W_2^1(\Omega_1)}dt+\int\limits_{\Omega_2}\sum_{i=1}^n(\Phi_i^{(2)}(x,h))^2dx$$
$$\leqslant\frac{2C_2}{\gamma}\int\limits_0^h\int\limits_{\Omega_2}(w^{(2)}(x,t))^2dx\,dt+\frac{2C_4}{\gamma}\int\limits_0^h\int\limits_{\Omega_2}\sum_{i=1}^n[\Phi_i^{(2)}(x,h)-\Phi_i^{(2)}(x,t)]^2dx\,dt$$
$$\leqslant\frac{2C_2}{\gamma}\int\limits_0^h\int\limits_{\Omega_2}(w^{(2)}(x,t))^2dx\,dt+\frac{4C_4}{\gamma}\int\limits_0^h\int\limits_{\Omega_2}\sum_{i=1}^n(\Phi_i^{(2)}(x,t))^2dx\,dt\tag{18}$$
$$+\frac{4C_4h}{\gamma}\int\limits_{\Omega_2}\sum_{i=1}^n(\Phi_i^{(2)}(x,h))^2dx.$$

Let $h=\gamma/8C_4$. Inequality (18) implies the inequality

$$\frac{dy(h)}{dh}\leqslant C_5y(h),\tag{19}$$

which holds for all $h \in [0, h_1]$, where $h_1 = \min(h_0, \bar{h})$,

$$y(h) = \int_0^h \int_{\Omega_2} (w^{(2)}(x, t))^2\, dx dt + \int_0^h \int_{\Omega_2} \sum_{i=1}^n (\Phi_i^{(2)}(x, t))^2\, dx dt,$$

and $C_5 = \max(4C_2/\gamma, 8C_4/\gamma)$. By (19),

$$\frac{d}{dh}(y(h)\, e^{-C_5 h}) \leqslant 0.$$

Integrating over h and taking into account that $y(0) = 0$, we find that $y(h)e^{-C_5 h} \leq 0$, whence $y(h) = 0$ for $h \in [0, h_1]$. From (18) it follows that $w^{(k)}(x, t) = 0$ for $(x, t) \in \Omega_k \times [0, h_1]$, $k = 1, 2$. The same reasoning can be repeated for the cylinders $\Omega_k \times [h_1, 2h_1]$, etc., until the cylinders $Q_T^{(k)}$ are exhausted. The theorem is proved.

Theorem 2. *Let conditions* a) *and* b) *of Theorem* 1 *be fulfilled, and let the function* $\partial^2 b^{(1)}/\partial t^2$ *be bounded and measurable in* $\overline{Q}_T^{(1)}$. *Then Problem* II *has no more than one generalized solution.*

The proof of this theorem is, in principle, carried out just like that of Theorem 1. As the arbitrary functions $\psi^{(k)}(x, t)$, $k = 1, 2$, in (14) written for the difference $w^{(k)} = u_1^{(k)} - u_2^{(k)}$, $k = 1, 2$, of two solutions of Problem II, we take the functions

$$\psi^{(1)}(x, t) = \begin{cases} \int_h^t d\tau \int_0^\tau w^{(1)}(x, \xi)\, d\xi, & 0 \leqslant t \leqslant h, \\ 0, & h \leqslant t \leqslant T, \end{cases}$$

$$\psi^{(2)}(x, t) = \begin{cases} \int_h^t w^{(2)}(x, \tau)\, d\tau, & 0 \leqslant t \leqslant h, \\ 0, & h \leqslant t \leqslant T. \end{cases}$$

Expressing everything using $\psi^{(k)}$, we come to an equality from which we can conclude that the $\psi^{(k)}$, and consequently also the $w^{(k)}$, are equal to zero.

Remark. In Theorems 1 and 2, one need make no assumptions about the smoothness of the boundary Γ, if, at the same time, one replaces the functions $\partial a_{ij}^{(2)}/\partial x_i$, $\partial a_i^{(2)}/\partial x_j$ and $\partial^2 b^{(2)}/\partial x_i \partial x_j$ by the functions $\partial a_i^{(2)}/\partial t$ and $\partial^2 b^{(2)}/\partial t \partial x_j$ in condition a). The proofs of the theorems are not changed in any essential way, except that the expression I must be transformed in the following manner:

$$I = -\int_{\Omega_2} \left(\frac{\partial b^{(2)}}{\partial x_j} a_{ij}^{(2)} + b^{(2)} a_i^{(2)}\right) \frac{\partial \psi^{(2)}}{\partial x_i} \psi^{(2)} dx \Big|_{t=0} - \int_0^h \int_{\Omega_2} \frac{\partial}{\partial t}\left(\frac{\partial b^{(2)}}{\partial x_j} a_{ij}^{(2)} + b^{(2)} a_i^{(2)}\right) \frac{\partial \psi^{(2)}}{\partial x_i} \psi^{(2)} dx dt - \int_0^h \int_{\Omega_2} \left(\frac{\partial b^{(2)}}{\partial x_j} a_{ij}^{(2)} + b^{(2)} a_i^{(2)}\right) \frac{\partial \psi^{(2)}}{\partial x_i} \cdot \frac{\partial \psi^{(2)}}{\partial t} dx dt.$$

These integrals are bounded in a manner analogous to the manner which was used in Theorem 1.

§3. Existence theorems

In this section the existence of generalized solutions of Problems I and II will be proved. To do this, we will use a method related to the Galerkin method.

Let us first look at Problem I. We will assume that the following conditions are fulfilled: a) the functions

$$b^{(k)},\ \frac{\partial b^{(k)}}{\partial x_j},\ \frac{\partial b^{(2)}}{\partial t},\ a_{ij}^{(k)},\ \frac{\partial a_{ij}^{(2)}}{\partial t},\ a_i^{(k)},\ a^{(k)}$$

are bounded and measurable in $\overline{Q}_T^{(k)}$, $k = 1, 2$, and $b^{(k)} \geq \beta = \text{const} > 0$; b) the functions $a_{ij}^{(1)}$, $a_i^{(1)}$, $a^{(1)}$ and $b^{(1)}$ do not depend on t; and c) for any function $v \in W_{2,0}^1(\Omega_1)$, where $W_{2,0}^1(\Omega_1)$ is the subspace of $W_2^1(\Omega_1)$ that consists of the functions that vanish on $S_1 \setminus \Gamma$, the inequality

$$\int_{\Omega_1} \left[b^{(1)} a_{ij}^{(1)} \frac{\partial v}{\partial x_i} \frac{\partial v}{\partial x_j} + \left(\frac{\partial b^{(1)}}{\partial x_j} a_{ij}^{(1)} + b^{(1)} a_i^{(1)} \right) \frac{\partial v}{\partial x_i} v + b^{(1)} a^{(1)} v^2 \right] dx \geqslant \nu \| v \|^2_{W_2^1(\Omega_1)} \tag{20}$$

is valid, where $\nu = \text{const} > 0$.

Let us introduce the following notation. Let $U(\Omega_1)$ be the subspace of $W_{2,0}^1(\Omega_1)$ consisting of the functions that satisfy the integral identity

$$\int_{\Omega_1} \left[b^{(1)} a_{ij}^{(1)} \frac{\partial v}{\partial x_i} \cdot \frac{\partial \Phi}{\partial x_j} + \left(\frac{\partial b^{(1)}}{\partial x_j} a_{ij}^{(1)} + b^{(1)} a_i^{(1)} \right) \frac{\partial v}{\partial x_i} \Phi + b^{(1)} a^{(1)} v \Phi \right] dx = 0 \tag{20a}$$

for any function $\Phi \in \overset{\circ}{W}{}_2^1(\Omega_1)$. Let $H(\Omega)$ be the subspace of $\overset{\circ}{W}{}_2^1(\Omega)$ that consists of all functions $v(x)$ that satisfy the integral identity (20a) for any function $\Phi \in \overset{\circ}{W}{}_2^1(\Omega_1)$. Let $\widetilde{\overset{\circ}{W}}{}_2^1(\Omega)$ be the subspace of $\overset{\circ}{W}{}_2^1(\Omega)$ that consists of all functions that vanish on Ω_2.

In what follows, we will make use of the following lemmas.

Lemma 1 (a refinement of well-known facts—see, for example, [7]). *The Hilbert space $W_{2,0}^1(\Omega_1)$ is the direct sum of the subspaces $\overset{\circ}{W}{}_2^1(\Omega_1)$ and $U(\Omega_1)$:*

$$W_{2,0}^1(\Omega_1) = \overset{\circ}{W}{}_2^1(\Omega_1) \dotplus U(\Omega_1).$$

Proof. It is sufficient to show that any element $u \in W_{2,0}^1(\Omega_1)$ can be represented in the form $u = v + w$, $v \in \overset{\circ}{W}{}_2^1(\Omega_1)$, $w \in U(\Omega_1)$, and that $\overset{\circ}{W}{}_2^1(\Omega_1) \cap U(\Omega_1) = \Theta$.

Let $u(x)$ be any element of $W_{2,0}^1(\Omega_1)$. From the function $u(x)$, construct the function $w(x)$ to be the generalized solution of the problem

$$\mathscr{L}_1^{(1)} w = 0,\quad w|_{S_1} = u|_{S_1} \tag{20b}$$

in $W_2^1(\Omega_1)$.

By virtue of (20) and the conditions imposed on the coefficients of the operator $\mathscr{L}_1^{(1)}$, this problem has a unique generalized solution in $W_2^1(\Omega_1)$ (see [8]). It is easy to see that $w \in U(\Omega_1)$. The difference $v = u - w$ is an element of $\overset{\circ}{W}{}_2^1(\Omega_1)$. The fact that the subspaces $\overset{\circ}{W}{}_2^1(\Omega_1)$ and $U(\Omega_1)$ can have only the zero element of $W_{2,0}^1(\Omega_1)$ in common follows from the uniqueness theorem for the generalized solution of problem (20b) in $W_2^1(\Omega_1)$. The lemma is proved.

In [3] it is shown that, if one looks at the functions of the space $H(\Omega)$ on the domain Ω_2 only, then the resulting space coincides with the subspace $W^1_{2,0}(\Omega_2)$ of $W^1_2(\Omega_2)$ that consists of the functions which vanish on $S_2\setminus\Gamma$. Thus the following lemma is valid.

Lemma 2. *The Hilbert space* $\overset{\circ}{W}{}^1_2(\Omega)$ *is the direct sum of the subspaces* $\tilde{\overset{\circ}{W}}{}^1_2(\Omega)$ *and* $H(\Omega)$:

$$\overset{\circ}{W}{}^1_2(\Omega) = \tilde{\overset{\circ}{W}}{}^1_2(\Omega) \dotplus H(\Omega).$$

To replace the function $u^{(1)}$ in (12), let us introduce the new function $v^{(1)} = u^{(1)} - \hat{u}^{(1)}$ where $\hat{u}^{(1)}$ is the generalized solution of the problem

$$\mathcal{L}_1^{(1)}\hat{u}^{(1)} = f_1^{(1)}(x,\ t),\quad \hat{u}^{(1)}\big|_{S_1} = 0 \tag{20c}$$

in $W^1_2(\Omega_1)$ for almost all $t\in[0,\ T]$. Then Problem I reduces to the problem of finding the element $\{v^{(1)},\ v^{(2)}\}\in V^1_0$ (V^1_0 is the subspace of V^1 that consists of all elements of V^1 for which $v^{(1)}\in U(\Omega_1)$ for almost all $t\in[0,\ T]$) that satisfies the integral identity

$$\begin{aligned}-\int\limits_{Q_T^{(2)}} \frac{\partial v^{(2)}}{\partial t}\cdot\frac{\partial}{\partial t}(b^{(2)}\psi^{(2)})\,dxdt + \sum_{k=1}^{2}\int_0^T \tilde{\mathcal{L}}_1^{(k)}(v^{(k)},\ \psi^{(k)})\,dt = \int\limits_{Q_T^{(1)}} b^{(1)}f_1^{(1)}\psi^{(1)}dxdt\\ + \int\limits_{Q_T^{(2)}} b^{(2)}f_3^{(2)}\psi^{(2)}dxdt + \int\limits_{\Omega_2} b^{(2)}\varphi_1^{(2)}\psi^{(2)}dx\,\big|_{t=0}\\ -\int_0^T \tilde{\mathcal{L}}_1^{(1)}(\hat{u}^{(1)},\ \psi^{(1)})\,dt - \int_0^T \tilde{\mathcal{L}}_1^{(2)}(\varphi_0^{(2)},\ \psi^{(2)})\,dt\end{aligned} \tag{20d}$$

for any $\psi(x,\ t) = \{\psi^{(1)}(x,\ t),\ \psi^{(2)}(x,\ t)\}$ in W^1_T and satisfies condition (13). By finding $u^{(1)}$, we automatically find $w = \{v^{(1)} + \hat{u}^{(1)},\ v^{(2)}\}$, the generalized solution of Problem I.

Let $\{\psi_k(x)\}$ be a basis of $H(\Omega)$. In [3] it is shown that $\{\psi_k(x)\}$ is a basis for $L_2(\Omega_2)$ also. For the sake of simplification that does not affect the proof, we will consider it to be orthonormalized in $L_2(\Omega_2)$ with the weight $b^{(2)}(x,\ 0)$; that is,

$$\int\limits_{\Omega_2} b^{(2)}(x,\ 0)\,\psi_m(x)\,\psi_n(x)\,dx = \delta_n^m,\quad n,\ m = 1,\ 2,\ \ldots.$$

The approximate solution $v_N(x,\ t) = \{v_N^{(1)}(x,\ t),\ v_N^{(2)}(x,\ t)\}$ will be sought in the form

$$v_N^{(1)}(x,\ t) = \sum_{m=1}^{N} c_m^N(t)\,\psi_m(x),\qquad v_N^{(2)}(x,\ t) = \sum_{m=1}^{N}\int_0^t c_m^N(\tau)\,d\tau\,\psi_m(x),$$

where the functions $c_m^N(t)$ are defined from the conditions

$$\begin{aligned}\int\limits_{\Omega_2} b^{(2)}(x,\ t)\frac{\partial^2 v_N^{(2)}}{\partial t^2}\psi_m dx + \sum_{k=1}^{2}\tilde{\mathcal{L}}_1^{(k)}(v_N^{(k)},\ \psi_m) - \int\limits_{\Omega_1} b^{(1)}f_1^{(1)}\psi_m dx - \int\limits_{\Omega_2} b^{(2)}f_3^{(2)}\psi_m dx\\ + \tilde{\mathcal{L}}_1^{(1)}(\hat{u}^{(1)},\ \psi_m) + \tilde{\mathcal{L}}_1^{(2)}(\varphi_0^{(2)},\ \psi_m) = 0\end{aligned} \tag{21}$$

and

$$c_m^N(0)=\int\limits_{\Omega_2} b^{(2)}(x,\ 0)\,\varphi_1^{(2)}(x)\,\psi_m(x)\,dx,\quad m=1,\ 2,\ \ldots,\ N. \tag{22}$$

Conditions (21) are a system of linear ordinary integro-differential equations of the form

$$\sum_{n=1}^{N} a_{mn}(t)\,\frac{dc_n^N(t)}{dt}+\sum_{n=1}^{N} A_{nm}(t)\,c_n^N(t)+\sum_{n=1}^{N} B_{nm}(t)\int\limits_0^t c_n^N(\tau)\,d\tau+F_m(t)=0. \tag{23}$$

Here $m=1,\ldots,N$, and the coefficients and the terms $F_m(t)$ are bounded and measurable on $[0,\ T]$. In order to verify the solvability of system (23) for the $dc_n^N(t)/dt$, it is sufficient to verify that the Gram determinant of any order N consisting of the elements

$$\int\limits_{\Omega_2} b^{(2)}(x,\ t)\,\psi_m(x)\,\psi_n(x)\,dx\equiv a_{mn}(t),$$

is nonzero for all $t\in[0,\ T]$. Indeed, if, for some $t_0\in[0,\ T]$ the Gram determinant of order N were zero, the system of functions

$$\{\sqrt{b^{(2)}(x,\ t_0)}\,\psi_m(x)\}$$

would be linearly independent, which is impossible because of the condition $b^{(2)}(x,\ t)\geq\beta=\text{const}>0$ and the linear independence of the system of functions $\{\psi_m(x)\}$. In view of all this, problems (23), (21) have a unique solution on $[0,\ T]$, and thus $v_N(x,\ t)$ is uniquely defined for every N. Let us show that the V^1 norms of the v_N are uniformly bounded; that is, that

$$\left\{\int\limits_{Q_T^{(2)}}\left(\frac{\partial v_N^{(2)}}{\partial t}\right)^2 dxdt+\sum_{k=1}^{2}\int\limits_{Q_T^{(k)}}\left[(v_N^{(k)})^2+\left(\frac{\partial v_N^{(k)}}{\partial x_i}\right)^2\right]dxdt\right\}^{1/2}\leqslant C_6$$

for all $N=1,2,\ldots$. In order to do this, multiply each of equations (21) by the corresponding $c_m^N(t)$. Sum the equalities thus obtained over all m from 1 to N and integrate the result over t from 0 to t_1. This yields

$$\int\limits_0^{t_1}\int\limits_{\Omega_2}\frac{\partial^2 v_N^{(2)}}{\partial t^2}\,b^{(2)}(x,\ t)\,\frac{\partial v_N^{(2)}}{\partial t}\,dxdt+\int\limits_0^{t_1}\mathscr{L}_1^{(1)}(v_N^{(1)},\ v_N^{(1)})\,dt+\int\limits_0^{t_1}\mathscr{L}_1^{(2)}\left(v_N^{(2)},\ \frac{\partial v_N^{(2)}}{\partial t}\right)dt$$
$$-\int\limits_0^{t_1}\int\limits_{\Omega_1} b^{(1)}f_1^{(1)}v_N^{(1)}\,dxdt-\int\limits_0^{t_1}\int\limits_{\Omega_2} b^{(2)}f_3^{(2)}\,\frac{\partial v_N^{(2)}}{\partial t}\,dxdt+\int\limits_0^{t_1}\mathscr{L}_1^{(1)}(\hat{u}^{(1)},\ v_N^{(1)})\,dt$$
$$+\int\limits_0^{t_1}\mathscr{L}_1^{(2)}\left(\varphi_0^{(2)},\ \frac{\partial v_N^{(2)}}{\partial t}\right)dt=0,$$

which can be transformed into the form

$$\frac{1}{2}\int_{\Omega_2} b^{(2)}\left(\frac{\partial v_N^{(2)}}{\partial t}\right)^2 dx\,|_{t=t_1} + \int_0^{t_1} \mathscr{L}_1^{(1)}(v_N^{(1)}, v_N^{(1)})\,dt + \frac{1}{2}\int_{\Omega_2} b^{(2)} a_{ij}^{(2)} \frac{\partial v_N^{(2)}}{\partial x_i}\frac{\partial v_N^{(2)}}{\partial x_j}\,dx\,|_{t=t_1}$$

$$= \frac{1}{2}\int_{\Omega_2} b^{(2)}\left(\frac{\partial v_N^{(2)}}{\partial t}\right)^2 dx\,|_{t=0} + \frac{1}{2}\int_0^{t_1}\int_{\Omega_2} \frac{\partial b^{(2)}}{\partial t}\left(\frac{\partial v_N^{(2)}}{\partial t}\right)^2 dx\,dt + \int_0^{t_1}\int_{\Omega_2}\left\{\frac{1}{2}\frac{\partial}{\partial t}(b^{(2)}a_{ij}^{(2)})\frac{\partial v_N^{(2)}}{\partial x_i}\frac{\partial v_N^{(2)}}{\partial x_j}\right.$$

$$\left. - \left[\left(\frac{\partial b^{(2)}}{\partial x_j} a_{ij}^{(2)} + b^{(2)} a_i^{(2)}\right)\frac{\partial v_N^{(2)}}{\partial x_i} + b^{(2)} a^{(2)} v_N^{(2)}\right]\frac{\partial v_N^{(2)}}{\partial t}\right\} dx\,dt - \int_{\Omega_2} b^{(2)} a_{ij}^{(2)} \frac{\partial \varphi_0^{(2)}}{\partial x_i}\cdot\frac{\partial v_N^{(2)}}{\partial x_j}\,dx\,|_{t=t_1} \tag{24}$$

$$+ \int_0^{t_1}\int_{\Omega_2}\left\{\frac{\partial}{\partial t}(b^{(2)}a_{ij}^{(2)})\cdot\frac{\partial \varphi_0^{(2)}}{\partial x_i}\cdot\frac{\partial v_N^{(2)}}{\partial x_j} - \left[\left(\frac{\partial b^{(2)}}{\partial x_j}a_{ij}^{(2)} + b^{(2)}a_i^{(2)}\right)\cdot\frac{\partial \varphi_0^{(2)}(x)}{\partial x_i} + b^{(2)}a^{(2)}\varphi_0^{(2)}\right]\frac{\partial v_N^{(2)}}{\partial t}\right\} dx\,dt$$

$$+ \int_0^{t_1}\int_{\Omega_1} b^{(1)} f_1^{(1)} v_N^{(1)}\,dx\,dt + \int_0^{t_1}\int_{\Omega_2} b^{(2)} f_3^{(2)} \frac{\partial v_N^{(2)}}{\partial t}\,dx\,dt - \int_0^{t_1} \mathscr{L}_1^{(1)}(\hat{u}^{(1)}, v_N^{(1)})\,dt.$$

By virtue of (20), we obtain from (24)

$$\frac{\beta}{2}\int_{\Omega_2}\left(\frac{\partial v_N^{(2)}}{\partial t}\right)^2 dx\,|_{t=t_1} + \nu\int_0^{t_1} \| v_N^{(1)} \|^2_{W_2^1(\Omega_1)}\,dt + \frac{\alpha\beta}{2}\int_{\Omega_2}\sum_{i=1}^n\left(\frac{\partial v_N^{(2)}}{\partial x_i}\right)^2 dx\,|_{t=t_1}$$

$$\leqslant \left\{M^2\left[(n+1)\sqrt{n} + \frac{2T+\sqrt{T}}{2}\right] + M\right\}\int_0^{t_1}\int_{\Omega_2}\left(\frac{\partial v_N^{(2)}}{\partial t}\right)^2 dx\,dt$$

$$+ \frac{M^2}{2}[3n + (n+1)\sqrt{n}]\int_0^{t_1}\int_{\Omega_2}\sum_{i=1}^n\left(\frac{\partial v_N^{(2)}}{\partial x_i}\right)^2 dx\,dt + \frac{\nu}{2}\int_0^{t_1}\| v_N^{(1)} \|^2_{W_2^1(\Omega_1)}\,dt$$

$$+ \frac{\alpha\beta}{4}\int_{\Omega_2}\sum_{i=1}^n\left(\frac{\partial v_N^{(2)}}{\partial x_i}\right)^2 dx\,|_{t=t_1} + \frac{M^2}{\nu}\int_0^{t_1}\int_{\Omega_1}(f_1^{(1)})^2\,dx\,dt + \frac{M}{2}\int_0^{t_1}\int_{\Omega_2}(f_3^{(2)})^2\,dx\,dt$$

$$+ \frac{M^2}{4}\left[\frac{4M^2n^2}{\alpha\beta} + 2nT + 2(n+1)\sqrt{n}\,T\right]\int_{\Omega_2}\sum_{i=1}^n\left(\frac{\partial \varphi_0^{(2)}}{\partial x_i}\right)^2 dx + \frac{M^2\sqrt{T}}{2}\int_{\Omega_2}(\varphi_0^{(2)})^2\,dx$$

$$+ \frac{M}{2}\int_{\Omega_2}(\varphi_1^{(2)}(x))^2\,dx + \frac{M^4[n^2 + 2(n+1)^2 n]}{\nu}\int_0^{t_1}\| \hat{u}^{(1)} \|^2_{W_2^1(\Omega_1)}\,dt,$$

where M is an upper bound for the functions

$$b^{(k)},\ \left|\frac{\partial b^{(k)}}{\partial x_j}\right|,\ \left|\frac{\partial b^{(2)}}{\partial t}\right|,\ |a_{ij}^{(k)}|,\ \left|\frac{\partial a_{ij}^{(2)}}{\partial t}\right|,\ |a_i^{(k)}|,\ |a^{(k)}|,$$

$i, j = 1, \ldots, n$, in $\overline{Q}_T^{(k)}$, $k = 1, 2$.

This inequality implies the inequality

$$\int_{\Omega_2}\left(\frac{\partial v_N^{(2)}}{\partial t}\right)^2 dx\,|_{t=t_1} + \int_0^{t_1}\| v_N^{(1)} \|^2_{W_2^1(\Omega_1)}\,dt + \int_{\Omega_2}\sum_{i=1}^n\left(\frac{\partial v_N^{(2)}}{\partial x_i}\right)^2 dx\,|_{t=t_1}$$

$$\leqslant C_7\left\{\int_0^{t_1}\int_{\Omega_2}\left(\frac{\partial v_N^{(2)}}{\partial t}\right)^2 dx\,dt + \int_0^{t_1}\int_{\Omega_2}\sum_{i=1}^n\left(\frac{\partial v_N^{(2)}}{\partial x_i}\right)^2 dx\,dt\right\} \tag{25}$$

$$+ C_8\left[\| f_1^{(1)} \|^2_{L_2(Q_T^{(1)})} + \| f_3^{(2)} \|^2_{L_2(Q_T^{(2)})} + \| \varphi_0^{(2)} \|^2_{W_2^1(\Omega_2)} + \| \varphi_1^{(2)} \|^2_{L_2(\Omega_2)}\right],$$

where the constants C_7 and C_8 depend only on M, n, T, α, β and ν. Here we have made use of the inequality

$$\int_0^{t_1} \| \hat{u}^{(1)} \|^2_{W_2^1(\Omega_1)} dt \leqslant \frac{M^2}{\nu^2} \int_0^{t_1} \| f_1^{(1)} \|^2_{L_2(\Omega_1)} dt, \tag{25a}$$

which, by virtue of (20), is valid for the generalized solution of problem (20c) in $W_2^1(\Omega_1)$ (see [8]).

Denote

$$\int_0^{t_1} \int_{\Omega_2} \left[\left(\frac{\partial v_N^{(2)}}{\partial t} \right)^2 + \sum_{i=1}^n \left(\frac{\partial v_N^{(2)}}{\partial x_i} \right)^2 \right] dx dt$$

by $y(t_1)$. From (25) it follows that $dy(t_1)/dt_1 \leq C_7 y(t_1) + C_9$, where

$$C_9 = C_8 \left[\| f_1^{(1)} \|^2_{L_2(Q_T^{(1)})} + \| f_3^{(2)} \|^2_{L_2(Q_T^{(2)})} + \| \varphi_0^{(2)} \|^2_{W_2^1(\Omega_2)} + \| \varphi_1^{(2)} \|^2_{L_2(\Omega_2)} \right].$$

Hence $y(t_1) \leq (e^{C_7 t_1} - 1)/C_7 C_9$. Substituting this bound into (25), we obtain

$$\begin{aligned} &\int_{\Omega_2} \left(\frac{\partial v_N^{(2)}}{\partial t} \right)^2 dx \Big|_{t=t_1} + \int_0^{t_1} \| v_N^{(1)} \|^2_{W_2^1(\Omega_1)} dt + \int_{\Omega_2} \sum_{i=1}^n \left(\frac{\partial v_N^{(2)}}{\partial x_i} \right)^2 dx \Big|_{t=t_1} \\ &\leqslant C_8 e^{C_7 T} \left[\| f_1^{(1)} \|^2_{L_2(Q_T^{(1)})} + \| f_3^{(2)} \|^2_{L_2(Q_T^{(2)})} + \| \varphi_0^{(2)} \|^2_{W_2^1(\Omega_2)} + \| \varphi_1^{(2)} \|^2_{L_2(\Omega_2)} \right], \end{aligned} \tag{26}$$

which is valid for all $t_1 \in [0, T]$. Inequality (26) and the bound

$$\| v_N^{(2)} \|^2_{L_2(Q_T^{(2)})} \leqslant T^2 \left\| \frac{\partial v_N^{(2)}}{\partial t} \right\|^2_{L_2(Q_T^{(2)})}$$

imply the inequality

$$\begin{aligned} &\int_0^T \| v_N^{(1)} \|^2_{W_2^1(\Omega_1)} dt + \int_{Q_T^{(2)}} \left[(v_N^{(2)})^2 + \sum_{i=1}^n \left(\frac{\partial v_N^{(2)}}{\partial x_i} \right)^2 + \left(\frac{\partial v_N^{(2)}}{\partial t} \right)^2 \right] dx dt \\ &\leqslant C_8 e^{C_7 T} (T^3 + T + 1) \left[\| f_1^{(1)} \|^2_{L_2(Q_T^{(1)})} + \| f_3^{(2)} \|^2_{L_2(Q_T^{(2)})} + \| \varphi_0^{(2)} \|^2_{W_2^1(\Omega_2)} + \| \varphi_1^{(2)} \|^2_{L_2(\Omega_2)} \right], \end{aligned} \tag{27}$$

the desired result.

By virtue of (27), out of the sequence $\{v_N^{(1)}, v_N^{(2)}\}$, one can select a subsequence $\{v_{N_l}^{(1)}, v_{N_l}^{(2)}\}$, $l = 1, 2, \ldots$, that converges weakly to $\{v^{(1)}, v^{(2)}\}$ in the V^1 norm. By virtue of a well-known property of weak convergence, inequality (27) remains valid for the limit element $\{v^{(1)}, v^{(2)}\}$, and $\{v^{(1)}, v^{(2)}\}$ is an element of V_0^1. Let us verify that the element $v(x, t) = \{v^{(1)}(x, t), v^{(2)}(x, t)\}$ satisfies (20d). In order to do this, multiply each of equations (21) by a corresponding smooth function $d_m(t)$ that is equal to zero at $t = T$. Sum the equalities thus obtained over all m from 1 to $N' \leq N$ and integrate the result over t from 0 to T. After integrating by parts in the first term, we obtain

$$-\int_{Q_T^{(2)}} \frac{\partial v_{N_l}^{(2)}}{\partial t} \frac{\partial}{\partial t}(b^{(2)}\psi_{N'}^{(2)})\,dxdt + \sum_{k=1}^{2}\int_0^T \mathscr{L}_1^{(1)}(v_{N_l}^{(1)},\ \psi_{N'}^{(1)})\,dt + \int_0^T \mathscr{L}_1^{(1)}(\hat{u}^{(1)},\ \psi_{N'}^{(1)})\,dt$$

$$-\int_{Q_T^{(1)}} b^{(1)}f_1^{(1)}\psi_{N'}^{(1)}\,dxdt - \int_{Q_T^{(2)}} b^{(2)}f_3^{(2)}\psi_{N'}^{(2)}\,dxdt - \int_{\Omega_2} b^{(2)}\frac{\partial v_{N_l}^{(2)}}{\partial t}\psi_{N'}^{(2)}\,dx\,\Big|_{t=0} \tag{28}$$

$$+\int_0^T \mathscr{L}_1^{(2)}(\varphi_0^{(2)},\ \psi_{N'}^{(2)})\,dt = 0,$$

where

$$\psi_{N'}^{(k)}(x,\ t) = \sum_{m=1}^{N'} d_m(t)\,\psi_m(x),\ k = 1,\ 2.$$

In this equality one can take the limit as $l \to \infty$, assuming $\psi_{N'}^{(k)}$ to be fixed, and, by this process, arrive at (28) for $v = \{v^{(1)},\ v^{(2)}\}$ (instead of $\{v_{N_l}^{(1)},\ v_{N_l}^{(2)}\}$). Indeed, in view of the fact mentioned above that $\{v_{N_l}^{(1)},\ v_{N_l}^{(2)}\}$ converges weakly to $\{v^{(1)},\ v^{(2)}\}$, one can take the limit as $l \to \infty$ in all of the volume integrals. The integral

$$\int_{\Omega_2} b^{(2)}\frac{\partial v_{N_l}^{(2)}}{\partial t}\psi_{N'}^{(2)}\,dx\,\Big|_{t=0}.$$

has the limit

$$\int_{\Omega_2} b^{(2)}(x,\ 0)\,\varphi_1^{(2)}(x)\,\psi_{N'}^{(2)}(x,\ 0)\,dx,$$

since

$$\int_{\Omega_2} b^{(2)}(x,\ 0)\left[\varphi_1^{(2)}(x) - \frac{\partial v_{N_l}^{(2)}}{\partial t}\Big|_{t=0}\right]^2 dx \to 0 \quad \text{as} \quad l \to \infty.$$

Thus we have proved that the limit element $v(x,\ t) = \{v^{(1)}(x,\ t),\ v^{(2)}(x,\ t)\}$ satisfies (20d) for all $\psi_{N'}(x,\ t) = \{\psi_{N'}^{(1)}(x,\ t),\ \psi_{N'}^{(2)}(x,\ t)\}$ of the form

$$\left\{\sum_{m=1}^{N'} d_m(t)\,\psi_m(x),\ \sum_{m=1}^{N'} d_m(t)\,\psi_m(x)\right\},$$

the $d_m(t)$ and the $\psi_m(x)$ having the properties mentioned above. But these elements $\psi_{N'}$ form a dense set in the Hilbert space $W_{T,0}^1$ consisting of all elements ψ of the space W_T^1 that belong to $H(\Omega)$ for almost all $t \in [0,\ T]$. In view of this, (20d) holds for all $\psi \in W_{T,0}^1$.

Now use Lemma 2 concerning the decomposition of the Hilbert space $\mathring{W}_2^1(\Omega)$ into the direct sum of the subspaces $\tilde{W}_2^1(\Omega)$ and $H(\Omega)$, and note that (20d) holds for all $\psi \in W_T^1$ that belong to $\tilde{W}_2^1(\Omega)$ for almost all $t \in [0,\ T]$. Consequently

$$w(x,\ t) = \{v^{(1)}(x,\ t) + \hat{u}^{(1)}(x,\ t),\ v^{(2)}(x,t)\}$$

is the generalized solution of Problem I.

Also, inequality (27) for the limit element $\{v^{(1)},\ v^{(2)}\}$ and inequality (25a) imply

$$\int_0^T \|u^{(1)}\|^2_{W_2^1(\Omega_1)}dt + \int_{Q_T^{(2)}} \left[(v^{(2)})^2 + \sum_{i=1}^n \left(\frac{\partial v^{(2)}}{\partial x_i}\right)^2 + \left(\frac{\partial v^{(2)}}{dt}\right)^2\right] dxdt$$
$$\leqslant C_{10}\left[\|f_1^{(1)}\|^2_{L_2(Q_T^{(1)})} + \|f_3^{(2)}\|^2_{L_2(Q_T^{(2)})} + \|\varphi_0^{(2)}\|^2_{W_2^1(\Omega_2)} + \|\varphi_1^{(2)}\|^2_{L_2(\Omega_2)}\right]. \quad (28a)$$

Thus we have proved the following theorem.

Theorem 3. *Assume that conditions* a)–c) *are fulfilled and that* $f_1^{(1)}(x, t) \in L_2(Q_T^{(1)})$, $f_3^{(2)}(x, t) \in L_2(Q_T^{(2)})$, $\phi_0^{(2)}(x) \in \overset{\circ}{W}{}_2^1(\Omega_2)$, *and* $\phi_1^{(2)}(x) \in L_2(\Omega_2)$. *Then Problem* I *has a generalized solution in* V^1. *Inequality* (28a) *is valid for this solution.*

Let us now consider Problem II. We will assume that the functions

$$b^{(k)},\ \frac{\partial b^{(k)}}{\partial x_j},\ \frac{\partial b^{(k)}}{\partial t},\ a_{ij}^{(k)},\ \frac{\partial a_{ij}^{(2)}}{\partial t},\ a_i^{(k)},\ a^{(k)}$$

are bounded and measurable in $\overline{Q}_T^{(k)}$, $k = 1, 2$, and that $b^{(k)}(x, t) \geq \beta = \mathrm{const} > 0$.

Let $\{\psi_n\}$ be a fundamental system of functions in $\overset{\circ}{W}{}_2^1(\Omega)$. Assume that

$$\int_\Omega b(x, 0)\psi_n(x)\psi_m(x)\,dx = \delta_n^m, \qquad n, m = 1, 2, \ldots,$$

where $b(x, t) = b^{(1)}(x, t)$ for $(x, t) \in Q_T^{(1)}$ and $b(x, t) = b^{(2)}(x, t)$ for $(x, t) \in Q_T^{(2)}$.

We will seek an approximate solution $w_N(x, t) = \{u_N^{(1)}(x, t), v_N^{(2)}(x, t)\}$ of Problem II in the form

$$u_N^{(1)}(x, t) = \sum_{n=1}^N c_n^N(t)\psi_n(x),\quad v_N^{(2)}(x, t) = \sum_{n=1}^N \int_0^t c_n^N(\tau)\,d\tau \cdot \psi_n(x),$$

where the functions $c_n^N(t)$ are defined by the equations

$$\int_{\Omega_1} \frac{\partial u_N^{(1)}}{\partial t} b^{(1)}(x, t)\psi_m(x)\,dx + \int_{\Omega_2} \frac{\partial^2 v_N^{(2)}}{\partial t^2} b^{(2)}(x, t)\psi_m(x)\,dx + \mathscr{L}_1^{(1)}(u_N^{(1)}, \psi_m)$$
$$+ \mathscr{L}_1^{(2)}(v_N^{(2)}, \psi_m) - \int_{\Omega_1} b^{(1)} f_2^{(1)}\psi_m dx - \int_{\Omega_2} b^{(2)} f_3^{(2)}\psi_m dx + \tilde{\mathscr{L}}_1^{(2)}(\varphi_0^{(2)}, \psi_m) \quad (29)$$
$$= 0,\ m = 1, 2, \ldots, N,$$

and the initial conditions

$$c_n^N(0) = \int_{\Omega_1} b^{(1)}(x, 0)\varphi_0^{(1)}(x)\psi_n(x)\,dx + \int_{\Omega_2} b^{(2)}(x, 0)\varphi_1^{(2)}(x)\psi_n(x)\,dx, n = 1, 2, \ldots, N. \quad (30)$$

These equations can be written in the form

$$\sum_{n=1}^N b_{mn}(t)\frac{dc_n^N(t)}{dt} + \sum_{n=1}^N A_{nm}(t)c_n^N(t) + \sum_{n=1}^N B_{nm}(t)\int_0^t c_n^N(\tau)\,d\tau + F_m(t) = 0, \quad (31)$$
$$m = 1, 2, \ldots, N,$$

where

$$b_{mn}(t)=\int_{\Omega} b(x,\ t)\,\psi_n(x)\,\psi_m(x)\,dx.$$

Just as in the proof of the solvability of Problem I, one can show that the Gram determinants of any order N consisting of the $b_{mn}(t)$ are nonzero for all t in $[0,\ T]$. Hence system (31) can be solved for the $dc_n^N(t)/dt$. Since the functions $A_{nm}(t)$, $B_{nm}(t)$ and $F_m(t)$ are bounded and measurable on $[0,\ T]$, problem (31), (30) has a unique solution on $[0,\ T]$, and thus the $w^N(x,\ t)$ are uniquely defined for any N. Let us show that the V^1 norms of the w^N are uniformly bounded with respect to N. In order to do this, multiply each of equalities (29) by the corresponding $c_m^N(t)$. Sum the resulting equalities over all m from 1 to N and integrate over t from 0 to t_1. This yields the equality

$$\int_0^{t_1}\int_{\Omega_1} b^{(1)}(x,t)\frac{\partial u_N^{(1)}}{\partial t}u_N^{(1)}dxdt+\int_0^{t_1}\int_{\Omega_2} b^{(2)}(x,t)\frac{\partial^2 v_N^{(2)}}{\partial t^2}\cdot\frac{\partial v_N^{(2)}}{\partial t}dxdt+\int_0^{t_1}\mathscr{L}_1^{(1)}(u_N^{(1)},u_N^{(1)})\,dt$$
$$+\int_0^{t_1}\mathscr{L}_1^{(2)}\left(v_N^{(2)},\ \frac{\partial v_N^{(2)}}{\partial t}\right)dt-\int_0^{t_1}\int_{\Omega_1} b^{(1)}f_2^{(1)}u_N^{(1)}dxdt-\int_0^{t_1}\int_{\Omega_2} b^{(2)}f_3^{(2)}\frac{\partial v_N^{(2)}}{\partial t}dxdt$$
$$+\int_0^{t_1}\mathscr{L}_1^{(2)}\left(\varphi_0^{(2)},\ \frac{\partial v_N^{(2)}}{\partial t}\right)dt=0,$$

which can be transformed into the form

$$\frac{1}{2}\int_{\Omega_1} b^{(1)}(u_N^{(1)})^2\,dx\,\Big|_{t=t_1}+\frac{1}{2}\int_{\Omega_2} b^{(2)}\left(\frac{\partial v_N^{(2)}}{\partial t}\right)^2 dx\,\Big|_{t=t_1}+\int_0^{t_1}\int_{\Omega_1} b^{(1)}a_{ij}^{(1)}\frac{\partial u_N^{(1)}}{\partial x_i}\cdot\frac{\partial u_N^{(1)}}{\partial x_j}\,dxdt$$
$$+\frac{1}{2}\int_{\Omega_2} b^{(2)}a_{ij}^{(2)}\frac{\partial v_N^{(2)}}{\partial x_i}\cdot\frac{\partial v_N^{(2)}}{\partial x_j}\,dx\,\Big|_{t=t_1}=\frac{1}{2}\int_{\Omega_1} b^{(1)}(u_N^{(1)})^2\,dx\,\Big|_{t=0}+\frac{1}{2}\int_{\Omega_2} b^{(2)}\left(\frac{\partial v_N^{(2)}}{\partial t}\right)^2 dx\,\Big|_{t=0}$$
$$+\frac{1}{2}\int_0^{t_1}\int_{\Omega_1}\frac{\partial b^{(1)}}{\partial t}(u_N^{(1)})^2\,dxdt+\frac{1}{2}\int_0^{t_1}\int_{\Omega_2}\frac{\partial b^{(2)}}{\partial t}\left(\frac{\partial v_N^{(2)}}{\partial t}\right)^2 dxdt+\int_0^{t_1}\int_{\Omega_2}\left\{\frac{1}{2}\frac{\partial}{\partial t}(b^{(2)}a_{ij}^{(2)})\frac{\partial v_N^{(2)}}{\partial x_i}\frac{\partial v_N^{(2)}}{\partial x_j}\right.$$
$$\left.-\left[\left(\frac{\partial b^{(2)}}{\partial x_j}a_{ij}^{(2)}+b^{(2)}a_i^{(2)}\right)\frac{\partial v_N^{(2)}}{\partial x_i}+b^{(2)}a^{(2)}v_N^{(2)}\right]\frac{\partial v_N^{(2)}}{\partial t}\right\}dxdt-\int_{\Omega_2} b^{(2)}a_{ij}^{(2)}\frac{\partial\varphi_0^{(2)}}{\partial x_i}\frac{\partial v_N^{(2)}}{\partial x_j}\,dx\,\Big|_{t=t_1}$$
$$+\int_0^{t_1}\int_{\Omega_2}\left\{\frac{\partial}{\partial t}(b^{(2)}a_{ij}^{(2)})\frac{\partial\varphi_0^{(2)}}{\partial x_i}\frac{\partial v_N^{(2)}}{\partial x_j}-\left[\left(\frac{\partial b^{(2)}}{\partial x_j}a_{ij}^{(2)}+b^{(2)}a_i^{(2)}\right)\frac{\partial\varphi_0^{(2)}(x)}{\partial x_i}+b^{(2)}a^{(2)}\varphi_0^{(2)}\right]\frac{\partial v_N^{(2)}}{\partial t}\right\}dxdt$$
$$+\int_0^{t_1}\int_{\Omega_1} b^{(1)}f_2^{(1)}u_N^{(1)}dxdt+\int_0^{t_1}\int_{\Omega_2} b^{(2)}f_3^{(2)}\frac{\partial v_N^{(2)}}{\partial t}dxdt$$
$$-\int_0^{t_1}\int_{\Omega_1}\left[\left(\frac{\partial b^{(1)}}{\partial x_j}a_{ij}^{(1)}+b^{(1)}a_i^{(1)}\right)\frac{\partial u_N^{(1)}}{\partial x_i}u_N^{(1)}+b^{(1)}a^{(1)}(u_N^{(1)})^2\right]dxdt.$$

In the same manner as in Theorem 3, we can derive from this equality the bound that is of interest to us:

$$\|w_N\|^2_{V^1} \equiv \int_0^T \|u_N^{(1)}\|^2_{W_2^1(\Omega_1)}\,dt + \int_{Q_T^{(2)}} \left[(v_N^{(2)})^2 + \sum_{i=1}^n \left(\frac{\partial v_N^{(2)}}{\partial x_i}\right)^2 + \left(\frac{\partial v_N^{(2)}}{\partial t}\right)^2 \right] dx\,dt$$
$$\leqslant C_{11}\left[\|f_2^{(1)}\|^2_{L_2(Q_T^{(1)})} + \|f_3^{(2)}\|^2_{L_2(Q_T^{(2)})} + \|\varphi_0^{(1)}\|^2_{L_2(\Omega_1)} + \|\varphi_1^{(2)}\|^2_{L_2(\Omega_2)} + \|\varphi_0^{(2)}\|^2_{W_2^1(\Omega_2)}\right], \tag{32}$$

where the constant C_{11} depends only on α, β, n, T and M. Here M is an upper bound for the functions

$$b^{(k)}, \quad \left|\frac{\partial b^{(k)}}{\partial x_j}\right|, \quad \left|\frac{\partial b^{(k)}}{\partial t}\right|, \quad |a_{ij}^{(k)}|, \quad \left|\frac{\partial a_{ij}^{(2)}}{\partial t}\right|, \quad |a_i^{(k)}|, \quad |a^{(k)}|,$$

$i, j = 1, \ldots, n$, on $\overline{Q}_T^{(k)}$, $k = 1, 2$.

The rest is proved just as in Theorem 3; that is, we multiply (29) by $d_m(t)$, where the $d_m(t)$ are continuously differentiable functions for $t \in [0, T]$ and satisfy the condition $d_m(T) = 0$. Sum the equalities thus obtained over m from 1 to $N' \leq N$, integrate over t from 0 to T, and then integrate the first two integrals by parts. Doing this, we obtain

$$-\int_{Q_T^{(1)}} u_N^{(1)} \frac{\partial}{\partial t}(b^{(1)}\psi_{N'}^{(1)})\,dx\,dt - \int_{Q_T^{(2)}} \frac{\partial v_N^{(2)}}{\partial t} \cdot \frac{\partial}{\partial t}(b^{(2)}\psi_{N'}^{(2)})\,dx\,dt + \int_0^T [\mathscr{L}_1^{(1)}(u_N^{(1)}, \psi_{N'}^{(1)})$$
$$+ \mathscr{L}_1^{(2)}(v_{N'}^{(2)}\psi_{N'}^{(2)})]\,dt - \int_{Q_T^{(1)}} b^{(1)} f_2^{(1)} \psi_{N'}^{(1)}\,dx\,dt - \int_{Q_T^{(2)}} b^{(2)} f_3^{(2)} \psi_{N'}^{(2)}\,dx\,dt \tag{33}$$
$$-\int_{\Omega_1} u_N^{(1)} b^{(1)} \psi_{N'}^{(1)}\,dx\,\Big|_{t=0} - \int_{\Omega_2} \frac{\partial v_N^{(2)}}{\partial t} b^{(2)} \psi_{N'}^{(2)}\,dx\,\Big|_{t=0} + \int_0^T \mathscr{L}_1^{(2)}(\varphi_0^{(2)}, \psi_{N'}^{(2)})\,dt = 0,$$

where

$$\psi_{N'}^{(k)}(x, t) = \sum_{m=1}^{N'} d_m(t)\,\psi_m(x), \quad k = 1, 2.$$

By virtue of (32), from the sequence $\{u_N^{(1)}, v_N^{(2)}\}$, we can extract a subsequence $\{u_{N_l}^{(1)}, v_{N_l}^{(2)}\}$ that converges weakly in V^1 to $\{u^{(1)}, v^{(2)}\}$ and can take the limit as $N_l \to \infty$ in (33) for $u_N^{(1)} = u_{N_l}^{(1)}$ and $v_N^{(2)} = v_{N_l}^{(2)}$. Doing this, we obtain (15) for $\psi = \psi_{N'} \equiv \{\psi_{N'}^{(1)}, \psi_{N'}^{(2)}\}$. In order to verify that the limit $\{u^{(1)}, v^{(2)}\}$ is the generalized solution of Problem II, it is sufficient to take the limit as $N' \to \infty$ in this identity.

Let us formulate this result as a theorem.

Theorem 4. *Let the functions $b^{(k)}$, $\partial b^{(k)}/\partial x_j$, $\partial b^{(k)}/\partial t$, $a_{ij}^{(k)}$, $\partial a_{ij}^{(2)}/\partial t$, $a_i^{(k)}$ and $a^{(k)}$ be bounded and measurable on $\overline{Q}_T^{(k)}$, $k = 1, 2$. Assume that $b^{(k)}(x, t) \geq \beta = \text{const} > 0$. Let $f_2^{(1)}(x, t) \in L_2(Q_T^{(1)})$, $f_3^{(2)}(x, t) \in L_2(Q_T^{(2)})$, $\phi_0^{(1)}(x) \in L_2(\Omega_1)$, $\phi_0^{(2)}(x) \in \overset{\circ}{W}{}_2^1(\Omega_2)$, and $\phi_1^{(2)}(x) \in L_2(\Omega_2)$. Then Problem* II *has a generalized solution in V^1, and for this solution*

$$\|w\|^2_{V^1} \equiv \|u^{(1)}\|^2_{W_2^{1,0}(Q_T^{(1)})} + \|v^{(2)}\|^2_{W_2^{1,1}(Q_T^{(2)})} \leqslant C_{11}\Big[\|f_2^{(1)}\|^2_{L_2(Q_T^{(1)})}$$
$$+ \|f_3^{(2)}\|^2_{L_2(Q_T^{(2)})} + \|\varphi_0^{(1)}\|^2_{L_2(\Omega_1)} + \|\varphi_1^{(2)}\|^2_{L_2(\Omega_2)} + \|\varphi_0^{(2)}\|^2_{W_2^1(\Omega_2)}\Big].$$

Denote by V^2 the Banach space of all elements of V^1 that have a finite norm

$$\|v\|_{V^2} = \operatorname*{ess\,sup}_{0 \leq t \leq T} \left(\left\| \frac{\partial v^{(2)}}{\partial t} \right\|_{L_2(\Omega_2)} + \|v^{(2)}\|_{W_2^1(\Omega_2)} \right) + \left(\int_0^T \|v^{(1)}\|^2_{W_2^1(\Omega_1)} dt \right)^{1/2},$$

and by $\hat{V}^2$ the Banach space of all elements of V^1 that have a finite norm

$$\|v\|_{\hat{V}^2} = \operatorname*{ess\,sup}_{0 \leq t \leq T} \left(\|v^{(1)}\|_{L_2(\Omega_1)} + \left\| \frac{\partial v^{(2)}}{\partial t} \right\|_{L_2(\Omega_2)} + \|v^{(2)}\|_{W_2^1(\Omega_2)} \right) + \left(\int_0^T \|v^{(1)}\|^2_{W_2^1(\Omega_1)} dt \right)^{1/2}.$$

We will assume that the $b^{(k)}$ do not depend on t. The following proposition holds.

Theorem 3′. *Let the hypotheses of Theorem 3 be fulfilled. Then Problem I has a generalized solution in* V^2 *for any* $f_1^{(1)}(x, t) \in L_2(Q_T^{(1)})$, $f_3^{(2)}(x, t) \in L_2(Q_T^{(2)})$, $\phi_0^{(2)}(x) \in \overset{\circ}{W}{}_2^1(\Omega_2)$, *and* $\phi_1^{(2)}(x) \in L_2(\Omega_2)$. *For this solution the inequality*

$$\|w\|^2_{V^2} \leq \tilde{C} \left[\|f_1^{(1)}\|^2_{L_2(Q_T^{(1)})} + \|f_3^{(2)}\|^2_{L_2(Q_T^{(2)})} + \|\varphi_0^{(2)}\|^2_{W_2^1(\Omega_2)} + \|\varphi_1^{(2)}\|^2_{L_2(\Omega_2)} \right] \tag{33a}$$

is valid.

Proof. Since for any element of V^1 the inequality

$$\operatorname*{ess\,sup}_{0 \leq t \leq T} \|v^{(2)}(x, t)\|_{W_2^1(\Omega_2)} \leq \hat{C} \|v\|_{V^1}$$

(see §1) is valid, in order to prove the theorem, it is sufficient to show that the norm

$$\operatorname*{ess\,sup}_{0 \leq t \leq T} \left\| \frac{\partial v^{(2)}(x, t)}{\partial t} \right\|_{L_2(\Omega_2)}$$

is finite. This is done in the same manner as in [9], Chapter III, §4. More specifically, in Theorem 3 it is proved that out of the sequence $\{v_N^{(1)}, v_N^{(2)}\}$ one can select a subsequence $\{v_{N_l}^{(1)}, v_{N_l}^{(2)}\}$, $l = 1, 2, \ldots$, that converges weakly in V^1 to $\{v^{(1)}, v^{(2)}\}$. Furthermore, from (26) it follows that the functions

$$c_m^{N_l}(t) = \int_{\Omega_2} b^{(2)}(x) \frac{\partial v_{N_l}^{(2)}}{\partial t} \psi_m(x)\, dx$$

are uniformly bounded. Let us show that, for fixed m and arbitrary $N_l \geq m$, they are equicontinuous on $[0, T]$. Indeed, from (21) we obtain

$$|c_m^{N_l}(t + \Delta t) - c_m^{N_l}(t)| \leq \int_t^{t+\Delta t} \left[\sum_{k=1}^{2} |\mathcal{L}_1^{(k)}(v_{N_l}^{(k)}, \psi_m)| + \left| \int_{\Omega_1} b^{(1)} f_1^{(1)} \psi_m dx \right| + \left| \int_{\Omega_2} b^{(2)} f_3^{(2)} \psi_m dx \right| + |\tilde{\mathcal{L}}_1^{(1)}(\hat{u}^{(1)}, \psi_m)| + |\tilde{\mathcal{L}}_1^{(2)}(\varphi_0^{(2)}, \psi_m)| \right] dt,$$

whence, using (27), it follows that

$$|c_m^{N_l}(t + \Delta t) - c_m^{N_l}(t)| \leq C' \sqrt{\Delta t},$$

where the constant C' does not depend on N_l; that is, the $c_m^{N_l}(t)$ are equicontinuous with respect to t. By the usual diagonal process, we can select a subsequence N_{l_p}, $p = 1, 2, \ldots$, for which the functions $c_m^{N_{l_p}}(t)$ converge uniformly on $[0, T]$ to some continuous function $c_m(t)$ for every $m = 1, 2, \ldots$. The functions $c_m(t)$, $m = 1, 2, \ldots$, determine

$$\frac{\partial v^{(2)}}{\partial t} = \sum_{m=1}^{\infty} c_m(t)\,\psi_m(x).$$

The $\partial v_{N_{l_p}}^{(2)}/\partial t$ converge weakly in $L_2(\Omega_2)$ to this function uniformly with respect to $t \in [0, T]$. Indeed, for any function $\psi^{(2)}(x)$ in $L_2(\Omega_2)$,

$$I = \int_{\Omega_2} b^{(2)}(x)\left(\frac{\partial v_{N_{l_p}}^{(2)}}{\partial t} - \frac{\partial v^{(2)}}{\partial t}\right)\psi^{(2)}(x)\,dx = \sum_{n=1}^{s} a_n \int_{\Omega_2} b^{(2)}(x)$$

$$\times\left(\frac{\partial v_{N_{l_p}}^{(2)}}{\partial t} - \frac{\partial v^{(2)}}{\partial t}\right)\psi_n(x)\,dx + \int_{\Omega_2} b^{(2)}(x)\left(\frac{\partial v_{N_{l_p}}^{(2)}}{\partial t} - \frac{\partial v^{(2)}}{\partial t}\right)\sum_{n=s+1}^{\infty} a_n\psi_n(x)\,dx, \tag{33b}$$

where

$$a_n = \int_{\Omega_2} b^{(2)}(x)\,\psi^{(2)}(x)\,\psi_n(x)\,dx.$$

Since

$$\left|\int_{\Omega_2} b^{(2)}(x)\left(\frac{\partial v_{N_{l_p}}^{(2)}}{\partial t} - \frac{\partial v^{(2)}}{\partial t}\right)\sum_{n=s+1}^{\infty} a_n\psi_n(x)\,dx\right| \leqslant C''\left(\sum_{n=s+1}^{\infty} a_n^2\right)^{1/2} \equiv C''Q(s),$$

where C'' does not depend on N_{l_p}, by choosing s sufficiently large, one can make $C''Q(s)$ less than any previously given $\epsilon > 0$. As was proved above, for fixed s the first sum in (33b) is less than ϵ for all $t \in [0, T]$, if N_{l_p} is taken to be sufficiently large. Thus $|I|$ can be made less than 2ϵ for all $t \in [0, T]$. We have hereby shown that the $\partial v_{N_{l_p}}^{(2)}/\partial t$, $p = 1, 2, \ldots$, converge to $\partial \tilde{v}^{(2)}/\partial t$ weakly in $L_2(\Omega_2)$ uniformly with respect to $t \in [0, T]$. By virtue of a well-known property of weak convergence and by Theorem 3, inequality (26) remains valid for the limit element $\{v^{(1)}, v^{(2)}\}$. This inequality, together with (25a), yields (33a) for the limit element $\{v^{(1)}, v^{(2)}\}$. Hence $\{u^{(1)}, v^{(2)}\}$ is an element of V^2. The theorem is proved.

An analogous theorem holds for Problem II.

Theorem 4′. *Let the hypotheses of Theorem 4 be fulfilled. Then Problem* II *has a generalized solution in* $\hat{V}^2$ *for any* $f_2^{(1)}(x, t) \in L_2(Q_T^{(1)})$, $f_3^{(2)}(x, t) \in L_2(Q_T^{(2)})$, $\phi_0^{(1)}(x) \in L_2(\Omega_1)$, $\phi_0^{(2)}(x) \in \overset{\circ}{W}{}_2^1(\Omega_2)$, *and* $\phi_1^{(2)}(x) \in L_2(\Omega_2)$. *For this solution,*

$$\|w\|_{\hat{V}^2}^2 \leqslant \hat{C}\left[\|f_2^{(1)}\|_{L_2(Q_T^{(1)})}^2 + \|f_3^{(2)}\|_{L_2(Q_T^{(2)})}^2 + \|\varphi_0^{(1)}\|_{L_2(\Omega_1)}^2 + \|\varphi_1^{(2)}\|_{L_2(\Omega_2)}^2 + \|\varphi_0^{(2)}\|_{W_2^1(\Omega_2)}^2\right].$$

The proof is analogous to that of Theorem 3′.

§4. The differential properties of generalized solutions

Without specifically stating so each time, we will assume that the closure of one of the domains Ω_k, $k = 1, 2$, say, for example, Ω_1, is contained in the interior of the domain Ω, and that the surfaces S and Γ are twice continuously differentiable.

Let us first look at Problem I. Assume that the following conditions are fulfilled.

$$\text{a)}\ \operatorname*{ess\,sup}_{(x,\,t)\in Q_T^{(k)}} \left| b^{(k)}, \frac{\partial b^{(k)}}{\partial x_j}, a_{ij}^{(k)}, \frac{\partial a_{ij}^{(k)}}{\partial x_i}, a_i^{(k)}, a^{(k)} \right| \leqslant M,$$

where $k = 1, 2$, and

$$\operatorname*{ess\,sup}_{(x,\,t)\in Q_T^{(2)}} \left| \frac{\partial b^{(2)}}{\partial t}, \frac{\partial^2 b^{(2)}}{\partial t^2}, \frac{\partial^2 b^{(2)}}{\partial t \partial x_j}, \frac{\partial a_{ij}^{(2)}}{\partial t}, \frac{\partial^2 a_{ij}^{(2)}}{\partial t^2}, \frac{\partial^2 a_{ij}^{(2)}}{\partial t \partial x_j}, \frac{\partial a_i^{(2)}}{\partial t}, \frac{\partial a^{(2)}}{\partial t} \right| \leqslant M;$$

b) $\varphi_0^{(2)}(x) \in \mathring{W}_2^2(\Omega_2)$, $\varphi_1^{(2)}(x) \in \mathring{W}_2^1(\Omega_2)$,

$$f_1^{(1)}(x, t) \in W_2^{0,1}(Q_T^{(1)}), \qquad f_3^{(2)}(x, t) \in W_2^{0,1}(Q_T^{(2)}),$$

where $W_2^{0,1}(Q_T^{(k)})$ is the Hilbert space with the scalar product

$$(u, v)_{W_2^{0,1}(Q_T^{(k)})} = \int_{Q_T^{(k)}} (uv + u_t v_t)\, dx dt, \; k = 1, 2.$$

In addition, let the function $f_1^{(1)}(x, 0)$ be such that the solution of the problem $L_1^{(1)} \hat{u}^{(1)} = f_1^{(1)}(x, 0)$, $\hat{u}^{(1)}|_{S_1} = 0$, belongs to the space $\mathring{W}_2^2(\Omega_1)$. The following proposition holds.

Lemma 3. *Let the hypotheses of Theorem 3 and conditions* a) *and* b) *be fulfilled. Then the generalized solution of Problem* I *has generalized derivatives* $\partial u^{(1)}/\partial t$ *and* $\partial^2 u^{(1)}/\partial t \partial x_i$ *in* $L_2(Q_T^{(1)})$, *and* $\partial^2 v^{(2)}/\partial t^2$ *and* $\partial^2 v^{(2)}/\partial t \partial x_i$ *in* $L_2(Q_T^{(2)})$. *For this solution the bound*

$$\begin{aligned} &\int_0^T \left[\left\| \frac{\partial u^{(1)}}{\partial t} \right\|_{W_2^1(\Omega_1)}^2 + \left\| \frac{\partial^2 v^{(2)}}{\partial t^2} \right\|_{L_2(\Omega_2)}^2 + \left\| \frac{\partial^2 v^{(2)}}{\partial t \partial x_i} \right\|_{L_2(\Omega_2)}^2 \right] dt \\ &\quad \leqslant C_{12} \left[\| f_1^{(1)} \|_{W_2^{0,1}(Q_T^{(1)})}^2 + \| f_3^{(2)} \|_{W_2^{0,1}(Q_T^{(2)})}^2 + \| \varphi_0^{(2)} \|_{W_2^2(\Omega_2)}^2 + \| \varphi_1^{(2)} \|_{W_2^1(\Omega_2)}^2 \right] \end{aligned} \tag{34}$$

holds, where the constant C_{12} *depends only on* M, n, T, α, β *and* ν.

Proof. We differentiate (21) with respect to t and obtain

$$\begin{aligned} &\int_{\Omega_2} \frac{\partial^3 v_N^{(2)}}{\partial t^3} b^{(2)}(x, t) \psi_m dx + \int_{\Omega_2} \frac{\partial^2 v_N^{(2)}}{\partial t^2} \frac{\partial b^{(2)}}{\partial t} \psi_m dx + \sum_{k=1}^{2} \mathscr{L}_1^{(k)} \left(\frac{\partial v_N^{(k)}}{\partial t}, \psi_m \right) \\ &\quad + \mathscr{L}_{1t}^{(2)}(v_N^{(2)}, \psi_m) - \int_{\Omega_1} b^{(1)} \frac{\partial f_1^{(1)}}{\partial t} \psi_m dx - \int_{\Omega_2} \frac{\partial b^{(2)}}{\partial t} f_3^{(2)} \psi_m dx \\ &\quad - \int_{\Omega_2} b^{(2)} \frac{\partial f_3^{(2)}}{\partial t} \psi_m dx + \mathscr{L}_1^{(1)} \left(\frac{\partial \hat{u}^{(1)}}{\partial t}, \psi_m \right) + \mathscr{L}_{1t}^{(2)}(\varphi_0^{(2)}, \psi_m) = 0, \end{aligned} \tag{35}$$

where

$$\mathscr{L}_{1t}^{(2)}(v, \psi_m) = \int_{\Omega_2} \Big[\frac{\partial}{\partial t}(b^{(2)} a_{ij}^{(2)}) \frac{\partial v}{\partial x_i} \cdot \frac{\partial \psi_m}{\partial x_j} + \frac{\partial}{\partial t}\Big(\frac{\partial b^{(2)}}{\partial x_j} a_{ij}^{(2)} + b^{(2)} a_i^{(2)} \Big) \frac{\partial v}{\partial x_i} \psi_m + \frac{\partial}{\partial t}(b^{(2)} a^{(2)}) v \psi_m \Big] dx.$$

Multiplying (35) by dc_m^N/dt, summing over m, and integrating over t from 0 to t_1, we find that

$$\int_0^{t_1}\int_{\Omega_2} b^{(2)}(x, t) \frac{\partial^3 v_N^{(2)}}{\partial t^3} \cdot \frac{\partial^2 v_N^{(2)}}{\partial t^2} dx dt + \int_0^{t_1}\int_{\Omega_2} \frac{\partial b^{(2)}}{\partial t} \Big(\frac{\partial^2 v_N^{(2)}}{\partial t^2} \Big)^2 dx dt$$
$$+ \int_0^{t_1} \mathscr{L}_1^{(1)} \Big(\frac{\partial v_N^{(1)}}{\partial t}, \frac{\partial v_N^{(1)}}{\partial t} \Big) dt + \int_0^{t_1} \mathscr{L}_1^{(2)} \Big(\frac{\partial v_N^{(2)}}{\partial t}, \frac{\partial^2 v_N^{(2)}}{\partial t^2} \Big) dt$$
$$+ \int_0^{t_1} \mathscr{L}_{1t}^{(2)} \Big(v_N^{(2)}, \frac{\partial^2 v_N^{(2)}}{\partial t^2} \Big) dt - \int_0^{t_1}\int_{\Omega_1} b^{(1)} \frac{\partial f_1^{(1)}}{\partial t} \cdot \frac{\partial v_N^{(1)}}{\partial t} dx dt$$
$$- \int_0^{t_1}\int_{\Omega_2} \frac{\partial b^{(2)}}{\partial t} f_3^{(2)} \frac{\partial^2 v_N^{(2)}}{\partial t^2} dx dt - \int_0^{t_1}\int_{\Omega_2} b^{(2)} \frac{\partial f_3^{(2)}}{\partial t} \cdot \frac{\partial^2 v_N^{(2)}}{\partial t^2} dx dt$$
$$+ \int_0^{t_1} \mathscr{L}_1^{(1)} \Big(\frac{\partial \hat{u}^{(1)}}{\partial t}, \frac{\partial v_N^{(1)}}{\partial t} \Big) dt + \int_0^{t_1} \mathscr{L}_{1t}^{(2)} \Big(\varphi_0^{(2)}, \frac{\partial^2 v_N^{(2)}}{\partial t^2} \Big) dt = 0.$$

This equality can be transformed into

$$\frac{1}{2} \int_{\Omega_2} b^{(2)} \Big(\frac{\partial^2 v_N^{(2)}}{\partial t^2} \Big)^2 dx \Big|_{t=t_1} + \int_0^{t_1} \mathscr{L}_1^{(1)} \Big(\frac{\partial v_N^{(1)}}{\partial t}, \frac{\partial v_N^{(1)}}{\partial t} \Big) dt$$
$$+ \frac{1}{2} \int_{\Omega_2} b^{(2)} a_{ij}^{(2)} \frac{\partial^2 v_N^{(2)}}{\partial t \partial x_i} \cdot \frac{\partial^2 v_N^{(2)}}{\partial t \partial x_j} dx \Big|_{t=t_1}$$
$$= \frac{1}{2} \int_{\Omega_2} \Big[b^{(2)} \Big(\frac{\partial^2 v_N^{(2)}}{\partial t^2} \Big)^2 + b^{(2)} a_{ij}^{(2)} \frac{\partial^2 v_N^{(2)}}{\partial t \partial x_i} \frac{\partial^2 v_N^{(2)}}{\partial t \partial x_j} \Big] dx \Big|_{t=0}$$
$$+ \frac{1}{2} \int_0^{t_1}\int_{\Omega_2} \Big\{ \frac{\partial}{\partial t}(b^{(2)} a_{ij}^{(2)}) \frac{\partial^2 v_N^{(2)}}{\partial t \partial x_i} \cdot \frac{\partial^2 v_N^{(2)}}{\partial t \partial x_j} -$$
$$- \Big[\Big(\frac{\partial b^{(2)}}{\partial x_j} a_{ij}^{(2)} + b^{(2)} a_i^{(2)} \Big) \frac{\partial^2 v_N^{(2)}}{\partial t \partial x_i} + b^{(2)} a^{(2)} \frac{\partial v_N^{(2)}}{\partial t} \Big] \frac{\partial^2 v_N^{(2)}}{\partial t^2} - \frac{\partial b^{(2)}}{\partial t} \Big(\frac{\partial^2 v_N^{(2)}}{\partial t^2} \Big)^2 \Big\} dx dt$$
$$- \int_0^{t_1} \mathscr{L}_{1t}^{(2)} \Big(v_N^{(2)}, \frac{\partial^2 v_N^{(2)}}{\partial t^2} \Big) dt + \int_0^{t_1}\int_{\Omega_1} b^{(1)} \frac{\partial f_1^{(1)}}{\partial t} \cdot \frac{\partial v_N^{(1)}}{\partial t} dx dt$$
$$+ \int_0^{t_1}\int_{\Omega_2} \Big(\frac{\partial b^{(2)}}{\partial t} f_3^{(2)} + b^{(2)} \frac{\partial f_3^{(2)}}{\partial t} \Big) \frac{\partial^2 v_N^{(2)}}{\partial t^2} dx dt$$
$$- \int_0^{t_1} \mathscr{L}_1^{(1)} \Big(\frac{\partial \hat{u}^{(1)}}{\partial t}, \frac{\partial v_N^{(1)}}{\partial t} \Big) dt - \int_0^{t_1} \mathscr{L}_{1t}^{(2)} \Big(\varphi_0^{(2)}, \frac{\partial^2 v_N^{(2)}}{\partial t^2} \Big) dt = \sum_{l=1}^{7} N_l. \quad (36)$$

It is obvious that

$$|N_1| \leqslant \frac{M}{2} \int\limits_{\Omega_2} \left(\frac{\partial^2 v_N^{(2)}}{\partial t^2} \right)^2 dx \bigg|_{t=0} + \frac{M^2 n}{2} \int\limits_{\Omega_2} \sum_{i=1}^{n} \left(\frac{\partial^2 v_N^{(2)}}{\partial t \partial x_i} \right)^2 dx \bigg|_{t=0}$$

and

$$|N_2| \leqslant \frac{M^2 (2n + (n+1)\sqrt{n}}{2} \int\limits_0^{t_1} \int\limits_{\Omega_2} \sum_{i=1}^{n} \left(\frac{\partial^2 v_N^{(2)}}{\partial t \partial x_i} \right)^2 dx dt$$

$$+ \frac{M^2((n+1)\sqrt{n}+1) + M}{2} \int\limits_0^{t_1} \int\limits_{\Omega_2} \left(\frac{\partial^2 v_N^{(2)}}{\partial t^2} \right)^2 dx dt + \frac{M^2}{2} \int\limits_0^{t_1} \int\limits_{\Omega_2} \left(\frac{\partial v_N^{(2)}}{\partial t} \right)^2 dx dt.$$

Here N_3 and N_7 can be bounded in analogous ways. For example, let us bound N_3. In order to do this, transform it in the following manner:

$$N_3 = -\int\limits_{\Omega_2} \frac{\partial}{\partial t} (b^{(2)} a_{ij}^{(2)}) \frac{\partial v_N^{(2)}}{\partial x_i} \cdot \frac{\partial^2 v_N^{(2)}}{\partial t \partial x_j} dx \,\Big|_{t=0}^{t=t_1}$$

$$+ \int\limits_0^{t_1} \int\limits_{\Omega_2} \left[\frac{\partial}{\partial t} (b^{(2)} a_{ij}^{(2)}) \frac{\partial^2 v_N^{(2)}}{\partial t \partial x_i} \frac{\partial^2 v_N^{(2)}}{\partial t \partial x_j} + \frac{\partial^2}{\partial t^2} (b^{(2)} a_{ij}^{(2)}) \frac{\partial^2 v_N^{(2)}}{\partial t \partial x_j} \cdot \frac{\partial v_N^{(2)}}{\partial x_i} \right] dx dt$$

$$- \int\limits_0^{t_1} \int\limits_{\Omega_2} \left[\frac{\partial}{\partial t} \left(\frac{\partial b^{(2)}}{\partial x_j} a_{ij}^{(2)} + b^{(2)} a_i^{(2)} \right) \frac{\partial v_N^{(2)}}{\partial x_i} \cdot \frac{\partial^2 v_N^{(2)}}{\partial t^2} + \frac{\partial}{\partial t} (b^{(2)} a^{(2)}) v_N^{(2)} \frac{\partial^2 v_N^{(2)}}{\partial t^2} \right] dx dt.$$

We have

$$|N_3| \leqslant \varepsilon_1 \int\limits_{\Omega_2} \sum_{i=1}^{n} \left(\frac{\partial^2 v_N^{(2)}}{\partial t \partial x_i} \right)^2 dx \,\Big|_{t=t_1} + \frac{M^4 n^2}{\varepsilon_1} \int\limits_{\Omega_2} \sum_{i=1}^{n} \left(\frac{\partial v_N^{(2)}}{\partial x_i} \right)^2 dx \,\Big|_{t=t_1}$$

$$+ 4M^2 n \int\limits_0^{t_1} \int\limits_{\Omega_2} \sum_{i=1}^{n} \left(\frac{\partial^2 v_N^{(2)}}{\partial t \partial x_i} \right)^2 dx dt + M^2 [(n+1)\sqrt{n}+1] \int\limits_0^{t_1} \int\limits_{\Omega_2} \left(\frac{\partial^2 v_N^{(2)}}{\partial t^2} \right)^2 dx dt$$

$$+ M^2 [(n+1)\sqrt{n} + 2n] \int\limits_0^{t_1} \| v_N^{(2)} \|^2_{W_2^1(\Omega_2)} dt.$$

Also

$$|N_4| \leqslant \varepsilon_2 \int\limits_0^{t_1} \int\limits_{\Omega_1} \left(\frac{\partial v_N^{(1)}}{\partial t} \right)^2 dx dt + \frac{M^2}{4\varepsilon_2} \int\limits_0^{t_1} \int\limits_{\Omega_1} \left(\frac{\partial f_1^{(1)}}{\partial t} \right)^2 dx dt,$$

$$|N_5| \leqslant M \int\limits_0^{t_1} \int\limits_{\Omega_2} \left(\frac{\partial^2 v_N^{(2)}}{\partial t^2} \right)^2 dx dt + \frac{M}{2} \int\limits_0^{t_1} \int\limits_{\Omega_2} \left[(f_3^{(2)})^2 + \left(\frac{\partial f_3^{(2)}}{\partial t} \right)^2 \right] dx dt.$$

In order to bound N_6, we make use of the inequality

$$\int\limits_0^{t_1} \left\| \frac{\partial \hat{u}^{(1)}}{\partial t} \right\|^2_{W_2^1(\Omega_1)} dt \leqslant \frac{M^2}{\nu^2} \int\limits_0^{t_1} \left\| \frac{\partial f_1^{(1)}}{\partial t} \right\|^2_{L_2(\Omega_1)} dt, \tag{37}$$

which, by virtue of (20), is valid for the generalized solution of the problem $L_1^{(1)} w = \partial f_1^{(1)}/\partial t$, $w|_{S_1} = 0$ in $W_2^1(\Omega_1)$, where $w = \partial \hat{u}^{(1)}/\partial t$ (see [7]). We obtain

$$|N_6| \leqslant \varepsilon_3 \int_0^{t_1} \left\| \frac{\partial v_N^{(1)}}{\partial t} \right\|^2_{W_2^1(\Omega_1)} dt + \frac{M^6 n}{4\varepsilon_3 \nu^2} (2(n+1)^2 + n) \int_0^{t_1} \left\| \frac{\partial f_1^{(1)}}{\partial t} \right\|^2_{L_2(\Omega_1)} dt.$$

Let $\epsilon_1 = \alpha\beta/8$ and $\epsilon_2 = \epsilon_3 = \nu/4$. Then (36) and the bounds we have obtained imply the inequality

$$\begin{aligned} &\int_{\Omega_2} \left(\frac{\partial^2 v_N^{(2)}}{\partial t^2} \right)^2 dx \,|_{t=t_1} + \int_{\Omega_2} \sum_{i=1}^n \left(\frac{\partial^2 v_N^{(2)}}{\partial t \partial x_i} \right)^2 dx \,|_{t=t_1} + \int_0^{t_1} \left\| \frac{\partial v_N^{(1)}}{\partial t} \right\|^2_{W_2^1(\Omega_1)} dt \\ &\qquad \leqslant C_{13} \left[\int_0^{t_1} \int_{\Omega_2} \left(\frac{\partial^2 v_N^{(2)}}{\partial t^2} \right)^2 dx dt + \int_0^{t_1} \int_{\Omega_2} \sum_{i=1}^n \left(\frac{\partial^2 v_N^{(2)}}{\partial t \partial x_i} \right)^2 dx dt \right] \\ &\qquad + C_{14} \left[\int_{\Omega_2} \left(\frac{\partial^2 v_N^{(2)}}{\partial t^2} \right)^2 dx \,|_{t=0} + \int_{\Omega_2} \sum_{i=1}^n \left(\frac{\partial^2 v_N^{(2)}}{\partial t \partial x_i} \right)^2 dx \,|_{t=0} \right] \qquad (38) \\ &\qquad + C_{15} \left[\| f_1^{(1)} \|^2_{W_2^{0,1}(Q_T^{(1)})} + \| f_3^{(2)} \|^2_{W_2^{0,1}(Q_T^{(2)})} + \| \varphi_0^{(2)} \|^2_{W_2^1(\Omega_2)} + \| \varphi_1^{(2)} \|^2_{L_2(\Omega_2)} \right], \end{aligned}$$

where the constants C_{13}, C_{14} and C_{15} depend only on M, n, T, α, β and ν. Here we have made use of (20) and (26).

It is easy to obtain a bound for the quantity

$$I = \int_{\Omega_2} \left(\frac{\partial^2 v_N^{(2)}}{\partial t^2} \right)^2 dx \,|_{t=0} + \int_{\Omega_2} \sum_{i=1}^n \left(\frac{\partial^2 v_N^{(2)}}{\partial t \partial x_i} \right)^2 dx \,|_{t=0}$$

from the initial data. Indeed, extend the function $\phi_1^{(2)}(x)$ by zero onto the whole domain Ω. Since $\phi_1^{(2)}(x) \in \overset{\circ}{W}{}_2^1(\Omega_2)$ (by virtue of the hypotheses of the lemma), it is easy to see that $\phi_1(x) \in H(\Omega)$, where $\phi_1(x) = 0$ for $x \in \Omega_1$ and $\phi_1(x) = \phi_1^{(2)}(x)$ for $x \in \Omega_2$. Hence, as the first of the basis functions $\psi_k(x)$, one can take the function

$$\psi_1(x) = \frac{\varphi_1(x)}{\left\| \sqrt{b^{(2)}(x, 0)}\, \varphi_1^{(2)}(x) \right\|_{L_2(\Omega_2)}}.$$

The initial conditions (22) take on the form

$$c_1^N(0) = \left\| \sqrt{b^{(2)}(x, 0)}\, \varphi_1^{(2)}(x) \right\|_{L_2(\Omega_2)}, \quad c_m^N(0) = 0, \qquad m = 2, 3, \ldots, N,$$

whence it follows that $v_N^{(1)}|_{t=0} = 0$. From (21) at $t = 0$, we obtain

$$\begin{aligned} &\int_{\Omega_2} b^{(2)} \frac{\partial^2 v_N^{(2)}}{\partial t^2} \psi_m dx \,|_{t=0} + \tilde{\mathscr{L}}_1^{(1)}(\hat{u}^{(1)}, \psi_m) - \int_{\Omega_1} b^{(1)} f_1^{(1)} \psi_m dx \\ &\qquad + \tilde{\mathscr{L}}_1^{(2)}(\varphi_0^{(2)}, \psi_m) - \int_{\Omega_2} b^{(2)} f_3^{(2)} \psi_m dx = 0. \qquad (39) \end{aligned}$$

From the hypotheses of the lemma, it follows that

$$\tilde{\mathscr{L}}_1^{(1)}(\hat{u}^{(1)}, \psi_m) = \int_{\Omega_1} b^{(1)} f_1^{(1)} \psi_m dx \,|_{t=0}.$$

Since $\phi_0^{(2)}(x) \in \overset{\circ}{W}{}_2^2(\Omega_2)$, equality (39) implies that

$$\int\limits_{\Omega_2} b^{(2)} \frac{\partial^2 v_N^{(2)}}{\partial t^2} \sum_{m=1}^{N} \frac{dc_m^N(t)}{dt} \psi_m(x)\, dx\,|_{t=0} = \int\limits_{\Omega_2} b^{(2)} \left(\frac{\partial^2 v_N^{(2)}}{\partial t^2}\right)^2 dx\,|_{t=0}$$

$$= \int\limits_{\Omega_2} b^{(2)} (-L_1^{(2)}\varphi_0^{(2)} + f_3^{(2)}) \frac{\partial^2 v_N^{(2)}}{\partial t^2}\, dx,$$

whence

$$\int\limits_{\Omega_2} \left(\frac{\partial^2 v_N^{(2)}}{\partial t^2}\right)^2 dx\,|_{t=0} \leqslant \frac{2M^4 n(n^2+1)}{\beta^2} \| \varphi_0^{(2)}(x) \|^2_{W_2^2(\Omega_2)} + \frac{2M^2}{\beta^2} \int\limits_{\Omega_2} [f_3^{(2)}(x,0)]^2\, dx. \quad (40)$$

Taking into account that

$$\left.\frac{\partial^2 v_N^{(2)}}{\partial t \partial x_i}\right|_{t=0} = \frac{\partial \varphi_1^{(2)}}{\partial x_i},$$

we obtain from (40) the bound of interest to us:

$$I \leqslant C_{16} \left[\| \varphi_0^{(2)}(x) \|^2_{W_2^2(\Omega_2)} + \| \varphi_1^{(2)} \|^2_{W_2^1(\Omega_2)} + \| f_3^{(2)} \|^2_{W_2^{0,1}(Q_T^{(2)})}\right]. \quad (41)$$

Just as in Theorem 3, from (38) and (41) we can obtain

$$\int_0^T \left[\left\| \frac{\partial^2 v_N^{(2)}}{\partial t^2} \right\|^2_{L_2(\Omega_2)} + \sum_{k=1}^{2} \left\| \frac{\partial v_N^{(k)}}{\partial t} \right\|^2_{W_2^1(\Omega_k)} \right] dt$$

$$\leqslant C_{17} \left[\| f_1^{(1)} \|^2_{W_2^{0,1}(Q_T^{(1)})} + \| f_3^{(2)} \|^2_{W_2^{0,1}(Q_T^{(2)})} + \| \varphi_0^{(2)} \|^2_{W_2^2(\Omega_2)} + \| \varphi_1^{(2)} \|^2_{W_2^1(\Omega_2)} \right]. \quad (42)$$

In (42), one can take the limit as $N \to \infty$ and conclude that the limit element $\{v^{(1)}, v^{(2)}\}$ has generalized derivatives $\partial v^{(1)}/\partial t$ and $\partial^2 v^{(1)}/\partial t \partial x_i$ and $L_2(Q_T^{(1)})$ and $\partial^2 v^{(2)}/\partial t^2$ and $\partial^2 v^{(2)}/\partial t \partial x_i$ in $L_2(Q_T^{(2)})$, and that (42) is valid for it. This is done in the same manner as in Theorem 3. Inequality (34) for the limit element $\{v^{(1)}, v^{(2)}\}$ follows from (37) and (42). The lemma is proved.

Lemma 4. *Let the hypotheses of Lemma 3 be fulfilled. Then the generalized solution of Problem* I *has generalized derivatives* $\partial^2 u^{(1)}/\partial x_i \partial x_j$ *in* $L_2(Q_T^{(1)})$ *and* $\partial^2 v^{(2)}/\partial x_i \partial x_j$ *in* $L_2(Q_T^{(2)})$, *and*

$$\int_0^T [\| u^{(1)} \|^2_{W_2^2(\Omega_1)} + \| v^{(2)} \|^2_{W_2^2(\Omega_2)}]\, dt$$

$$\leqslant C_{18} \left[\| f_1^{(1)} \|^2_{W_2^{0,1}(Q_T^{(1)})} + \| f_3^{(2)} \|^2_{W_2^{0,1}(Q_T^{(2)})} + \| \varphi_0^{(2)} \|^2_{W_2^2(\Omega_2)} + \| \varphi_1^{(2)} \|^2_{W_2^1(\Omega_2)} \right]. \quad (43)$$

Proof. The existence of square integrable generalized derivatives $\partial^2 u^{(1)}/\partial x_i \partial x_j$ inside the cylinder $Q_T^{(1)}$ and of square integrable generalized derivatives $\partial^2 v^{(2)}/\partial x_i \partial x_j$ inside $Q_T^{(2)}$ and near S_T can be established just as in [8] and [10]. Hence, in order to prove the lemma, it is sufficient to demonstrate the existence of the second derivatives of $u^{(1)}$ and $v^{(2)}$ with respect to the x_i near Γ_T.

In the integral identity (12), integrate by parts, transferring the derivative with respect to t from $b^{(2)}\psi^{(2)}$ to $\partial v^{(2)}/\partial t$. This is possible, since, by virtue of Lemma 3, the generalized solution of Problem I has the generalized derivative

$\partial^2 v^{(2)}/\partial t^2$ in $L_2(Q_T^{(2)})$. Also, in the identity obtained, set $\psi^{(k)}(x, t) = f(t)\psi^{(k)}(x)$, where $f(t) \in L_2(0, T)$ and where

$$\psi(x) = \begin{cases} \psi^{(1)}(x), & x \in \Omega_1, \\ \psi^{(2)}(x), & x \in \Omega_2, \end{cases}$$

is any function in $\overset{\circ}{W}{}_2^1(\Omega)$. We then have

$$\int_0^T \Big[\int_{\Omega_2} b^{(2)} \frac{\partial^2 v^{(2)}}{\partial t^2} \psi^{(2)} dx + \mathscr{L}_1^{(1)}(u^{(1)}, \psi^{(1)}) + \mathscr{L}_1^{(2)}(v^{(2)}, \psi^{(2)}) - \int_{\Omega_1} b^{(1)} f_1^{(1)} \psi^{(1)} dx$$
$$- \int_{\Omega_2} b^{(2)} f_3^{(2)} \psi^{(2)} dx + \mathscr{L}_1^{(2)}(\varphi_0^{(2)}, \psi^{(2)}) \Big] f(t)\, dt = 0.$$

This identity implies that, at almost all t,

$$\int_{\Omega_2} b^{(2)} \frac{\partial^2 v^{(2)}}{\partial t^2} \psi^{(2)} dx + \mathscr{L}_1^{(1)}(u^{(1)}, \psi^{(1)}) + \mathscr{L}_1^{(2)}(v^{(2)}, \psi^{(2)}) + \mathscr{L}_1^{(2)}(\varphi_0^{(2)}, \psi^{(2)})$$
$$- \int_{\Omega_1} b^{(1)} f_1^{(1)} \psi^{(1)} dx - \int_{\Omega_2} b^{(2)} f_3^{(2)} \psi^{(2)} dx = 0. \tag{44}$$

The surface Γ is twice continuously differentiable, and therefore can be covered by a finite number of overlapping domains, each of which can be mapped onto a part of the plane by a one-to-one, twice continuously differentiable transformation $y = y(x)$. In this transformation, none of the required properties of equations (1) and (2) are destroyed, and the equations retain the same form. Hence, without loss of generality, we will assume that, in the x-coordinates, one of these $(n-1)$-dimensional domains $\Gamma_1 \subset \Gamma$ is already in the plane $x_n = 0$.

As $\psi(x)$, take the function

$$\psi(x) = \begin{cases} \left[\zeta(x) \sum_{l=1}^{n-1} \frac{\partial^2}{\partial x_l^2} (\zeta u_\rho^{(1)}) \right]_\rho, & x \in \Omega_1, \\ \left[\zeta(x) \sum_{l=1}^{n-1} \frac{\partial^2}{\partial x_l^2} \left(\zeta \frac{\partial v_\rho^{(2)}}{\partial t} \right) \right]_\rho, & x \in \Omega_2, \end{cases}$$

where $\zeta(x)$ is the twice continuously differentiable cut-off function given by

$$\zeta(x) = \begin{cases} 1, & x \in Q' = \Gamma_1' \times \left[-\frac{\delta}{2} \leqslant x_n \leqslant \frac{\delta}{2} \right], \\ 0, & x \notin Q'' = \Gamma_1'' \times \left[-\frac{3\delta}{4} \leqslant x_n \leqslant \frac{3\delta}{4} \right], \end{cases} \quad \Gamma_1 \supset \Gamma_1'' \supset \Gamma_1'.$$

Let $Q^{(1)} = \Gamma_1 \times [-\delta \leq x_n < 0] \subset \Omega_1$, $Q^{(2)} = \Gamma_1 \times [0 < x_n \leq \delta] \subset \Omega_2$. The symbol f_ρ designates the average of f over $x' = (x_1, \ldots, x_{n-1})$,

$$f_\rho(x) = \frac{1}{\rho^{n-1}} \int_{|x'-y'| \leqslant \rho} \omega\left(\frac{|x'-y'|}{\rho} \right) f(y', x_n)\, dy'.$$

Take the radius ρ to be less than the distance from Q'' to S. This is possible, since $\psi(x) \in \overset{\circ}{W}{}_2^1(\Omega)$. Substitute $\psi(x)$ into (44). Integrate the equality thus obtained over t from 0 to t_1 and transform it as follows:

$$\sum_{l=1}^{n-1}\sum_{i,j=1}^{n}\int_0^{t_1}\int_{\Omega_1} b^{(1)}a_{ij}^{(1)}\frac{\partial^2\zeta u_\rho^{(1)}}{\partial x_i\partial x_l}\frac{\partial^2\zeta u_\rho^{(1)}}{\partial x_j\partial x_l}\,dxdt + \frac{1}{2}\sum_{l=1}^{n-1}\sum_{i,j=1}^{n}\int_{\Omega_2} b^{(2)}a_{ij}^{(2)}\frac{\partial^2\zeta v_\rho^{(2)}}{\partial x_i\partial x_l}\cdot\frac{\partial^2\zeta v_\rho^{(2)}}{\partial x_j\partial x_l}\,dx\,|_{t=t_1} = R(t_1), \tag{45}$$

where

$$\begin{aligned}
R(t_1) = {}& \sum_{l=1}^{n-1}\sum_{i,j=1}^{n}\int_0^{t_1}\int_{\Omega_1}\left(b^{(1)}a_{ij}^{(1)}\frac{\partial u^{(1)}}{\partial x_i}\right)_\rho\frac{\partial\zeta}{\partial x_j}\cdot\frac{\partial^2}{\partial x_l^2}(\zeta u_\rho^{(1)})\,dxdt \\
&+\sum_{l=1}^{n-1}\sum_{i,j=1}^{n}\int_0^{t_1}\int_{\Omega_1}\left(b^{(1)}a_{ij}^{(1)}\frac{\partial^2\zeta u_\rho^{(1)}}{\partial x_i\partial x_l} - \zeta b^{(1)}a_{ij}^{(1)}\frac{\partial^2 u_\rho^{(1)}}{\partial x_i\partial x_l}\right)\frac{\partial^2\zeta u_\rho^{(1)}}{\partial x_j\partial x_l}\,dxdt \\
&+\sum_{l=1}^{n-1}\sum_{i,j=1}^{n}\int_0^{t_1}\int_{\Omega_1}\zeta\left[b^{(1)}a_{ij}^{(1)}\frac{\partial^2 u_\rho^{(1)}}{\partial x_i\partial x_l} - \frac{\partial}{\partial x_l}\left(b^{(1)}a_{ij}^{(1)}\frac{\partial u^{(1)}}{\partial x_i}\right)_\rho\right]\frac{\partial^2\zeta u_\rho^{(1)}}{\partial x_j\partial x_l}\,dxdt \\
&-\sum_{l=1}^{n-1}\sum_{i,j=1}^{n}\int_0^{t_1}\int_{\Omega_1}\frac{\partial\zeta}{\partial x_l}\left(b^{(1)}a_{ij}^{(1)}\frac{\partial u^{(1)}}{\partial x_i}\right)_\rho\cdot\frac{\partial^2\zeta u_\rho^{(1)}}{\partial x_j\partial x_l}\,dxdt \\
&+\int_0^{t_1}\int_{\Omega_1}\left[\sum_{i=1}^{n}\left(\sum_{j=1}^{n}\frac{\partial b^{(1)}}{\partial x_j}a_{ij}^{(1)} + b^{(1)}a_i^{(1)}\right)\frac{\partial u^{(1)}}{\partial x_i} + b^{(1)}a^{(1)}u^{(1)}\right]\left[\zeta\sum_{l=1}^{n-1}\frac{\partial^2}{\partial x_l^2}(\zeta u_\rho^{(1)})\right]_\rho dxdt \\
&+\sum_{l=1}^{n-1}\sum_{i,j=1}^{n}\int_0^{t_1}\int_{\Omega_2}\left(b^{(2)}a_{ij}^{(2)}\frac{\partial v^{(2)}}{\partial x_i}\right)_\rho\frac{\partial\zeta}{\partial x_j}\cdot\frac{\partial^2}{\partial x_l^2}\left(\zeta\frac{\partial v_\rho^{(2)}}{\partial t}\right)dxdt + \\
&+\frac{1}{2}\sum_{l=1}^{n-1}\sum_{i,j=1}^{n}\int_0^{t_1}\int_{\Omega_2}\frac{\partial b^{(2)}a_{ij}^{(2)}}{\partial t}\cdot\frac{\partial^2\zeta v_\rho^{(2)}}{\partial x_i\partial x_l}\frac{\partial^2\zeta v_\rho^{(2)}}{\partial x_j\partial x_l}\,dxdt \\
&+\sum_{l=1}^{n-1}\sum_{i,j=1}^{n}\int_0^{t_1}\int_{\Omega_2} b^{(2)}a_{ij}^{(2)}\left[\frac{\partial^2\zeta v_\rho^{(2)}}{\partial x_i\partial x_l} - \zeta\frac{\partial^2 v_\rho^{(2)}}{\partial x_i\partial x_l}\right]\cdot\frac{\partial^3(\zeta v_\rho^{(2)})}{\partial t\partial x_j\partial x_l}\,dxdt \\
&+\sum_{l=1}^{n-1}\sum_{i,j=1}^{n}\int_0^{t_1}\int_{\Omega_2}\zeta\left[b^{(2)}a_{ij}^{(2)}\frac{\partial^2 v_\rho^{(2)}}{\partial x_i\partial x_l} - \frac{\partial}{\partial x_l}\left(b^{(2)}a_{ij}^{(2)}\frac{\partial v^{(2)}}{\partial x_i}\right)_\rho\right]\cdot\frac{\partial^3(\zeta v_\rho^{(2)})}{\partial t\partial x_j\partial x_l}\,dxdt \\
&-\sum_{l=1}^{n-1}\sum_{i,j=1}^{n}\int_0^{t_1}\int_{\Omega_2}\frac{\partial\zeta}{\partial x_l}\left(b^{(2)}a_{ij}^{(2)}\frac{\partial v^{(2)}}{\partial x_i}\right)_\rho\cdot\frac{\partial^3(\zeta v_\rho^{(2)})}{\partial t\partial x_j\partial x_l}\,dxdt \\
&+\int_0^{t_1}\int_{\Omega_2}\left[\sum_{i=1}^{n}\left(\sum_{j=1}^{n}\frac{\partial b^{(2)}}{\partial x_j}a_{ij}^{(2)} + b^{(2)}a_i^{(2)}\right)\frac{\partial v^{(2)}}{\partial x_i} + b^{(2)}a^{(2)}v^{(2)}\right] \\
&\times\left[\zeta\sum_{l=1}^{n-1}\frac{\partial^2}{\partial x_l^2}\left(\zeta\frac{\partial v_\rho^{(2)}}{\partial t}\right)\right]_\rho dxdt + \int_0^{t_1}\int_{\Omega_2} b^{(2)}\frac{\partial^2 v^{(2)}}{\partial t^2}\left[\zeta\sum_{l=1}^{n-1}\frac{\partial^2}{\partial x_l^2}\left(\zeta\frac{\partial v_\rho^{(2)}}{\partial t}\right)\right]_\rho dxdt \\
&+\int_0^{t_1}\mathscr{L}_1^{(2)}\left(\varphi_0^{(2)}, \left[\zeta\sum_{l=1}^{n-1}\frac{\partial^2}{\partial x_l^2}\left(\zeta\frac{\partial v_\rho^{(2)}}{\partial t}\right)\right]_\rho\right)dt - \int_0^{t_1}\int_{\Omega_1} b^{(1)}f_1^{(1)}\left[\zeta\sum_{l=1}^{n-1}\frac{\partial^2}{\partial x_l^2}(\zeta u_\rho^{(1)})\right]_\rho dxdt \\
&-\int_0^{t_1}\int_{\Omega_2} b^{(2)}f_3^{(2)}\left[\zeta\sum_{l=1}^{n-1}\frac{\partial^2}{\partial x_l^2}\left(\zeta\frac{\partial v_\rho^{(2)}}{\partial t}\right)\right]_\rho dxdt \equiv \sum_{m=1}^{15} R_m(t_1).
\end{aligned}$$

Let us show that

$$|R(t_1)| \leqslant \frac{\alpha\beta}{2} \sum_{l=1}^{n-1} \sum_{i=1}^{n} \int_0^{t_1} \left\| \frac{\partial^2 \zeta u_\rho^{(1)}}{\partial x_i \partial x_l} \right\|_{L_2(\Omega_1)}^2 dt + \frac{\alpha\beta}{4} \sum_{l=1}^{n-1} \sum_{i=1}^{n} \left\| \frac{\partial^2 \zeta v_\rho^{(2)}}{\partial x_i \partial x_l} \right\|_{L_2(\Omega_2)}^2 \Bigg|_{t=t_1}$$

$$+ C_{19} \sum_{l=1}^{n-1} \sum_{i=1}^{n} \int_0^{t_1} \left\| \frac{\partial^2 \zeta v_\rho^{(2)}}{\partial x_i \partial x_l} \right\|_{L_2(\Omega_2)}^2 dt \tag{46}$$

$$+ C_{20}(\alpha, \beta) \left[\| f_1^{(1)} \|_{W_2^{0,1}(Q_T^{(1)})}^2 + \| f_3^{(2)} \|_{W_2^{0,1}(Q_T^{(2)})}^2 + \| \varphi_0^{(2)} \|_{W_2^2(\Omega_2)}^2 + \| \varphi_1^{(2)} \|_{W_2^1(\Omega_2)}^2 \right].$$

Indeed, as is easy to verify, for $R_1(t_1)$, $R_2(t_1)$, $R_4(t_1)$ and $R_5(t_1)$ the bounds

$$|R_m(t_1)| \leqslant \varepsilon_1 \sum_{l=1}^{n-1} \sum_{i=1}^{n} \int_0^{t_1} \left\| \frac{\partial^2 \zeta u_\rho^{(1)}}{\partial x_i \partial x_l} \right\|_{L_2(\Omega_1)}^2 dt + C_{21}(\varepsilon_1) \int_0^{t_1} \| u^{(1)} \|_{W_2^1(\Omega_1)}^2 dt, \tag{47}$$
$$m = 1, 2, 4, 5.$$

are valid. Let us now bound $R_3(t_1)$. We have

$$b^{(1)} a_{ij}^{(1)} \frac{\partial^2 u_\rho^{(1)}}{\partial x_i \partial x_l} - \frac{\partial}{\partial x_l} \left(b^{(1)} a_{ij}^{(1)} \frac{\partial u^{(1)}}{\partial x_i} \right)_\rho = \left(b^{(1)} a_{ij}^{(1)} \frac{\partial^2 u_\rho^{(1)}}{\partial x_i \partial x_l} - \frac{\partial}{\partial x_l} b^{(1)} a_{ij}^{(1)} \frac{\partial u_\rho^{(1)}}{\partial x_i} \right)$$
$$+ \left[\frac{\partial}{\partial x_l} b^{(1)} a_{ij}^{(1)} \frac{\partial u_\rho^{(1)}}{\partial x_i} - \frac{\partial}{\partial x_l} \left(b^{(1)} a_{ij}^{(1)} \frac{\partial u^{(1)}}{\partial x_i} \right)_\rho \right].$$

It is easy to see that

$$\int_0^{t_1} \sum_{l=1}^{n-1} \sum_{i,j=1}^{n} \left\| b^{(1)} a_{ij}^{(1)} \frac{\partial^2 u_\rho^{(1)}}{\partial x_i \partial x_l} - \frac{\partial}{\partial x_l} b^{(1)} a_{ij}^{(1)} \frac{\partial u_\rho^{(1)}}{\partial x_i} \right\|_{L_2(\Omega_1)}^2 dt \leqslant C_{22} \int_0^{t_1} \| u^{(1)} \|_{W_2^1(\Omega_1)}^2 dt. \tag{48}$$

Also,

$$\frac{\partial}{\partial x_l} b^{(1)} a_{ij}^{(1)} \frac{\partial u_\rho^{(1)}}{\partial x_i} - \frac{\partial}{\partial x_l} \left(b^{(1)} a_{ij}^{(1)} \frac{\partial u^{(1)}}{\partial x_i} \right)_\rho = \frac{1}{\rho^{n-1}} \int_{|x'-y'| \leqslant \rho} [b^{(1)}(x, t) a_{ij}^{(1)}(x, t)$$
$$- b^{(1)}(y', x_n, t) a_{ij}^{(1)}(y', x_n, t)] \cdot \frac{\partial \omega\left(\frac{|x'-y'|}{\rho}\right)}{\partial x_l} \cdot \frac{\partial u^{(1)}}{\partial y_i} dy' \tag{49}$$
$$+ \frac{\partial (b^{(1)} a_{ij}^{(1)})}{\partial x_l} \cdot \frac{1}{\rho^{n-1}} \int_{|x'-y'| \leqslant \rho} \omega\left(\frac{|x'-y'|}{\rho}\right) \frac{\partial u^{(1)}}{\partial y_i} dy'.$$

Bounding the norm of the right side of (49) and using the bound

$$|b^{(1)}(x, t) a_{ij}^{(1)}(x, t) - b^{(1)}(y', x_n, t) a_{ij}^{(1)}(y', x_n, t)| \leqslant 2M^2 \sqrt{n-1}\, |x' - y'|$$

and the boundedness of the averaging operator with the kernel

$$\frac{|x'-y'|}{\rho^{n-1}} \frac{\partial \omega\left(\frac{|x'-y'|}{\rho}\right)}{\partial x_l}$$

in $L_2(\Omega_1)$, we obtain a bound of type (48) for the left side of (49). From this it is easy to obtain a bound of type (47) for $R_3(t_1)$. The quantities $R_6(t_1)$, $R_8(t_1)$, $R_9(t_1)$, $R_{10}(t_1)$ and $R_{11}(t_1)$ can be bounded in the same manner (to bound $R_9(t_1)$ it is necessary only to repeat the reasoning carried out when bounding $R_3(t_1)$). For example, let us bound $R_6(t_1)$. In order to do this, transform it by integration by parts as follows:

$$R_6(t_1)=\sum_{l=1}^{n-1}\sum_{i,\,j=1}^{n}\int\limits_{\Omega_2}\left(b^{(2)}a_{ij}^{(2)}\frac{\partial v^{(2)}}{\partial x_i}\right)_\rho\cdot\frac{\partial\zeta}{\partial x_j}\cdot\frac{\partial^2(\zeta v_\rho^{(2)})}{\partial x_l^2}\,dx\Big|_{t=t_1}$$
$$-\sum_{l=1}^{n-1}\sum_{i,\,j=1}^{n}\int\limits_0^{t_1}\int\limits_{\Omega_2}\frac{\partial}{\partial t}\left(b^{(2)}a_{ij}^{(2)}\frac{\partial v^{(2)}}{\partial x_i}\right)_\rho\cdot\frac{\partial\zeta}{\partial x_j}\cdot\frac{\partial^2(\zeta v_\rho^{(2)})}{\partial x_l^2}\,dxdt.$$

It is easy to see that

$$|R_6(t_1)|\leqslant\varepsilon_2\sum_{l=1}^{n-1}\sum_{i=1}^{n}\left\|\frac{\partial^2(\zeta v_\rho^{(2)})}{\partial x_i\partial x_l}\right\|_{L_2(\Omega_2)}^2\Big|_{t=t_1}+C_{23}(\varepsilon_2)\|v^{(2)}(x,\ t_1)\|_{W_2^1(\Omega_2)}^2$$
$$+C_{24}\sum_{l=1}^{n-1}\sum_{i=1}^{n}\int\limits_0^{t_1}\left\|\frac{\partial^2\zeta v_\rho^{(2)}}{\partial x_i\partial x_l}\right\|_{L_2(\Omega_2)}^2dt+C_{25}\int\limits_0^{t_1}\left\|\frac{\partial v^{(2)}}{\partial t}\right\|_{W_2^1(\Omega_2)}^2dt.$$

It is obvious that

$$|R_7(t_1)|\leqslant C_{26}\int\limits_0^{t_1}\sum_{l=1}^{n-1}\sum_{i=1}^{n}\left\|\frac{\partial^2(\zeta v_\rho^{(2)})}{\partial x_i\partial x_l}\right\|_{L_2(\Omega_2)}^2dt.$$

Let us now bound $R_{12}(t_1)$. We have

$$R_{12}(t_1)=-\sum_{l=1}^{n-1}\int\limits_0^{t_1}\int\limits_{\Omega_2}\frac{\partial\zeta}{\partial x_l}\cdot\left(b^{(2)}\frac{\partial^2v^{(2)}}{\partial t^2}\right)_\rho\cdot\frac{\partial^2\zeta v_\rho^{(2)}}{\partial x_l\partial t}\,dxdt$$
$$-\frac{1}{2}\sum_{l=1}^{n-1}\int\limits_{\Omega_2}b^{(2)}\left(\frac{\partial^2\zeta v_\rho^{(2)}}{\partial x_l\partial t}\right)^2dx\Big|_{t=0}^{t=t_1}+\frac{1}{2}\sum_{l=1}^{n-1}\int\limits_0^{t_1}\int\limits_{\Omega_2}\frac{\partial b^{(2)}}{\partial t}\left(\frac{\partial^2\zeta v_\rho^{(2)}}{\partial t\partial x_l}\right)^2dxdt+$$
$$+\sum_{l=1}^{n-1}\int\limits_0^{t_1}\int\limits_{\Omega_2}\left[b^{(2)}\frac{\partial^3\zeta v_\rho^{(2)}}{\partial t^2\partial x_l}-\zeta b^{(2)}\frac{\partial^3v_\rho^{(2)}}{\partial t^2\partial x_l}\right]\cdot\frac{\partial^2\zeta v_\rho^{(2)}}{\partial t\partial x_l}\,dxdt$$
$$+\sum_{l=1}^{n-1}\int\limits_0^{t_1}\int\limits_{\Omega_2}\zeta\left[b^{(2)}\frac{\partial^3v_\rho^{(2)}}{\partial t^2\partial x_l}-\frac{\partial}{\partial x_l}\left(b^{(2)}\frac{\partial^2v^{(2)}}{\partial t^2}\right)_\rho\right]\frac{\partial^2\zeta v_\rho^{(2)}}{\partial x_l\partial t}\,dxdt.$$

Using reasoning analogous to that used above, we obtain

$$|R_{12}(t_1)|\leqslant C_{27}\int\limits_0^{t_1}\left\|\frac{\partial^2v^{(2)}}{\partial t^2}\right\|_{L_2(\Omega_2)}^2dt+C_{28}\int\limits_0^{t_1}\left\|\frac{\partial v^{(2)}}{\partial t}\right\|_{W_2^1(\Omega_2)}^2dt$$
$$+C_{29}\left\|\frac{\partial v^{(2)}}{\partial t}\right\|_{W_2^1(\Omega_2)}^2\Big|_{t=t_1}+C_{30}\|\varphi_1^{(2)}\|_{W_2^1(\Omega_2)}^2.$$

Note that, by virtue of Lemma 3,

$$\frac{\partial w}{\partial t}=\left\{\frac{\partial u^{(1)}}{\partial t},\ \frac{\partial v^{(2)}}{\partial t}\right\}\in V^1.$$

Hence

$$\operatorname*{ess\,sup}_{0\leqslant t\leqslant T}\left\|\frac{\partial v^{(2)}(x,\ t)}{\partial t}\right\|_{W_2^1(\Omega_2)}^2\leqslant C_{31}\left\|\frac{\partial w}{\partial t}\right\|_{V^1}^2\leqslant C_{32}\Big[\|f_1^{(1)}\|_{W_2^{0,1}(Q_T^{(1)})}^2$$
$$+\|f_3^{(2)}\|_{W_2^{0,1}(Q_T^{(2)})}^2+\|\varphi_0^{(2)}\|_{W_2^2(\Omega_2)}^2+\|\varphi_1^{(2)}\|_{W_2^1(\Omega_2)}^2\Big]$$

(cf. §1). Let us proceed to bounding $R_{13}(t_1)$. We have

$$R_{13}(t_1)=\int\limits_{\Omega_2}\Big\{-\frac{\partial}{\partial x_j}\Big(b^{(2)}a_{ij}^{(2)}\frac{\partial\varphi_0^{(2)}}{\partial x_i}\Big)_\rho+\Big[\Big(\frac{\partial b^{(2)}}{\partial x_j}a_{ij}^{(2)}+b^{(2)}a_i^{(2)}\Big)\frac{\partial\varphi_0^{(2)}}{\partial x_i}\Big]_\rho$$

$$+(b^{(2)}a^{(2)}\varphi_0^{(2)})_\rho\Big\}\cdot\zeta\sum_{l=1}^{n-1}\frac{\partial^2\zeta v_\rho^{(2)}}{\partial x_l^2}dx\Big|_{t=t_1}-\int\limits_0^{t_1}\int\limits_{\Omega_2}\Big\{-\frac{\partial}{\partial x_j}\Big(\frac{\partial(b^{(2)}a_{ij}^{(2)})}{\partial t}\cdot\frac{\partial\varphi_0^{(2)}}{\partial x_i}\Big)_\rho$$

$$+\Big[\frac{\partial}{\partial t}\Big(\frac{\partial b^{(2)}}{\partial x_j}a_{ij}^{(2)}+b^{(2)}a_i^{(2)}\Big)\frac{\partial\varphi_0^{(2)}}{\partial x_i}\Big]_\rho+\Big[\frac{\partial(b^{(2)}a^{(2)})}{\partial t}\varphi_0^{(2)}\Big]_\rho\Big\}\zeta\sum_{l=1}^{n-1}\frac{\partial^2\zeta v_\rho^{(2)}}{\partial x_l^2}dxdt.$$

Hence

$$|R_{13}(t_1)|\leqslant\varepsilon_3\sum_{l=1}^{n-1}\sum_{i=1}^{n}\Big\|\frac{\partial^2\zeta v_\rho^{(2)}}{\partial x_i\partial x_l}\Big\|_{L_2(\Omega_2)}^2\Big|_{t=t_1}+C_{33}(\varepsilon_3)\|\varphi_0^{(2)}\|_{W_2^2(\Omega_2)}^2$$
$$+C_{34}\int\limits_0^{t_1}\sum_{l=1}^{n-1}\sum_{i=1}^{n}\Big\|\frac{\partial^2\zeta v_\rho^{(2)}}{\partial x_i\partial x_l}\Big\|_{L_2(\Omega_2)}^2dt.$$

It is easy to verify that

$$|R_{14}(t_1)|\leqslant\varepsilon_4\int\limits_0^{t_1}\sum_{l=1}^{n-1}\sum_{i=1}^{n}\Big\|\frac{\partial^2\zeta u_\rho^{(1)}}{\partial x_i\partial x_l}\Big\|_{L_2(\Omega_1)}^2dt+C_{35}(\varepsilon_4)\|f_1^{(1)}\|_{L_2(Q_T^{(1)})}^2,$$

$$|R_{15}(t_1)|\leqslant\varepsilon_5\sum_{l=1}^{n-1}\sum_{i=1}^{n}\Big\|\frac{\partial^2\zeta v_\rho^{(2)}}{\partial x_i\partial x_l}\Big\|_{L_2(\Omega_2)}^2\Big|_{t=t_1}+C_{36}\int\limits_0^{t_1}\sum_{l=1}^{n-1}\sum_{i=1}^{n}\Big\|\frac{\partial^2\zeta v_\rho^{(2)}}{\partial x_i\partial x_l}\Big\|_{L_2(\Omega_2)}^2dt$$
$$+C_{37}(\varepsilon_5)\|f_3^{(2)}\|_{W_2^{0,1}(Q_T^{(2)})}^2.$$

Combining these inequalities and using (27) and (34), already proved, we obtain (46).

(45) and (46) imply the inequality

$$\sum_{l=1}^{n-1}\sum_{i=1}^{n}\int\limits_0^{t_1}\Big\|\frac{\partial^2\zeta u_\rho^{(1)}}{\partial x_i\partial x_l}\Big\|_{L_2(\Omega_1)}^2dt+\sum_{l=1}^{n-1}\sum_{i=1}^{n}\Big\|\frac{\partial^2\zeta v_\rho^{(2)}}{\partial x_i\partial x_l}\Big\|_{L_2(\Omega_2)}^2\Big|_{t=t_1}$$
$$\leqslant C_{38}\sum_{l=1}^{n-1}\sum_{i=1}^{n}\int\limits_0^{t_1}\Big\|\frac{\partial^2\zeta v_\rho^{(2)}}{\partial x_i\partial x_l}\Big\|_{L_2(\Omega_2)}^2dt+C_{39}\big[\|f_1^{(1)}\|_{W_2^{0,1}(Q_T^{(1)})}^2$$
$$+\|f_3^{(2)}\|_{W_2^{0,1}(Q_T^{(2)})}^2+\|\varphi_0^{(2)}\|_{W_2^2(\Omega_2)}^2+\|\varphi_1^{(2)}\|_{W_2^1(\Omega_2)}^2\big].$$

Using the same reasoning as in the proof of Theorem 3, we obtain

$$\sum_{l=1}^{n-1}\sum_{i=1}^{n}\int\limits_0^{T}\Big[\Big\|\frac{\partial^2\zeta u_\rho^{(1)}}{\partial x_i\partial x_l}\Big\|_{L_2(\Omega_1)}^2+\Big\|\frac{\partial^2\zeta v_\rho^{(2)}}{\partial x_i\partial x_l}\Big\|_{L_2(\Omega_2)}^2\Big]dt$$
$$\leqslant C_{40}\Big[\|f_1^{(1)}\|_{W_2^{0,1}(Q_T^{(1)})}^2+\|f_3^{(2)}\|_{W_2^{0,1}(Q_T^{(2)})}^2+\|\varphi_0^{(2)}\|_{W_2^2(\Omega_2)}^2+\|\varphi_1^{(2)}\|_{W_2^1(\Omega_2)}^2\Big],\quad(50)$$

where C_{40} does not depend on ρ.

From this we conclude that, near Γ_T, all generalized second derivatives except $\partial^2u^{(1)}/\partial x_n^2$ and $\partial^2v^{(2)}/\partial x_n^2$ exist and that they satisfy an inequality of type (50). But then it follows from (1) and (2) that $\partial^2u^{(1)}/\partial x_n^2$ and $\partial^2v^{(2)}/\partial x_n^2$ also exist and satisfy the same inequality. The lemma is proved.

The following theorem ensues from Lemmas 4 and 3.

Theorem 5. *Let the hypotheses of Lemma 3 be fulfilled. Then the generalized solution of Problem* I *has generalized derivatives*

$$\frac{\partial u^{(1)}}{\partial t},\ \frac{\partial^2 u^{(1)}}{\partial t \partial x_i},\ \frac{\partial^2 u^{(1)}}{\partial x_i \partial x_j} \in L_2(Q_T^{(1)})$$

and

$$\frac{\partial^2 v^{(2)}}{\partial t^2},\ \frac{\partial^2 v^{(2)}}{\partial t \partial x_i},\ \frac{\partial^2 v^{(2)}}{\partial x_i \partial x_j} \in L_2(Q_T^{(2)}).$$

It satisfies equations (1) *and* (2) *almost everywhere in* $Q_T^{(1)}$ *and* $Q_T^{(2)}$, *respectively, and it satisfies conditions* (3) *and* (8) *in the sense of the Sobolev imbedding theorems. Moreover,*

$$\int_0^T \Big[\Big\| \frac{\partial u^{(1)}}{\partial t} \Big\|^2_{W_2^1(\Omega_1)} + \Big\| \frac{\partial v^{(2)}}{\partial t} \Big\|^2_{W_2^1(\Omega_2)} + \sum_{i,j=1}^n \Big\| \frac{\partial^2 u^{(1)}}{\partial x_i \partial x_j} \Big\|^2_{L_2(\Omega_1)} + \sum_{i,j=1}^n \Big\| \frac{\partial^2 v^{(2)}}{\partial x_i \partial x_j} \Big\|^2_{L_2(\Omega_2)}$$
$$+ \Big\| \frac{\partial^2 v^{(2)}}{\partial t^2} \Big\|^2_{L_2(\Omega_2)} \Big] dt \leqslant C_{41} \big[\| f_1^{(1)} \|^2_{W_2^{0,1}(Q_T^{(1)})} + \| f_3^{(2)} \|^2_{W_2^{0,1}(Q_T^{(2)})}$$
$$+ \| \varphi_0^{(2)} \|^2_{W_2^2(\Omega_2)} + \| \varphi_1^{(2)} \|^2_{W_2^1(\Omega_2)} \big].$$

An analogous theorem holds for Problem II.

Theorem 6. *Let the hypotheses of Theorem 4 be fulfilled, and assume that the given problems satisfy the following conditions:*

a) *The functions*

$$b^{(k)},\ \frac{\partial b^{(k)}}{\partial x_j},\ \frac{\partial b^{(k)}}{\partial t},\ \frac{\partial^2 b^{(2)}}{\partial t^2},\ \frac{\partial^2 b^{(k)}}{\partial t \partial x_j},\qquad a_{ij}^{(k)},\ \frac{\partial a_{ij}^{(k)}}{\partial t},\ \frac{\partial a_{ij}^{(k)}}{\partial x_i},\ \frac{\partial^2 a_{ij}^{(2)}}{\partial t^2},\ \frac{\partial^2 a_{ij}^{(2)}}{\partial t \partial x_j},$$
$$a_i^{(k)},\ \frac{\partial a_i^{(k)}}{\partial t},\ a^{(k)},\ \frac{\partial a^{(k)}}{\partial t}$$

are bounded and measurable in $Q_T^{(k)}$, $k = 1, 2$.

b) $f_2^{(1)}(x, t) \in W_2^{0,1}(Q_T^{(1)}),\ f_3^{(2)}(x, t) \in W_2^{0,1}(Q_T^{(2)}),$

$$\varphi_0^{(1)}(x) \in \mathring{W}_2^2(\Omega_1),\ \varphi_0^{(2)}(x) \in \mathring{W}_2^2(\Omega_2),\ \varphi_1^{(2)}(x) \in \mathring{W}_2^1(\Omega_2).$$

Then the generalized solution of Problem II *has generalized derivatives*

$$\frac{\partial u^{(1)}}{\partial t},\ \frac{\partial^2 u^{(1)}}{\partial t \partial x_i},\ \frac{\partial^2 u^{(1)}}{\partial x_i \partial x_j} \in L_2(Q_T^{(1)})$$

and

$$\frac{\partial^2 v^{(2)}}{\partial t^2},\ \frac{\partial^2 v^{(2)}}{\partial t \partial x_i},\ \frac{\partial^2 v^{(2)}}{\partial x_i \partial x_j} \in L_2(Q_T^{(2)}),$$

It satisfies equations (6) *and* (2) *almost everywhere in* $Q_T^{(1)}$ *and* $Q_T^{(2)}$, *respectively, and also satisfies conditions* (7), (3) *and* (8). *For this solution, the following inequality is valid:*

$$\int_0^T \Big[\Big\| \frac{\partial u^{(1)}}{\partial t} \Big\|^2_{W_2^1(\Omega_1)} + \Big\| \frac{\partial v^{(2)}}{\partial t} \Big\|^2_{W_2^1(\Omega_2)} + \sum_{i,j=1}^n \Big(\Big\| \frac{\partial^2 u^{(1)}}{\partial x_i \partial x_j} \Big\|^2_{L_2(\Omega_1)} + \Big\| \frac{\partial^2 v^{(2)}}{\partial x_i \partial x_j} \Big\|^2_{L_2(\Omega_2)} \Big)$$
$$+ \Big\| \frac{\partial^2 v^{(2)}}{\partial t^2} \Big\|^2_{L_2(\Omega_2)} \Big] dt \leqslant C_{42} \big[\| f_2^{(1)} \|^2_{W_2^{0,1}(Q_T^{(1)})} + \| f_3^{(2)} \|^2_{W_2^{0,1}(Q_T^{(2)})}$$
$$+ \| \varphi_0^{(1)} \|^2_{W_2^2(\Omega_1)} + \| \varphi_0^{(2)} \|^2_{W_2^2(\Omega_2)} + \| \varphi_1^{(2)} \|^2_{W_2^1(\Omega_2)} \big].$$

§5. Problem I with coefficients that depend on t

Let us consider the case when the coefficients of the elliptic operator in equation (1) of Problem I depend on t. Without loss of generality, we will assume that $\phi_0^{(2)}(x) = \phi_1^{(2)}(x) \equiv 0$.

Let us introduce the following notation. Let W be the Hilbert space with the scalar product

$$(u, v)_W = \sum_{k=1}^{2} \int_{Q_T^{(k)}} u^{(k)} v^{(k)}\, dxdt;$$

Let V_0^1 be the Hilbert space consisting of all elements of V^1 that satisfy the condition $u^{(2)}|_{t=0} = 0$. Let W_T^{-1} be the space of linear functionals on W_T^1.

The space of generalized functions corresponds to the space W_T^{-1}. It is isometric to W_T^1. Every linear functional $l(u)$ on W_T^1 can be represented in a unique way by the formula $l(u) = (\alpha, u)_w$, where $\alpha = \{\alpha^{(1)}, \alpha^{(2)}\}$ is some fixed element of W_T^{-1} (see [4]).

Let $D(A)$ be the set of all elements $u = \{u^{(1)}, u^{(2)}\}$ in $W - L_2(Q_T^{(1)}) \times L_2(Q_T^{(2)})$ that have all generalized derivatives up to and including second order with respect to the x_i in $L_2(Q_T^{(1)})$ and $L_2(Q_T^{(2)})$, respectively, that have generalized derivatives $\partial^2 u^{(2)}/\partial t^2$ and $\partial^2 u^{(2)}/\partial t \partial x_i$ in $L_2(Q_T^{(2)})$, and that satisfy the conditions

$$u^{(k)}\big|_{S_T^{(k)} \setminus \Gamma_T} = 0,\ k = 1, 2,\ u^{(1)}\big|_{\Gamma_T} = \frac{\partial u^{(2)}}{\partial t}\Big|_{\Gamma_T},\ b^{(1)} \frac{\partial u^{(1)}}{\partial N^{(1)}}\Big|_{\Gamma_T} = b^{(2)} \frac{\partial u^{(2)}}{\partial N^{(2)}}\Big|_{\Gamma_T}, \tag{51}$$

$$u^{(2)}\big|_{t=0} = 0,\quad u_t^{(2)}\big|_{t=0} = 0. \tag{52}$$

Define the operator A on this set in the following manner:

$$Au = \{\mathcal{L}_1^{(1)} u^{(1)},\ (x, t) \in Q_T^{(1)};\ \mathcal{L}_3^{(2)} u^{(2)},\ (x, t) \in Q_T^{(2)}\}. \tag{53}$$

Let us now deal with the construction of an extension of the operator A, which is defined on $D(A)$. On the elements $u = \{u^{(1)}, u^{(2)}\}$ of V_0^1, define the operator $\tilde{A}$ by

$$(\tilde{A}u, \psi)_W = -\int_{Q_T^{(2)}} \frac{\partial u^{(2)}}{\partial t} \cdot \frac{\partial (b^{(2)}\psi^{(2)})}{\partial t}\, dxdt + \sum_{k=1}^{2} \int_0^T \tilde{\mathcal{L}}_1^{(k)} (u^{(k)}, \psi^{(k)})\, dt,\ \psi \in W_T^1. \tag{54}$$

Defining the norm of $\tilde{A}u$ to be the norm of the functional $(\tilde{A}u, \psi)_W$ on W_T^1, we can write

$$\|\tilde{A}u\|_{W_T^{-1}} = \sup_{\psi \in W_T^1} \frac{|(\tilde{A}u, \psi)_W|}{\|\psi\|_{W_T^1}}.$$

If

$$\operatorname*{ess\,sup}_{(x, t) \in Q_T^{(k)}} \left| b^{(k)}, \frac{\partial b^{(2)}}{\partial t}, \frac{\partial b^{(k)}}{\partial x_j}, a_{ij}^{(k)}, a_i^{(k)}, a^{(k)} \right| \leqslant M,$$

then the functional $(\tilde{A}u, \psi)_W$ is bounded, since $|(\tilde{A}u, \psi)_W| \le C_{43} \|u\|_{V^1} \|\psi\|_{W_T^1}$, where C_{43} depends only on M. It is easy to see that the operator $\tilde{A}$ is an extension of A.

Lemma 5. *Assume that the following conditions are satisfied:*

a) $\operatorname*{ess\,sup}_{(x,\,t)\in Q_T^{(k)}} \left| b^{(k)}, \frac{\partial b^{(k)}}{\partial x_j}, \frac{\partial b^{(2)}}{\partial t}, a_{ij}^{(k)}, \frac{\partial a_{ij}^{(2)}}{\partial t}, a_i^{(k)}, a^{(k)} \right| \leqslant M,$

where $k = 1, 2$.

b) *For any function* $u^{(1)} \in W^1_{2,0}(\Omega_1)$ *the inequality*

$$\int_{\Omega_1} \left\{ b^{(1)} a_{ij}^{(1)} \frac{\partial u^{(1)}}{\partial x_i} \cdot \frac{\partial u^{(1)}}{\partial x_j} + \left(\frac{\partial b^{(1)}}{\partial x_j} a_{ij}^{(1)} + b^{(1)} a_i^{(1)} \right) \frac{\partial u^{(1)}}{\partial x_i} u^{(1)} + b^{(1)} a^{(1)} [u^{(1)}]^2 \right\} dx \geqslant \nu \| u^{(1)} \|^2_{W_2^1(\Omega_1)} \tag{55}$$

is valid, where $\nu = \text{const} > 0$.

Then the range $R(\widetilde{A})$ *of* $\widetilde{A}$ *is dense in* W_T^{-1}.

Proof. If $R(\widetilde{A})$ is not dense in W_T^{-1}, then an element $\psi = \{\psi^{(1)}, \psi^{(2)}\}$ of W_T^1 can be found such that, for any $u = \{u^{(1)}, u^{(2)}\} \in V_0^1$,

$$-\int_{Q_T^{(2)}} \frac{\partial u^{(2)}}{\partial t} \cdot \frac{\partial (b^{(2)} \psi^{(2)})}{\partial t} dx dt + \sum_{k=1}^{2} \int_0^T \mathscr{L}_1^{(k)} (u^{(k)}, \psi^{(k)})\, dt = 0. \tag{56}$$

As the $u = \{u^{(1)}, u^{(2)}\}$ in (56), take the element

$$u(x, t) = \begin{cases} 0, & t \in [0, h], \\ \hat{\psi}(x, t), & t \in [h, T], \end{cases} \tag{57}$$

where

$$\hat{\psi}(x, t) = \left\{ \psi^{(1)}(x, t), \int_h^t \psi^{(2)}(x, \tau)\, d\tau \right\}.$$

It is an admissible element, since $u(x, t) \in V_0^1$. Substituting $u(x, t)$ into (56) and integrating by parts, we transform the identity in the following manner:

$$\frac{1}{2} \int_{\Omega_2} b^{(2)} \left(\frac{\partial \Phi^{(2)}}{\partial t} \right)^2 dx \Big|_{t=h} + \int_h^T \mathscr{L}_1^{(1)} (\psi^{(1)}, \psi^{(1)})\, dt + \frac{1}{2} \int_{\Omega_2} b^{(2)} a_{ij}^{(2)} \frac{\partial \Phi^{(2)}}{\partial x_i} \cdot \frac{\partial \Phi^{(2)}}{\partial x_j} dx \Big|_{t=h}$$
$$= \frac{1}{2} \int_h^T \int_{\Omega_2} \frac{\partial b^{(2)}}{\partial t} \left(\frac{\partial \Phi^{(2)}}{\partial t} \right)^2 dx dt + \frac{1}{2} \int_h^T \int_{\Omega_2} \frac{\partial (b^{(2)} a_{ij}^{(2)})}{\partial t} \cdot \frac{\partial \Phi^{(2)}}{\partial x_i} \cdot \frac{\partial \Phi^{(2)}}{\partial x_j} dx dt$$
$$- \int_h^T \int_{\Omega_2} \frac{\partial (b^{(2)} a_{ij}^{(2)})}{\partial t} \cdot \frac{\partial \Phi^{(2)}(x, h)}{\partial x_i} \frac{\partial \Phi^{(2)}(x, t)}{\partial x_j} dx dt$$
$$- \int_h^T \int_{\Omega_2} \left\{ \left(\frac{\partial b^{(2)}}{\partial x_j} a_{ij}^{(2)} + b^{(2)} a_i^{(2)} \right) \frac{\partial}{\partial x_i} [\Phi^{(2)}(x, t) - \Phi^{(2)}(x, h)] \cdot \frac{\partial \Phi^{(2)}}{\partial t} + b^{(2)} a^{(2)} [\Phi^{(2)}(x, t) - \Phi^{(2)}(x, h)] \cdot \frac{\partial \Phi^{(2)}}{\partial t} \right\} dx dt,$$

where

$$\Phi^{(2)}(x, t) = \int_T^t \psi^{(2)}(x, \tau)\, d\tau.$$

This implies

$$\int_{\Omega_2} \left(\frac{\partial \Phi^{(2)}}{\partial t} \right)^2 dx \Big|_{t=h} + \sum_{i=1}^{n} \int_{\Omega_2} \left(\frac{\partial \Phi^{(2)}}{\partial x_i} \right)^2 dx \Big|_{t=h} + \int_h^T \| \psi^{(1)} \|^2_{W_2^1(\Omega_1)} dt$$
$$\leqslant C_{44} \int_h^T \left[\left\| \frac{\partial \Phi^{(2)}}{\partial t} \right\|^2_{L_2(\Omega_2)} + \sum_{i=1}^{n} \left\| \frac{\partial \Phi^{(2)}}{\partial x_i} \right\|^2_{L_2(\Omega_2)} \right] dt, \tag{58}$$

where C_{44} depends only on M, n, T, α, β and ν. This inequality implies that

$$-\frac{dy(h)}{dh} \leqslant C_{44} y(h),$$

where

$$y(h) = \int_h^T \left[\left\| \frac{\partial \Phi^{(2)}}{\partial t} \right\|^2_{L_2(\Omega_2)} + \sum_{i=1}^n \left\| \frac{\partial \Phi^{(2)}}{\partial x_i} \right\|^2_{L_2(\Omega_2)} \right] dt.$$

Therefore

$$-\frac{d}{dh}\left[y(h)\, e^{C_{44}h}\right] \leqslant 0.$$

Integrating over h and taking into account that $y(T) = 0$, we find that $y(h)e^{C_{44}h} \leq 0$, which implies that $y(h) = 0$. But then, since $\Phi^{(2)}(x, t) = \int^t \psi^{(2)}(x, \tau)\, d\tau$, we have $\psi^{(2)}(x, t) = 0$ in $Q_T^{(2)}$. From (58) it follows that also $\psi^{(1)}(x, t)^T = 0$ in $Q_T^{(1)}$. From this we conclude that the set $R(\widetilde{A})$ is dense in W_T^{-1}. The lemma is proved.

Lemma 6. *Let the hypothesis of Lemma 5 be fulfilled. Then*

$$\int_{Q_T^{(2)}} \left(\frac{\partial u^{(2)}}{\partial t}\right)^2 dx\,dt + \sum_{k=1}^{2} \int_{Q_T^{(k)}} \left[\sum_{i=1}^n \left(\frac{\partial u^{(k)}}{\partial x_i}\right)^2 + (u^{(k)})^2\right] dx\,dt$$
$$\leqslant C_{45}\left[\|\mathcal{L}_1^{(1)} u^{(1)}\|^2_{L_2(Q_T^{(1)})} + \|\mathcal{L}_3^{(2)} u^{(2)}\|^2_{L_2(Q_T^{(2)})}\right] \equiv C_{45}\|Au\|^2_W \tag{59}$$

for all $u(x, t) \in D(A)$.

Proof. We have

$$\frac{1}{2}\int_{\Omega_2} b^{(2)}\left(\frac{\partial u^{(2)}}{\partial t}\right)^2 dx\Big|_{t=t_1} + \frac{1}{2}\int_{\Omega_2} b^{(2)} a_{ij}^{(2)} \frac{\partial u^{(2)}}{\partial x_i}\cdot\frac{\partial u^{(2)}}{\partial x_j}\, dx\Big|_{t=t_1} + \int_0^{t_1} \widetilde{\mathcal{L}}_1^{(1)}(u^{(1)}, u^{(1)})\, dt$$
$$= \frac{1}{2}\int_0^{t_1}\int_{\Omega_2} \frac{\partial b^{(2)}}{\partial t}\left(\frac{\partial u^{(2)}}{\partial t}\right)^2 dx\,dt + \frac{1}{2}\int_0^{t_1}\int_{\Omega_2} \frac{\partial (b^{(2)} a_{ij}^{(2)})}{\partial t}\cdot\frac{\partial u^{(2)}}{\partial x_i}\cdot\frac{\partial u^{(2)}}{\partial x_j}\, dx\,dt$$
$$-\int_0^{t_1}\int_{\Omega_2}\left[\left(\frac{\partial b^{(2)}}{\partial x_j} a_{ij}^{(2)} + b^{(2)} a_i^{(2)}\right)\frac{\partial u^{(2)}}{\partial x_i} + b^{(2)} a^{(2)} u^{(2)}\right]\frac{\partial u^{(2)}}{\partial t}\, dx\,dt$$
$$+\int_0^{t_1}\int_{\Omega_1} b^{(1)} \mathcal{L}_1^{(1)} u^{(1)}\cdot u^{(1)}\, dx\,dt + \int_0^{t_1}\int_{\Omega_2} b^{(2)} \mathcal{L}_3^{(2)} u^{(2)}\cdot\frac{\partial u^{(2)}}{\partial t}\, dx\,dt.$$

This implies

$$\int_{\Omega_2}\left(\frac{\partial u^{(2)}}{\partial t}\right)^2 dx\Big|_{t=t_1} + \sum_{i=1}^n \int_{\Omega_2}\left(\frac{\partial u^{(2)}}{\partial x_i}\right)^2 dx\Big|_{t=t_1} + \int_0^{t_1}\|u^{(1)}(x, t)\|^2_{W_2^1(\Omega_1)}\, dt$$
$$\leqslant C_{46}\int_0^{t_1}\int_{\Omega_2}\left[\left(\frac{\partial u^{(2)}}{\partial t}\right)^2 + \sum_{i=1}^n\left(\frac{\partial u^{(2)}}{\partial x_i}\right)^2\right] dx\,dt$$
$$+ C_{47}\left[\|\mathcal{L}_1^{(1)} u^{(1)}\|^2_{L_2(Q_T^{(1)})} + \|\mathcal{L}_3^{(2)} u^{(2)}\|^2_{L_2(Q_T^{(2)})}\right].$$

Inequality (59) is derived from this inequality in exactly the same way as that used in proving Theorem 3. The lemma is proved.

Lemma 6 implies that the operator A has a closed extension $\overline{A}$, and that $\overline{A} \subseteq \tilde{A}$. Indeed, if a sequence of elements $\{u_n\}$ in $D(A)$ is such that $\|A(u_n - u_m)\|_W \to 0$ as $n, m \to \infty$, then

$$\left\| \frac{\partial (u_n^{(2)} - u_m^{(2)})}{\partial t} \right\|^2_{L_2(Q_T^{(2)})} + \sum_{k=1}^{2} \int_0^T \| u_n^{(k)} - u_m^{(k)} \|^2_{W_2^1(\Omega_k)} dt \to 0$$

(cf. (59)) and, certainly, then $\|u_n - u_m\|_W \to 0$. From this we can conclude that closing the operator $A\cdot$ is equivalent to closing the range $R(A)$ of A, and that the elements $u(x, t)$ by which $D(A)$ is completed belong to V_0^1. The following theorem ensues directly from this fact and from Lemmas 5 and 6.

Theorem 7. *The operator A has a closed extension $\overline{A}$ such that $R(\overline{A}) = \overline{R(A)} = W$, and $\overline{A}$ has a bounded inverse.*

Thus the unique solvability of the operator equation $\overline{A}u = \{f_1^{(1)}, f_3^{(2)}\}$ for any $f_1^{(1)}(x, t) \in L_2(Q_T^{(1)})$ and any $f_3^{(2)}(x, t) \in L_2(Q_T^{(2)})$ has been proved. The solution $u(x, t)$ of this equation belongs to V_0^1.

The author expresses his deep gratitude to Professor O. A. Ladyženskaja for her critical evaluation of the results of this article and for valuable suggestions.

BIBLIOGRAPHY

1. O. A. Ladyženskaja and L. Stupjalis, *Equations of mixed type*, Vestnik Leningrad. Univ. 20 (1965), no. 19, 38–46. (Russian) MR 35 #3275.

2. L. Stupjalis, *A spectral problem for equations of mixed type*, Vestnik Leningrad. Univ. 22 (1967), no. 13, 86–102. (Russian) MR 36 #4175.

3. O. A. Ladyženskaja and L. Stupjalis, *Boundary value problems for equations of mixed type*, Trudy Mat. Inst. Steklov. 116 (1971), 101–136 = Proc. Steklov Inst. Math. 116 (1971), 102–139 (1973).

4. L. Stupjalis, *Boundary value problems for elliptic-hyperbolic equations*, Trudy Mat. Inst. Steklov. 125 (1973), 211–229 = Proc. Steklov Inst. Math. 125 (1973), 200–217 (1975). MR 50 #789.

5. H. Cohen and S. I. Rubinow, *Some mathematical topics in biology*, Proc. Sympos. System Theory (New York, 1965), Microwave Res. Inst. Sympos. Ser., Vol. 15, Polytechnic Press, Polytechnic Inst. Brooklyn, Brooklyn, N. Y.; Wiley, New York, 1965, pp. 321–337. MR 41 #5077.

6. L. N. Slobodeckiĭ, *Generalized Sobolev spaces and their application to boundary value problems for partial differential equations*, Leningrad. Gos. Ped. Inst. Učen. Zap. 197 (1958), 54–112; English transl., Amer. Math. Soc. Transl. (2) 57 (1966), 207–276. MR 34 #3075.

7. M. I. Višik and O. A. Ladyženskaja, *Boundary value problems for partial differential equations and certain classes of operator equations*, Uspehi Mat. Nauk 11 (1956), no. 6 (72), 41–97; English transl., Amer. Math. Soc. Transl. (2) 10 (1958), 223–281. MR 20 #1091.

8. O. A. Ladyženskaja and N. N. Ural′ceva, *Linear and quasi-linear equations of elliptic type*, "Nauka", Moscow, 1964; English transl., Academic Press, New York, 1968. MR 35 #1955; 39 #5941.

9. O. A. Ladyženskaja, V. A. Solonnikov and N. N. Ural′ceva, *Linear and quasi-linear equations of parabolic type*, "Nauka", Moscow, 1967; English transl., Amer. Math. Soc., Providence, R. I., 1968. MR 39 #3159a, b.

10. O. A. Ladyženskaja, *The mixed problem for a hyperbolic equation*, GITTL, Moscow, 1953. (Russian) MR **17**, 160.

Translated by J. LAVERY

Trudy Mat. Inst. Steklov
127 (1975)

Proc. Steklov Inst. Math.
127 (1975)

THE GENERALIZED NEUMANN PROBLEM FOR SECOND-ORDER ELLIPTIC EQUATIONS IN THE PLANE

UDC 517.919

A. L. TRESKUNOV

Abstract. A boundary value problem with oblique derivative (with homogeneous boundary condition) is considered for a two-dimensional elliptic equation. In the case of a linear equation an a priori estimate of the solution is obtained in Morrey space. A corollary of this result is an a priori estimate of the Hölder constant for the first derivatives of the solutions of a quasilinear equation satisfying the indicated boundary condition. The existence of a classical solution of this problem is proved under continuity conditions on the coefficients and conditions of weak growth of the right side of the equation.

Bibliography: 3 titles.

In this paper we consider an elliptic equation of the form $au_{xx} + 2bu_{xy} + cu_{yy} = f$, $(x, y) \in \Omega$, where Ω is a bounded region with a sufficiently smooth boundary. For the case of the first boundary value problem the properties of solutions of this equation have been throughly studied for $(x, y) \in \Omega$ and for $(x, y) \in \bar{\Omega}$ under minimal assumptions on the coefficients and the right side. A thorough exposition of these results, with references, may be found in [1].

We shall be interested in the smoothness up to the boundary of the solution of the "problem with directional derivative" (the generalized Neumann problem) for this equation: suppose that on $\partial\Omega$ we are given a smooth vector field $\mathbf{l}$; the problem is to find a solution of the equation satisfying the condition $\partial u/\partial l|_{\partial\Omega} = 0$.

As a characteristic of $\mathbf{l}$ and $\partial\Omega$ we introduce below the $\mathbf{l}$-curvature $K_l(s)$ of the boundary at the point s, and we shall say that the boundary is $\mathbf{l}$-convex if $K_l(s) \geq 0$ for all s.

Under the condition of $\mathbf{l}$-convexity for a linear equation with bounded measurable coefficients and $f(x, y) \in L_{2+\epsilon}(\Omega)$ we shall show that the derivatives of the solutions of the boundary value problem with directional derivative belong to a certain Morrey space, and are therefore Hölder continuous up to the boundary.

If in fact we replace the requirement of $\mathbf{l}$-convexity by the condition $(\mathbf{l}, \mathbf{n}) \geq \delta > 0$, where $\mathbf{n}$ is the exterior normal to $\partial\Omega$, while δ is some number, then the solution of the above-mentioned boundary value problem for a quasilinear elliptic equation which satisfies certain standard restrictions (compare with [1]) exists and has continuous second derivatives.

AMS (MOS) subject classifications (1970). Primary 35J25, 35J60, 35B45.

For the study of the solutions of problems with directional derivatives we shall follow H. O. Cordes [2] and first obtain the necessary estimate for the simplest equation with constant coefficients—Poisson's equation; and the equation with variable coefficients will be treated as a "perturbation" of Poisson's equation.

Concepts and notation which are not defined in the text are standard, and can be found for example in [1]. The symbols C and δ with indices will designate constants that are defined in terms of known quantities.

I wish to express my deep appreciation to Professor N. N. Ural'ceva for her attention to this work and her valuable advice.

§1. An estimate for the solution of the generalized Neumann problem in the space $W_2^2(\Omega)$

Let Ω be a bounded region in R^2 and suppose that on $\partial\Omega$ a vector field $\mathbf{l}$ is given. We consider the generalized homogeneous Neumann problem for Poisson's equation in Ω:

$$\Delta u = f(x, y), \quad (x, y) \in \Omega, \tag{1}$$

$$\left.\frac{\partial u}{\partial l}\right|_{\partial\Omega} = 0. \tag{2}$$

Lemma 1. *Suppose that $\partial\Omega$ and $\mathbf{l}$ belong to the class C^2, that $f(x, y)$ belongs to $L_2(\Omega)$, and that $u(x, y)$ belongs to $W_2^2(\Omega)$. Then*

$$\iint_\Omega [u_{xx}^2 + 2u_{xy}^2 + u_{yy}^2]\,dxdy = \iint_\Omega f^2(x, y)\,dxdy - \int_{\partial\Omega} K_l(s)\,|\nabla u|^2\,ds, \tag{3}$$

where $K_l(s) = \lim_{\Delta s\to 0} \widehat{\mathbf{l}(s)\mathbf{l}(s+\Delta s)}/\Delta s$ is the $\mathbf{l}$-curvature of $\partial\Omega$, s is the arc length on $\partial\Omega$, and the sign of the angle is defined according to the usual geometric rule.

Proof. It is sufficient to prove that (3) holds for $u(x, y)$ in $C^3(\Omega)$. After proving this we can prove the lemma by a standard limit procedure. Therefore, let $u(x, y)$ belong to $C^3(\Omega)$.

By Green's formula

$$\iint_\Omega [u_{xx}^2 + 2u_{xy}^2 + u_{yy}^2]\,dxdy = \iint_\Omega (u_{xx} + u_{yy})^2\,dxdy + \int_{\partial\Omega} \left[\frac{1}{2}(u_x^2 + u_y^2)_n - \Delta u \cdot u_n\right] ds, \tag{4}$$

where $(\cdot)_n$ is the derivative in the direction of the exterior normal to $\partial\Omega$ at the point of integration.

We examine the expression under the integral sign in the last term, in a neighborhood of a fixed point on the boundary (we shall assume that this point corresponds to $s = s_0$). For convenience we introduce new rectangular coordinates with origin at this fixed point, such that the axis OY is in the direction of $\mathbf{l}(s_0)$. Since $I(s_0) \equiv [\frac{1}{2}(u_x^2 + u_y^2)_n - \Delta u \cdot u_n]_{s=s_0}$ is invariant under orthogonal transformations of the independent variables, we can keep the same formula in the new coordinates.

Using the boundary condition, we have

$$\begin{aligned} I(s_0) &= u_x (u_x)_n - u_{xx}u_n - u_{yy}u_n \\ &= u_x u_{xx} \cos nx + u_x u_{xy} \cos ny - u_{xx}u_x \cos nx - u_{yy}u_x \cos nx \\ &= u_x [(u_y)_x \cos ny - (u_y)_y \cos nx] = -u_x(s_0)(u_y)_s(s_0), \end{aligned}$$

where $(\cdot)_s$ is the derivative with respect to arc length on $\partial\Omega$, or, equivalently, in the direction of the tangent pointed toward increasing s. Naturally, $\cos ny = -\cos sx$ and $\cos nx = \cos sy$.

We compute $(u_y)_s(s_0)$ according to the definition:

$$(u_y)_s(s_0) = \lim_{\Delta s\to 0} \frac{u_y(s_0+\Delta s) - u_y(s_0)}{\Delta s} = \lim_{\Delta s\to 0} \frac{u_y(\hat{s_0} + \Delta s)}{\Delta s} =$$

$$= \lim_{\Delta s\to 0} \frac{u_t(s_0+\Delta s) \cos \widehat{\mathbf{t}(s_0+\Delta s)\, \mathbf{l}(s_0)}}{\Delta s} = u_x(s_0) \lim_{\Delta s\to 0} \frac{\sin \widehat{\mathbf{l}(s_0)\ \ \mathbf{l}(s_0+\Delta s)}}{\Delta s}$$

$$= u_x(s_0) K_l(s_0).$$

Here we have used the notation $\mathbf{t}(s)$ for a vector perpendicular to $\mathbf{l}(s)$, and we have also used the continuity of the first derivatives of $u(x, y)$ on the boundary.

Finally, $I(s) = -|\nabla u|^2 K_l(s)$. Substituting this into (4), we obtain the desired inequality (3).

Definition 1. A region Ω is called $\mathbf{l}$-*convex* if $K_l(s) \geq 0$ on $\partial\Omega$. (If $\mathbf{l} = \mathbf{n}$, then an $\mathbf{l}$-convex region is simply convex).

Corollary. *If a region* Ω *is* $\mathbf{l}$-*convex, then Lemma* 1 *yields an a priori estimate for the solution of the problem* (1), (2):

$$\iint_\Omega [u_{xx}^2 + 2u_{xy}^2 + u_{yy}^2]\,dxdy \leqslant \iint_\Omega f^2(x, y)\,dxdy.$$

Now suppose that we are given an elliptic equation in Ω

$$\mathscr{L}[u] \equiv a(x, y)u_{xx} + 2b(x, y)u_{xy} + c(x, y)u_{yy} = f(x, y), \tag{5}$$

where $a(x, y)$, $b(x, y)$ and $c(x, y)$ are bounded measurable functions in Ω satisfying the ellipticity condition

$$\nu(\xi^2 + \eta^2) \leqslant a\xi^2 + 2b\xi\eta + c\eta^2 \leqslant \mu(\xi^2 + \eta^2), \quad \nu > 0. \tag{6}$$

Theorem 1. *Suppose that* Ω *and* $\mathbf{l}$ *belong to the class* C^2, *and assume that* $f(x, y) \in L_2(\Omega)$ *and* $u(x, y) \in W_2^2(\Omega)$. *Then the following estimate holds for the solution of the boundary value problem* (5), (2):

$$\iint_\Omega [u_{xx}^2 + 2u_{xy}^2 + u_{yy}^2]\,dxdy \leqslant C_1 \iint_\Omega f^2 dxdy + C_2 \int_{\Omega} \frac{|K_l(s)| - K_l(s)}{2} |\nabla u|^2 ds,$$

in which C_1 *and* C_2 *depend only on* ν *and* μ *from* (6).

Proof. We write $a + c = T$. Then

$$\|\Delta u\|_{L_2(\Omega)} \leqslant \left\|\frac{\mathcal{L}[u]}{T}\right\|_{L_2(\Omega)} + \left\|\frac{\mathcal{L}[u]}{T} - \Delta u\right\|_{L_2(\Omega)}. \tag{7}$$

For the last term on the right side we have

$$\left\|\frac{\mathcal{L}[u]}{T} - \Delta u\right\|^2_{L_2(\Omega)} = \iint_\Omega \left[\left(\frac{a}{T} - 1\right)u_{xx} + 2\frac{b}{T}u_{xy} + \left(\frac{c}{T} - 1\right)u_{yy}\right]^2 dxdy$$

$$\leqslant \max_{(x,y)\in\Omega}\left[\left(\frac{a}{T} - 1\right)^2 + \left(\frac{b}{T}\right)^2 + \left(\frac{c}{T} - 1\right)^2\right] \iint_\Omega [u_{xx}^2 + 2u_{xy}^2 + u_{yy}^2]\, dxdy.$$

Let λ_1 and λ_2 be the eigenvalues of the matrix $\binom{a\ b}{b\ c}$. Then

$$\left(\frac{a}{T} - 1\right)^2 + 2\left(\frac{b}{T}\right)^2 + \left(\frac{c}{T} - 1\right)^2 = \left(\frac{\lambda_1}{\lambda_1 + \lambda_2} - 1\right)^2 + \left(\frac{\lambda_2}{\lambda_1 + \lambda_2} - 1\right)^2 = 1 - \frac{2\lambda_1\lambda_2}{(\lambda_1 + \lambda_2)^2}.$$

But since it follows from (6) that $\nu \leq \lambda_1, \lambda_2 \leq \mu$, we conclude that

$$\max_{(x,y)\in\Omega}\left[\left(\frac{a}{T} - 1\right)^2 + \left(\frac{b}{T}\right)^2 + \left(\frac{c}{T} - 1\right)^2\right] \leqslant 1 - \frac{2\mu\nu}{(\nu + \mu)^2}.$$

This, together with (7) and Lemma 1, gives

$$\iint_\Omega [u_{xx}^2 + 2u_{xy}^2 + u_{yy}^2]\, dxdy + \int_\Omega K_l(s)\,|\nabla u|^2\, ds \leqslant \left(1 + \frac{1}{2\varepsilon}\right)\iint_\Omega \frac{f^2}{T^2}\, dxdy$$

$$+ \left(1 + \frac{\varepsilon}{2}\right)\left(1 - \frac{2\nu\mu}{(\nu + \mu)^2}\right)\iint_\Omega [u_{xx}^2 + 2u_{xy}^2 + u_{yy}^2]\, dxdy$$

for any $\epsilon > 0$. Choosing ϵ so that

$$\left(1 + \frac{\varepsilon}{2}\right)\left(1 - \frac{2\nu\mu}{(\nu + \mu)^2}\right) < 1,$$

we obtain the conclusion of the theorem after some obvious manipulations.

Corollary. *If, in addition to the conditions of Theorem* 1, *the region* Ω *is* l-*convex and* $K_l(s) \not\equiv 0$, *it then follows that the solution of the boundary value problem* (5), (2) *satisfies the a priori estimate*

$$\|u - \bar{u}\|_{W_2^2(\Omega)} \leqslant C_3 \|f\|_{L_2(\Omega)}, \tag{8}$$

where $\bar{u} = (1/\text{meas}\ \Omega)\iint_\Omega u(x, y)\,dxdy$.

§2. An a priori estimate for the norm of the solution of the generalized Neumann problem in the space $C^{1,\alpha}(\Omega)$

We shall assume throughout the following that the region Ω and the field l satisfy the conditions of the corollary to Theorem 1.

Definition 2. The *Morrey space* $L_{2,\gamma}(\Omega)$, $\gamma \leq 2$, is the collection of functions from $L_2(\Omega)$ for which the norm

$$\|u\|_{L_{2,\gamma}(\Omega)} \equiv \|u\|_{2,\gamma,\Omega} \equiv \max_{(x_0, y_0)\in\Omega} \sqrt{\int_0^R r^{-1-\gamma} \|u\|^2_{L_2(\Omega_r(x_0, y_0))}\, dr}$$

is finite, where R is the diameter of Ω; $\Omega_r(x_0, y_0) = \Omega \cap K_r(x_0, y_0)$; $K_r(x_0, y_0)$ is the disk of radius r with center at (x_0, y_0).

Lemma 2. *Suppose that* $f(x, y) \in L_{r,\gamma}(\Omega)$. *Then for a solution of the problem* (1), (2) *belonging to the space* $W_2^2(\Omega)$ *the following estimate holds*:

$$\sqrt{1-\delta\gamma}\,\|\sqrt{u_{xx}^2+2u_{xy}^2+u_{yy}^2}\|_{2,\gamma,\Omega} \leqslant \|f\|_{2,\gamma,\Omega} + C_4\|\sqrt{u_{xx}^2+2u_{xy}^2+u_{yy}^2}\|_{2,\Omega} \tag{9}$$

with $\delta \geq \pi\sqrt{2}$ *and* $\gamma < 1/\delta$.

Proof. We first consider a strictly interior disk $K_r(x_0, y_0)$. Denoting $C_r(x_0, y_0) = \partial K_r(x_0, y_0)$, we apply Green's formula (cf. [1], Chapter IX, §6) to obtain

$$\iint\limits_{K_r(x_0, y_0)} [u_{xx}^2+2u_{xy}^2+u_{yy}^2]\,dxdy = \iint\limits_{K_r(x_0, y_0)} f^2 dxdy - 2\int\limits_{C_r(x_0, y_0)} u_x \frac{\partial u_y}{\partial s}\,ds.$$

We estimate the last integral using Cauchy's and Wirtinger's inequalities, and the notation $\bar{u}_x = (1/2\pi)\int_0^{2\pi} u_x(re^{i\phi})d\phi$:

$$\left|\int\limits_{C_r(x_0, y_0)} u_x \frac{\partial u_y}{\partial s}\,ds\right| = \left|\int_0^{2\pi} u_x(re^{i\varphi}) \frac{\partial u_y(re^{i\varphi})}{\partial \varphi}\,d\varphi\right|$$

$$= \left|\int_0^{2\pi} [u_x(re^{i\varphi}) - \bar{u}_x] \frac{\partial u_y}{\partial \varphi}\,d\varphi\right| \leqslant \sqrt{\int_0^{2\pi}\left(\frac{\partial u_x}{\partial \varphi}\right)^2 d\varphi \cdot \int_0^{2\pi}\left(\frac{\partial u_y}{\partial \varphi}\right)^2 d\varphi}$$

$$\leqslant \frac{r}{2}\int\limits_{C_r(x_0, y_0)} [u_{xx}^2+2u_{xy}^2+u_{yy}^2]\,ds.$$

Substituting this estimate into the preceding equation, we obtain for almost all r

$$\varphi(r) \leqslant r\varphi'(r) + F(r), \tag{10a}$$

where

$$\varphi(r) \equiv \iint\limits_{\Omega_r(x_0, y_0)} [u_{xx}^2+2u_{xy}^2+u_{yy}^2]\,dxdy, \quad F(r) \equiv \iint\limits_{\Omega_r(x_0, y_0)} f^2 dxdy.$$

A disk $K_r(x_0, y_0)$ in general position can be treated analogously. In this case, using the computations in Lemma 1, we obtain

$$\varphi(r) = F(r) - 2\int\limits_{C_r(x_0, y_0)\cap\Omega} u_x \frac{\partial u_y}{\partial s}\,ds - \int\limits_{K_r(x_0, y_0)\cap\partial\Omega} K_l(s)\,|\nabla u|^2\,ds$$

$$\leqslant F(r) - 2\int\limits_{C_r(x_0, y_0)\cap\Omega} u_x \frac{\partial u_y}{\partial s}\,ds.$$

In order to estimate the last integral, we assume that the coordinate system has been chosen in such a way that the axis OX is directed parallel to Γ at one of the points of $C_r(x_0, y_0) \cap \partial\Omega$, and therefore at this point $u_x = 0$. The symbol ϕ_0 ($\leq 2\pi$) will denote the radian measure of the arc $C_r \cap \Omega$.

Using Cauchy's inequality and the fact that

$$\int_0^{\varphi_0} v^2(\varphi)\,d\varphi \leqslant \frac{\varphi_0^2}{2}\int_0^{\varphi_0} [v'(\varphi)]^2\,d\varphi,$$

$v(0) = 0$, we obtain

$$\left| \int_{C_r(x_0, y_0)\cap\Omega} u_x \frac{\partial u_y}{\partial s} ds \right| = \left| \int_0^{\varphi_0} u_x \frac{\partial u_y}{\partial \varphi} d\varphi \right| \leqslant \frac{\varphi_0}{\sqrt{2}} \sqrt{\int_0^{\varphi_0} \left(\frac{\partial u_x}{\partial \varphi}\right)^2 d\varphi \int_0^{\varphi_0} \left(\frac{\partial u_y}{\partial \varphi}\right)^2 d\varphi}$$

$$\leqslant \frac{\pi}{\sqrt{2}} r \int_{C_r(x_0, y_0)\cap\Omega} [u_{xx}^2 + 2u_{xy}^2 + u_{yy}^2] ds = \frac{\pi}{\sqrt{2}} r\varphi'(r).$$

Denoting $\pi\sqrt{2} = \delta$, we write the above inequality and (10a) in the form

$$\varphi(r) \leqslant \delta r \varphi'(r) + F(r), \tag{10b}$$

where (10b) is valid for all $(x_0, y_0) \in \Omega$ and $r \leq R$.

Let $\gamma < 1/\delta$. Then, multiplying (10) by $r^{-1-\gamma}$ and integrating from 0 to R, we obtain

$$\int_0^R r^{-1-\gamma} \varphi(r)\, dr \leqslant \delta \int_0^R r^{-1-\gamma} \varphi'(r)\, dr + \int_0^R r^{-1-\gamma} F(r)\, dr$$

$$\leqslant \delta \frac{\varphi(R)}{R^\gamma} + \delta\gamma \int_0^R r^{-1-\gamma} \varphi(r)\, dr + \int_0^R r^{-1-\gamma} F(r)\, dr,$$

and from this we obtain

$$(1 - \gamma\delta) \int_0^R r^{-1-\gamma} \varphi(r)\, dr \leqslant \frac{\delta}{R^\gamma} \varphi(R) + \int_0^R r^{-1-\gamma} F(r)\, dr,$$

which yields the desired inequality (9).

Theorem 2. *For a solution of the boundary value problem* (5), (2) *under the assumptions* (6) *and* $f(x, y) \in L_{2,\gamma}(\Omega)$, *the following a priori estimate holds*:

$$\left\| \sqrt{u_{xx}^2 + 2u_{xy}^2 + u_{yy}^2} \right\|_{2, \gamma, \Omega} \leqslant C_5 \| f \|_{2, \gamma, \Omega}, \tag{11}$$

if $\gamma < \sqrt{2}\,\mu\nu/\pi(\mu + \nu)^2$.

Proof. Here we essentially need only repeat the proof of Theorem 1. The sole distinction is that we need to estimate not the norm in $L_2(\Omega)$, but the norm in $L_{2,\gamma}(\Omega)$. In this case the computations in the proof of Theorem 1 yield

$$\| \Delta u \|_{L_{2,\gamma}(\Omega)}^2 \leqslant \left(1 + \frac{1}{2\varepsilon}\right) \left\| \frac{f}{T} \right\|_{L_{2,\gamma}}^2 + \left(1 + \frac{\varepsilon}{2}\right)\left(1 - \frac{2\nu\mu}{(\nu + \mu)^2}\right) \left\| \sqrt{u_{xx}^2 + 2u_{xy}^2 + u_{yy}^2} \right\|_{L_{2,\gamma}}.$$

Furthermore, using Lemma 2, the l-convexity of Ω and Theorem 1, we obtain

$$(1 - \delta\gamma) \left\| \sqrt{u_{xx}^2 + 2u_{xy}^2 + u_{yy}^2} \right\|_{2, \gamma, \Omega}^2 \leqslant C_4' \| f \|_{2, \Omega}^2$$

$$+ \left(1 + \frac{1}{2\varepsilon}\right) \left\| \frac{f}{T} \right\|_{2, \gamma, \Omega}^2 + \left(1 + \frac{\varepsilon}{2}\right)\left(1 - \frac{2\nu\mu}{(\nu + \mu)^2}\right) \left\| \sqrt{u_{xx}^2 + 2u_{xy}^2 + u_{yy}^2} \right\|_{2, \gamma, \Omega}.$$

By choosing ϵ sufficiently small, depending on $\gamma < \sqrt{2}\,\mu\nu/\pi(\mu + \nu)^2$, ν and μ, and using the fact that in a finite region $\|f\|_{2,\Omega} \leq C_6 \|f\|_{2,\gamma,\Omega}$, we obtain (11).

Remark. If we carry out the computations in the proof of Lemma 2 and Theorem 2 more carefully, we can obtain the conclusion of Theorem 2 with $\gamma < 2\nu/\mu$.

A similar result for estimates at interior points was obtained by Talenti in [3].

Corollary. It follows from Theorem 2 and a lemma of Morrey ([1], Chapter II, §4, Lemma 4.1) that the solution of the problem (5), (2) belongs to the space $C^{1,\gamma/2}(\Omega)$, which constitutes the principle result of the preceding arguments.

§3. An a priori estimate for the solution of boundary value problem (2) for a quasilinear equations and an existence theorem

We consider the boundary value problem for an equation

$$a(x, y, u, u_x, u_y)u_{xx} + 2b(x, y, u, u_x, u_y)u_{xy} + c(x, y, u, u_x, u_y)u_{yy} = f(x, y, u, u_x, u_y), \quad (12)$$

which satisfies the ellipticity condition (6) and the condition

$$|f(x, y, u, u_x, u_y)| \leqslant \delta_1[u_x^2 + u_y^2] + C(1 + u_x^2 + u_y^2)^{1-\varepsilon}, \quad (13)$$

where δ_1 is sufficiently small and $\epsilon > 0$, uniformly with respect to the argument of the function f.

We place one more condition on the field $\mathbf{l}$:

$$(\mathbf{n}, \mathbf{l}) \geqslant \delta_2 > 0. \quad (14)$$

Lemma 3. *Suppose that the field* $\mathbf{l}$ *satisfies condition* (14). *Then for certain constants* C_6 *and* C_7, *depending on* δ_2, *the smoothness characteristic of* $\partial\Omega$, $\mathbf{l}$ *and the size of* Ω, *the following estimate will hold for any* $u(x, y) \in W_2^2(\Omega)$:

$$\iint_\Omega |\nabla u|^4\, dxdy \leqslant C_6 \max_\Omega u^2(x, y) \iint_\Omega [u_{xx}^2 + 2u_{xy}^2 + u_{yy}^2]\, dxdy + C_7 \max_\Omega u^4(x, y).$$

Proof. Suppose that there exists a positive definite smooth matrix-valued function $\binom{A\ B}{B\ C}$ with lower bound λ and upper bound Λ such that on the boundary of Ω the conormal $\mathbf{N}$, determined with respect to this matrix, is colinear with $\mathbf{l}$, and therefore

$$\frac{\partial u}{\partial N} \equiv (Au_x + Bu_y)\cos nx + (Bu_x + Cu_y)\cos ny = \frac{\partial u}{\partial l} = 0.$$

Using the inequality $\lambda|\nabla u|^2 \leq Au_x^2 + 2Bu_xu_y + Cu_y^2$ and applying Green's formula and Hölder's and Young's inequalities, we obtain

$$\begin{aligned}\iint_\Omega |\nabla u|^4\, dxdy &\leqslant \frac{1}{\lambda}\iint_\Omega [Au_x^2 + 2Bu_xu_y + Cu_y^2]\, dxdy \\ &= \frac{1}{\lambda}\int_{\partial\Omega} |\nabla u|^2 u \frac{\partial u}{\partial N}\, ds - \frac{1}{\lambda}\iint_\Omega 2u\{(u_xu_{xx} + u_yu_{yx})[Au_x + Bu_y] \\ &\quad + (u_xu_{xy} + u_yu_{yy})[Bu_x + Cu_y]\}\, dxdy \\ &\quad - \frac{1}{\lambda}\iint_\Omega u|\nabla u|^2(Au_{xx} + 2Bu_{xy} + Cu_{yy})\, dxdy \\ &\quad - \frac{1}{\lambda}\iint_\Omega u|\nabla u|^2[A_xu_x + B_xu_y + B_yu_x + C_yu_y]\, dxdy \leqslant\end{aligned}$$

$$\leqslant C_6 \max_{\Omega} u^2(x, y) \iint_{\Omega} [u_{xx}^2 + 2u_{xy}^2 + u_{yy}^2]\, dx dy + C_7 \max_{\Omega} u^4(x, y),$$

since the boundary integral is 0.

In this inequality C_6 is determined by λ and Λ, while C_7 is also an upper bound for the derivatives of the elements of the matrix and the quantity meas Ω. Therefore the proof of the lemma reduces to the construction of the matrix described above.

At each point of the boundary of Ω we introduce a system of coordinates: the axis OX will be tangent, and the axis OY will be normal; $\mathbf{l} = l_x \mathbf{i} + l_y \mathbf{j}$. In this system of coordinates we define

$$\begin{pmatrix} A & B \\ B & C \end{pmatrix} = \begin{pmatrix} \frac{1}{l_y} & l_x \\ l_x & l_y \end{pmatrix}$$

and return to the old system of coordinates. It is clear that on the boundary the matrix function constructed satisfies the stated conditions.

Let δ_3 be the width of the strip in which through every point there passes exactly one continuation of the normal to $\partial\Omega$. At each point of this strip we define (in the special coordinate system)

$$\begin{pmatrix} A & B \\ B & C \end{pmatrix} = \begin{pmatrix} \frac{1}{l_y} & l_x \\ l_x & l_y \end{pmatrix} \frac{y + \delta_3}{\delta_3} - \frac{y}{\delta_3} \begin{pmatrix} 1 & 0 \\ 0 & 1 \end{pmatrix},$$

and outside the strip we define $\left(\begin{smallmatrix} A & B \\ B & C \end{smallmatrix}\right) = \left(\begin{smallmatrix} 1 & 0 \\ 0 & 1 \end{smallmatrix}\right)$.

After returning to the old system of coordinates, we obtain the desired matrix. This proves the lemma.

Lemma 4. *Assume that conditions* (6), (13) *and* (14) *hold, and that the solution of the boundary value problem* (12), (2) *satisfies the a priori estimate*

$$\max_{\Omega} |u(x, y)| \leqslant M. \tag{15}$$

Then for δ_1 *sufficiently small in condition* (13) *the following a priori estimate holds*:

$$\|u\|_{W_2^2(\Omega)} \leqslant C_8 \cdot M.$$

To prove the lemma it is sufficient to apply Theorem 1 and Lemma 3 to the boundary value problem (12), (2), and then use condition (15).

Corollary. *Under the conditions of Lemma* 4, $f(x, y, u, u_x, u_y) \in L_{2,\gamma}(\Omega)$ *for any* $\gamma < 2$.

In order to prove this, it is sufficient to use the Sobolev imbedding theorem and the fact that $L_p(\Omega) \subset L_{2,\gamma}(\Omega)$ for $p > p_0(\gamma)$.

The following theorem follows easily from the above corollary and Theorem 2.

Theorem 3. *Under the conditions of Lemma* 4, *a solution of the boundary value problem* (12), (2) *satisfies the a priori estimate* $|u|_{C^{1,\gamma/2}(\Omega)} \leq C_9 M$, *with* γ *as in Theorem* 2.

Using Theorem 3, it is not difficult to obtain an existence theorem for a classical solution of the problem (12), (2).

Theorem 4. *Let a, b, c and f from the equation* (12) *be Hölder continuous in all of their arguments for some exponent* $\alpha > 0$, *and assume that* $\partial f(x, y, u, 0, 0)/\partial u \geq \delta_4 > 0$ *for all* $(x, y) \in \Omega$ *and* $u \in R'$.

Then under the conditions of Theorem 3 *the boundary value problem* (12), (2) *has a solution* $u(x, y) \in C^{2,\alpha}(\Omega) \cap C^2(\bar{\Omega})$.

The proof of this theorem is analogous to the proof given in [1] for the existence of a solution for the Dirichlet problem.

BIBLIOGRAPHY

1. O. A. Ladyženskaja and N. N. Ural′ceva, *Linear and quasi-linear equations of elliptic type*, "Nauka", Moscow, 1964; English transl., Academic Press, New York, 1968. MR 35 #1955; 39 #5941.

2. Heinz Otto Cordes, *Über die erste Randwertaufgabe bei quasilinearen Differentialgleichungen zweiter Ordnung in mehr als zwei Variablen*, Math. Ann. **131** (1956), 278–312. MR 19, 961.

3. Giorgio Talenti, *Equazioni lineari ellittiche in due variabili*, Matematiche (Catania) 21 (1966), 339–376. MR 34 #4681.

Translated by J. MACKI